D0794020

A Philosophy of Time

To Joseph [illegible]
with all good
wishes

[illegible]

By Louis Aaron Reitmeister

PARADISE FOUND
IF TOMORROW COMES
GIST OF PHILOSOPHY
PHILOSOPHY OF LOVE
WHEN TOMORROW COMES
WHAT LIFE MEANS TO GREAT PHILOSOPHERS
THE CRISIS OF 1940: BEING AN APPEAL TO COMMON SENSE
THE NATURE OF POWER
THE GODS AND MY FRIENDS
BY THE WAY
A PHILOSOPHY OF TIME

In preparation:
THE PHILOSOPHY AND NATURE OF FRIENDSHIP
MY MOTHER—A BIOGRAPHY

Louis Aaron Reitmeister

A PHILOSOPHY OF TIME

1962
New York
The Citadel Press

FIRST EDITION

 Library of Congress Catalog Card Number 62-14811. Manufactured in the United States of America. Published by The Citadel Press, 222 Park Avenue South, New York 3, N. Y.

CONTENTS

TO
PAW-PAW

PREFATORY

A day is the price we pay for living twenty-four hours less. This essay is an attempt to make this payment wisely and happily.

LOUIS AARON REITMEISTER

Great Neck, N. Y.
January, 1962

The Bird of Time has but a little way
To flutter—and the Bird is on the Wing

OMAR KHAYYAM

ONE

THE *NATURE* OF TIME

TIME is a short word, a very ordinary often-used word, something that almost everyone believes has always existed and will always exist. It is a something taken for granted, like the very air we breathe, and it is usually looked at or observed when we have a train to make, a date to fill, a bill to pay, a show to see, money to get or a letter to send. It is part of the mechanics of a day and of so many days and months that make up a calendar, which is nice to look at and we even mark on its day our various appointments, hopes, dreams, and even our fears. It is very natural, very human, to feel it is an ever-present, ever-ready thing, always seemingly available because it is always here. The idea of a shortage of it never enters into our minds. It is nothing to stare at, to weigh as we would weigh potatoes, to buy or sell in the literal sense, although men have bought it and sold it as wares and money. It is not a gas or a substance, a subject or an object, a number or a sum, something about which we can express or impress our emotions. We cannot call it beautiful or ugly because we cannot see it or even imagine it to be of some form or spirit. We cannot say it is evil or it is good because a thing is evil or good if we can isolate it, identify it, separate it from us, look at it and judge it whether it contains purpose or intent, whether it operates of itself, consciously or unconsciously for good or evil, just like ourselves. Life is the occupant of Time and yet there is no one around to collect the rent. It has no color or shades of color, not even an opaqueness, odor or fragrance, to arouse our esthetic desires to translate it into some arrangement of softness, hardness, fineness, firmness, of strength or weakness, of farness or nearness. It has neither height nor depth and yet we seem to know that it is here. We cannot feel it, yet we know it surrounds us. We sense it but cannot touch it because there is nothing to touch.

Time is the most immaterial thing there is, and without it the material doesn't matter. It is something, yet of itself nothing, though everything that exists does so only within it. It is the most prized to those who want it, cruel to those who do not. The infant brings it with him; the dying leave it behind. It is the most obedient slave, for it lets you do with it what you please, yet it cups the universe in its palm and frames the limitless expanse of space. It is the most tolerant and patient, always new, yet the oldest of the old. It will forgive if we seek it; it will punish if we fear it; it will reward if we earn it. It is "Judgment Day" every day. Of itself it has no record or recognition. Out of it the buds appear, into it the leaves wither and disappear. The unborn are part of it, youth does not see it, the aged fear it, the dying are resigned to it, and the dead go back to it. But we do feel that *Time exists.* It must, we are sure, and so there must be also some way of describing it or trying to affix its nature in the same manner as we do other things around us. Such a simple thing, and yet many people would consider it laughable that any serious attempt is made to describe or try to analyze something which is so clearly elemental.

A word, like anything else, should be discernible and something of meaning, recognizable in experience, even if in thought, which is part of experience, if it has to have a justifiable reason to exist. This should be no restriction on the imagination or even on the fantasies of men, whether they be poet, painter, or peasant. Freud made it sufficiently clear that even dreams are not only part of human and other animal experiences but further, that they affect our conscious or wakened hours to a specific extent. To the primitive savage, dreams were as real to him as any other experience and he applied it as such. Even the Bible plainly states in many places that dreams were the means used by God to make his wishes known or to bring revelations, prophecies, etc., to his mortal emissaries on earth, which is merely a continuity of the primitive's mentality.

However, the words "forever," "evermore," "everlasting," "eternal," not discernible or recognizable in human experience or in natural phenomena, become completely meaningless, and the only justification for their existence as words can only mean, if anything, "states" of impossible perception. A poet looks at the horizon and beautifies it with words. While we can follow the horizon around the world, return to our starting point without touching it or reaching it, yet it is discernible and part of human experience. But the horizon is merely a view of distance and contains only space within distance, and its distance depends upon how high or low we view it from. Remove the perspective of specific distance and you remove or move the horizon, if that were possible. However, when the theologian tells us to believe in his "reality" of *eternity,* he is merely saying something without meaning because it is neither discernible in his experience nor in the experience of his worshippers. He is stating an *idea,* which is part of human experience, but the actual and recognizable perceptibility of the idea is beyond the human mind to identify its content, form, substance or length. In short, the idea is identified as a thought, and a thought is experience, but the *idea* of the idea is just nothing.

Eternity cannot be related to Time, as Time, in a recognizable state, is one of related measurement, but eternity, having no need of measurement, has no need for the co-existent factor of Time. We have often heard the expression that "eternity is timeless." It is true. There could be no time where there is no need of it. A thing without end, if it could be conceived, is a thing without measure. *Time is a measure* of a specific period between the beginning of something and the end of it. Measure is impossible at one end only. Nor is it conceivable in infinity. The value of life is known only when we know its limitations. We can never measure the value of life until we appreciate its finity, its certainties. Even if the exact date or marked beginning is not known and the exact length or duration is not known, as long as the something involved is one of known, estimated or assumed finiteness, it becomes a determinable factor requiring the co-existent of Time. Time has meaning in terms of the hour, not in terms of eternity.

Eternity is a word obviously created to indicate a something which we could never conceive, perceive, or understand. It's a word that makes Time completely unnecessary and meaningless, taking it out of measurement and reducing it to nothing. Theology attempts to make eternity appear as a definable factor, a quantum of certainty, a similar continuity of life on earth, as if "forever" could be reduced to months or years or any period conceivable to the mind of man. Theology attempts to make Time appear as if it were existent outside of anything conscious enough to identify it. It is doubtful whether the living creatures on earth other than man are even conscious of it or aware of it. Even if they are, it most probably is to a much lesser extent than our consciousness of it.

Time is only reducible to a term of measure within the *present* as the point of determination. The past and the future, to exist even in the mind, must of necessity be tied to the umbilical cord of the present, to which and from which it draws its identity, and from its identity its measure. Only from its measure can we determine its value and its meaning. If we keep this in mind as a *constant,* the present is the time we really possess or occupy. If we try to tie the past to the present to this extent we are trying to live in a non-existent past. If we tie the present to the future, to that extent we are trying to delude ourselves that we are living tomorrow today when tomorrow has no content of Time and without Time we do not live. To tie the past to the present is like dragging the anchor at the stern; by tying the future to the present we are dragging down the bow of our little ship. In the *present* of our compass we can more fully realize what we are constantly leaving behind our stern and we are also guiding ourselves to what lies ahead, but the boat can only be located, at any given point, in the present and in the present only.

Time does not exist either for the past or for the future. It doesn't exist for us when we sleep soundly; we only know it existed when we awaken and calculate that we slept so many hours of time by looking at a clock; we re-establish its existence by our consciousness. Yet one could say, "If Time exists when we are conscious to sense it, cannot we assume that it also exists even if we do not?" The answer to that is to try it.

We should hold in mind that Time is an invisible *intangible,* discernible and identifiable in reality only when it is screened against tangibles of measure and so becomes a measure in itself as long as the tangibles reflecting or revealing it are co-existents with it. If the universe did not exist but just space, why Time? If our earth did not exist, why Time on earth? If a consciousness did not exist to recognize the earth, the universe, or the individual himself, how could there be a "something" to recognize Time as a measure of all or any of the three?

As soon as Time is discernible it becomes finite.

Infinity and non-existence are concomitant factors or the sum of two nothingnesses. Time and existence are also two concomitant factors or the sum of two somethings. Now, of course, it would be silly, living in the world, seeing the stars, the moon and the sun, to say that if I died today that these things no longer will exist and therefore Time, as co-existent, also ceases to exist. We can assume that the universe and its Time continue to exist because we see other people dying while we are alive and the universe stays with us. But alas, that's as long as *we* do live. If I died today I doubt whether *I* can prove tomorrow that the universe or Time do exist. As far as I would be concerned, they couldn't exist because the Universe, Time and I, have become co-factors of nothingness. I could not sense Time because I have no live, conscious mind to sense it with. Sum total: No mind to discern, no Time to measure, no Universe to exist. Deduction: it appears to be that as long as I live and possess a conscious mind I can recognize the certainty of the Universe and the nature of Time as a relative co-existent factor so that my perspective not only becomes one of substance but also one of measure. The clocks's clicking is heard because *I* can hear it; the clock is seen because *I* can see it and the world it rests on. A doctor is operating on a patient who is under anesthesia. The patient is completely unconscious. The patient has no sense of time; the doctor most certainly has and he most probably is watching it very closely. Loren Eiseley remarks that "life and time bear some curious relationships to each other that is not shared by inanimate things. It is in the brain that this world opens."[1]

What does all this mean? To me it seems that the *I* and *Time* are of the same essence. The *I,* by being *I,* becomes Time as well, and if the *I* is non-existent then Time, so far as that *I* is concerned, also ceases to exist.

Then what *is* Time? An *identity of oneself,* discernible and recognizable by that self's consciousness of itself and its *related* identities of other existing things. I am conscious; I identify myself; I recognize in my own identity the finity of Time, just as finite as myself, no more, no less.

The metaphysician, in rebuttal, might claim: "If *Time* and the *I* are of the same essence and as God gave each human being a soul, which is presumably the *I* of the body, and which is eternal, then this soul is also attached or in essence the same as Time eternal."

Firstly this is a contention purely *a priori.* Before it can be attempted to prove that Time is eternal with the eternal soul or *I,* the theologian must prove that the soul or *I* is something existent apart or away from the body, that it actually exists after death, and the nature of its consciousness and expression after death of the body. The burden of proof is on the one who claims a proposition, not on the one who cannot affirm it or denies it because of the lack of such proof. The theologian will have to prove conclusively the identity of this spirit or soul, its form, its manner of communication, the nature of its manifestations, how it travels, where and why. *He cannot apply a natural proposition to support a supernatural proposition.* It is either natural or it isn't. If it isn't natural, then Time and Space are not co-existent factors. Albert Einstein, probably the greatest of all scientists, clearly differentiated between what is scientific and what is non-scientific, what is natural and what is supernatural, when he said: "I cannot imagine a God who rewards and punishes the objects of his creation, whose purposes are modeled after our own—a God, in short, who is but a reflection of human frailty. Neither can I believe that the individual survives the death of his body, although feeble souls harbor such thoughts through fear or ridiculous egotism."

Again we pause to inquire into the nature of Time. I have previously stated that *Time* and the *I* are of the same essence. Now we realize that they are of the same *value,* being a plural identity of one essence. But value is not a thing in itself; it is a something related to things outside of itself, a sort of by-product of an experience. A diamond ring suspended in space, unknown and unrelated, can scarcely be expected to have value. So it is with a person. Our sense of value depends on the nature of our relationships with things, people, periods, events, etc., outside of ourselves but which becomes part of our experience in being. Therefore, where in the first instance we recognize Time and the Self to be of the same essence in being and its nature of oneness, the value of this essence is determined by the *status* or *content* of the self and its relationships with external factors in the field which we can call experience. When Spencer tried to define life as a stage or series of relationships in which internal and external factors are involved, he was attempting to fuse together the constant manifestations being experienced in Time by any particular *state of being*. However, it is only a conscious state of being, or what we know to be a conscious state of being, that makes possible the recognition or discernibility of a state of Time and when this occurs Time is also recognized as a co-existent factor in its relative value to the self.

The relative value of Time to a lion who instinctively is stalking a zebra and is gauging the time element to bring the victim within striking distance, or the value of Time to a scientist counting off the seconds of his timer as he is doing his chemistry, or the value of Time to a pilot who is calculating to make his trip in so many hours and minutes, or the value of Time to a peon in a little Mexican village sleeping away the day in the shade of a tree, or the value of Time to a murderer who is waiting to take his last walk to the gallows, or the value of Time to a young man waiting to kiss his fiancée, or the value of Time to the Captain of a submarine in measuring his torpedo to the speed of his victim vessel, or the value of Time to a person who has heard the bad news of cancer and that he has only a few months to live, or the value of Time to an expectant mother in labor or the expectant father walking circles on a path of cigarette butts, or the value of Time to a dying man who strives to see yet another day—in all of these the Time essence may be the same but the relative value is different in each case as it affects each being in a state of different content and experience.

So, in plain words of common sense, what does all this mean? It means that Time is a thing of value because, firstly, it is of the same value as ourselves in being, which means we do not live apart from it but within it, a plural identity of one essence. Secondly, the identity of this essence, which consists of the *Self* and *Time,* is one of measure because of its finity in its relationship to other measures in time which are also finite. Thirdly, this self or essence has value, and the nature of its value depends on the status or content of the self and its particular relationships with things outside of itself in what we call the field of experience or existence. Now we should try to discern what is the nature of *value* so that it can be possibly identified and verified as a status or thing of specific measure and thus established as something of certainty and substance. In so doing we shall strive to find the *meaning* of value.

Value is a *qualitative* word and by/of itself is meaningless. It cannot indicate a status of being but merely a perspective, a *way* of measuring, not a measure. A beautiful large diamond is a god to worship and to wear by one woman who considers herself greater, richer, more outstanding, more important, more secure, than other women with smaller diamonds; this is her value of the stone. To another person, a jeweler, it represents the value of so many dollars and how much profit he can reap by selling it. To still another, the poet, it may be no more valuable than the beauty of a snow-flake. To still another it may mean the value of evil, the crimes that people would commit just to possess it. To the smuggler it represents the convenience of handling little things for big money. To the scientist it is merely a piece of purified carbon that can perform certain functions and uses. Yet the diamond itself remains the same, what it is and will always be so. Its value depends on its relationship to something else; of itself there is no value.

Let's take another example. We have here a very learned professor who believes sincerely that he is so wise that he seems too much for ordinary people to digest, that is, to appreciate his austere genius. So he hides away in some remote place where he hasn't any contact with people. He is supremely alone with his great brain. His knowledge is profound, his intellectual horizon wide and deep but he loves to think alone with the stars. What has happened here? No one can deny his right of privacy, but we cannot reasonably sustain that the value of his brain to other people outside of himself has increased by so doing. Even to himself it may be decreased because the *value potential* has been restricted, lessened, sublimated. The potential has been reduced from a maximum potential to a minimum potential limited to himself.

A monk, existing in some remote monastery on one of the Greek islands, is practically shut off from the world. No sex, not even any living symbol of femininity like a chicken or a pig is allowed. Alone he chastises himself daily, and each night he sleeps on straw just to prove to his God how much suffering he can endure before he receives his reward of softer cushions in heaven. He eats little, wants little, most of the time praying for his soul in his dungeon cell, waiting for God's call to hoist his spirit to the pearly gates. What has happened here? The value minimum has become obliterated because he has transferred whatever time-value potential (measure) he could have possibly possessed to an unknowable infinity (measureless and therefore meaningless) until his own being has become meaningless and of no value. He has wasted away his life trying to bribe his own shadow.

Value, therefore, is a *perspective* of kind and degree, varying with people, circumstances, purposes, places and events. Co-existent with the perspective of value by any person is *Time,* which is a co-existent of the person himself. It is a way of looking at things, a privilege we might say, an opinion or a way of life.

Perspective is a transient, illusory factor, being limited to our senses and, because of this, very often creates an element of confusion unless it finds itself founded on a reasonable discernibility of its true identity. Take *color,* for instance, which is an illusion of reality. The identity of color is born in perspective; perspective is born in relationship. Relationship being a transient, color becomes a transient factor, a temporary, momentary identity depending wholly on the relationship of the viewer and the viewed and what's between them. Measure itself serves to allay the illusion between things but cannot solve the measure of all things because of its finiteness and limitations. Perspective, thus born out of relationships, gives birth, in turn, to measure.

Law does not set any standards of value because of the nature of itself; it tries to standardize the means of conduct and procedure by which values should exhibit and express themselves in such a manner as to allow people the same opportunity of value-measurement under similar circumstances.

The theologies do not set standards of value because they do not trade with things of measure or finites. To "reward" a person for his value of conformance with a value of endlessness of either heaven or hell is to first confuse and then dilute the entire perspective of value into nothingness. "Man's quest for certainty is, in the last analysis, a quest for meaning. But the meaning lies buried within himself rather than in the void that he has vainly searched for portents since antiquity." [2]

The problem we face is how to be able to arrange our perspectives or of our focuses of things, including ourselves, so that the value-potential can be maximum reached instead of minimum reached. This is the test of wisdom and brings us through another door into another room where we find the task of finding out the *meaning* of value. When we can find the meaningfulness of value, we can then begin the road of trying to arrange some method of *verifying* the potential of one value against another in such a manner as to produce *judgments* that can activate the least possible number of regrets and bring about the greatest possible amount of happiness and general satisfaction.

Meaning is a concept of relationship, and because its nature or "body" rises out of relationship, it is a transient, something in flux, subject to change as do relationships; and their values change or become modified by time and event. To me it seems there are no meanings in terms of absolutes, dogmas, supernaturals, because these are hypothetical assumptions beyond the possibility of perception by man and as such are completely meaningless. A person states that God exists; this is an assumption that has not been verified. The man is entitled to his belief or idea and his right to throw his life away for it if he wants; he has the right to say that God has meaning to him. But I think he is mistaken. What he really believes is that he believes in a God, which means that the act of believing has meaning to him, not the God which he could not possibly perceive. This God has not been verified to him but he believes it because if he didn't he may get punished and if he believes he may get rewarded. Fear of punishment in hell or reward in heavenly bliss have meaning to him. Fear of ostracism from his group, society, position or party, if he does not believe, has meaning to him. The emotional satisfaction of being considered a good man of the neighborhood, has meaning to him when he goes to his church or temple. If he is a faker, like so many are, he will use the veneer of belief to show to his friends and associates that he is a God-fearing man and thus create a confidence-game so that he can cheat or swindle them; and such a veneer has meaning to him. The minister who gets a fat salary, a home, a car, looked up to with respect and reverence plus many handshakes on the Sabbath, whether he really believes or not, the maintenance of so comfortable an economic position has meaning to him. But God is not a discernment and is not recognizable, and as such, is meaningless. The theologian will write millions of words about the majesty of the Lord, morals, etc., assuming the proposition that an absolute exists. But if he wants to be *absolutely* honest the most he can really do is to hope that what he believes in, irrespective of the inability of the human mind to fully perceive his God in verifiable sense, is true in spite of his "worldly" limitation to reach it.

Before we wander away too far from our main premise—the nature of meaning—let us bear in mind that meaning, in its origin, is an individualistic event before it could possibly be, by aggregate of similar meanings, made into a social, economic or political event. Like Time, and because of it, meaning is also attached to the *I* and its relationships to other things. It is a *variant,* like value, and its nature, measure and comprehension depend on each particular event related to the *I,* including the self itself. Against what screen can we project anything to reveal its meaning if possible?

Experience. And experience only operates in a fallible field of imperfections, changeabilities, flexibilities, variations. What is this field? Nature. And nature is not supernatural, divine, or a God. The human mind has what we call *intelligence,* which being part of nature and experience, follows the parental pattern of being neither perfect nor harmonious, and such an intelligence, being a perspective of limitation, cannot conceive a perfection or a harmony without limitation.

What does all this mean? Simply this. If we really want to get down to Earth, let's stay there. As a matter of fact, we can't do otherwise. If we ever leave the Earth, it will be inside of rockets, perhaps, but never on the wings of angels.

Let us recapitulate on the identity of meaning: it is a *variant* and a *transient,* like ourselves, and subject to the impress of event, circumstance and condition at any particular time according to the value placed on it by the self or *I,* which, in turn, is subject to the same changes of event, heritage, environment and relationship. How can we measure this meaning? We can measure it by each person's scale of the value of Time (co-existent with the Self) in ratio to the thing or event about which the inquiry for meaning is directed.

To summarize, briefly, our initial venture into the land of a philosophy of Time, we have brought to light that Time does not exist by itself alone, that of itself it is nothing, without essence, measure, discernment or recognition. It is a created factor to a consciousness which becomes aware of it. It then becomes a co-existent factor of measure because it becomes verifiable and discernible. It is a measure because it is as finite as the consciousness that gave it birth. It is verifiable as the self is identifiable. A person completely insane may be conscious, but the consciousness is unable, because of its uncontrollability, to identify itself and thus cannot identify the existence of Time as a plural identity of itself.

As a result of establishing finites as measures in Time, infinities become meaningless and abstractions, mere words indicating "states" of incomprehensibilities. We will sketch, most briefly, an outline of the doctrine of a future life after death, from primitive cultures to modern times, to impress the point that if we are to be serious about obtaining the most satisfaction in our mortal life, we must seek it in the mortal world, of which we may know something, instead of trying to seek it in some other world about which we know nothing.

Further, we have seen that Time becomes a value in ratio to the value of the self, but to know value we must also sense or understand the meaning of value. This meaning is a variant and a transient like value itself, like Time, like the Self, and becomes a variable of relationship and perspective. As a result, there must be some kind of self-direction in guiding us through the various meanings of Time-Value if we are to make the most out of life with the least number of regrets and the most satisfaction. This calls for some analysis of the nature of experience, and the relationships of this experience to the Self and its Time. Here is the test of courage and intelligence. Courage to face up to realities as we find them; intelligence to translate and manifest these realities into a happy life. We shall now proceed to some discussion on the nature of intelligence and the nature of experience.

The nature of *Intelligence* is another subject, that, to properly analyze, would take at least a volume in itself. Here again we must restrict ourselves, reluctantly, to its skeletal form and frame, identity, and its relationship to Time. Intelligence indicates an action which is both a cause and an effect of cognition. In *Homo sapiens* this cognition is not only an identifiable manifestation of one's own existence, but it is also a discernible factor that this existence is part of the experience of the Self and the related experience which surrounds it. Time is the area of polarity within which and through which the circulating vessels of event manifest themselves constantly. "The prime function of philosophy," writes John Dewey, "is that of rationalizing the *possibilities* of experience,"[3] and Intelligence is the means by which these possibilities become known, activated, meaningful. Intelligence, whether it identifies itself as such or not, can only be a part of the natural realm, processed out of and into the objectivity of experience. "The true import of the doctrine of the operative or practical character of knowing, of intelligence, is objective."[4]

Intelligence is something of both degree and of kind, as varied as human beings are, both in the mind potential and in the myriad variations and changeabilities of their own experiences and the experiences of things and events in constant flux and impress with them. Even the hopes of paradise and the fears of hell are cognitions in a very mortal brain occurring or being experienced in a very mortal, earthly way. "There is nothing superhuman about the mind," writes Dr. Philip Eichler. "It is very human; nothing supernatural, just commonplace, and quite within physiologic bounds. Mentation is no more sublime and no more of a miracle than circulation or respiration."[5]

Spinoza, whom Miguel de Unamuno called "the most logical and consistent of athiests"[6] wrote that "the mind can imagine nothing, neither can it remember anything that is past, save during the continuance of the body."

It seems to me that a person can only *feel* the identity of experience as a constant factor within Time, and the identity of Time is dependent upon the mind to *sense* it by its own consciousness. This is not a denial that if I were to die today that experience and Time cease to exist because I cease to exist. It means that so far as we know, it ceases to exist for *me.* Ceasing to exist ceases also the mind potential to identify anything else. No mind to sense existence, no mind to sense Time. The whole idea of any value of Time is based upon what it reflects, like a mirror, the experiences appearing before it, and the life of any person is a particular reflection of this experience. The purpose of this essay is to attempt an evaluation of this experience, of different types of human experiences before this mirror of Time, and perhaps by these reflections we may possibly gather, not only the value of Time itself, its meaning and its measure, but the value, meaning, and measure of ourselves to ourselves and to others within this Time-Mirror which reflects our own identities and the experiences within which these identities are manifesting themselves.

Intelligence implies the ability to change, modify, accrete, variate itself, otherwise there would be no possibility of changing the nature and substance of judgments. It also indicates a status of imperfection at any point inasmuch as a perfection, if it could be even conceived, has no need of any change of itself or any modification and contains no possibility of self-criticism or self-analysis. To state that nature is perfect is to attempt to convey the idea that it contains no intelligence, purpose or design, as all these things imply a process of perfecting or imperfecting, according to anyone's particular perspective or sense of value concerning it. The real reason that Plato took time off to believe in the Perfect Being, the Ideal of Ultimate Reality and the perfection of its fixed, non-changeable Soul, is because he had someone else to do the cooking for him. Because Plato had his *Ideas,* the world stood still for two thousand years. While the metaphysician contemplated the Perfect Being (as if he could), the practical knowledge-seeking empiricist and instrumentalist gave man all the invention, progress, and the general cultural, social, economic and political advance, both good and bad, that we have today. Garrett Hardin puts it nicely: "Perfection is not a characteristic of the single individual or the unique act of Nature. . . . The generation of error is without end."[7] Condorcet expresses the same theme in just a different pose: "That Nature has set no limit to the perfecting of the human faculties, that the perfectibility of man is truly indefinite; that the progress of this perfectibility, henceforth independent of any power that might wish to arrest it, has no other limit than the duration of the globe on which Nature has placed us."[8]

Intelligence cannot be conceived of as being a moral process because it, being a method of finding judgments, decisions, viewpoints, or determinants of actions and judgments, can be applied for evil as well as for good, and here again the universality of anything as evil or good cannot be reasonably established because evil and good are perspectives, not things of universal certainty; it is merely anyone's particular opinion or pattern. In the course of history and during the rise and fall of social and political institutions, in the development and constant modifications, variabilities, erosions and accretions of social, economic, religious and political cultures, what was considered good and righteous by one group at one time was held to be evil and sinful by another. Even within each group or culture various individualistic perspectives and opinions hardly considered anything in the same identical or similar light. The religious books are loaded with never-ceasing expressions and exaltations about the righteous, the good and the moral, versus the sinful, the transgressions, the evils and the immoral. But what is exactly good and what is exactly bad, as a certain judgment before the god or gods, was, in reality, a judgment of a certain individual or group of individuals. Let's face it: There are in this world hundreds of different religions, each considered by each particular creed to be *the one* and only true religion regardless of its tolerance of other creeds. In almost every religion there are many subdivisions or offshoots, as we find in Christianity, Judaism, Hinduism, Buddhism, Islamism, etc. Which is the correct or true one? Intelligence, properly and fairly applied, can come only to one conclusion.

A man may have faith as a result of his intelligence but faith itself and by itself cannot lead to intelligence. It can lead to mental slavery, as we shall soon see later on in this book. The nature of intelligence, as evolved in man and the other animals, seems to be a natural endowment, not a metaphysical mysticism. It is a natural endowment with which we can *sense,* through reflections and judgments, the possible certainty and, as a result, the possible value of any particular thing or event. Inasmuch as any person is confined to his heritage and limited in his experience, his intelligence, and as a result, his judgments, can never be infallible. It can be correct or wrong, true or false, good or bad, etc. By training this intelligence, through knowledge and education, we encompass a wider range of experience. Our mental visibility becomes clearer and farther, and this brings us so much nearer to a better and a more true evaluation of things and experiences, because we are thus more able to identify them in terms of content and certainty. We have heard often the expression in the business world that "nothing can take the place of experience." This may be true, but we must hold in mind that an idiot, if he insists upon remaining one or cannot help himself being something else, will not be better by so many more years of experience, while one who applies his intelligence properly and constructively may encompass greater benefit within a shorter period of time. What does this mean in Time? It means that the one who has only the belief, through faith, that he is going to live forever, hardly lives at all because he is constantly forfeiting today for tomorrow, while the one who intelligently believes in what he can possibly know, in the certainties of experience, does not live forever, but he does desire to make the most of the time he really has.

If we grasp the evidence that life is a process of evolution, like all else around us, above and below us, we will grasp that everything within life is also in process—no goal, no destination—but just a process. This process is *experience.* Further, we will see more clearly that any expression in the history of man has been *processed,* not out of predetermined will or direction or purpose, but out of the nature of things themselves and which contains constantly the nature of flux and changeability. Nothing is static. A shark must move, otherwise it will suffocate. In a way, we move too, constantly. Life is a continual series of constant relationships of integration (birth) and disintegration (death or the absence of life). At every moment things within us and outside of us are going through this process of integration and disintegration, emergence and disappearance, birth and death, each effect becoming a cause of both. James Hutton wrote, "Nature lives in motion." [9]

Intelligence is the *movement* of our mental resources, operating only within earthly experience, subject to change and modification, as each day brings us so many more hours of experience. This keeps the door wide open to knowledge and gradually cultivates an intellectual humility for our limited knowledge and an enthusiastic gratitude for the almost unlimited panorama of experience that lies before us. That we can hardly be able to experience all of it is clearly established by the nature of ourselves and our life, but to blindfold our eyes in fear of something of which we have no knowledge or certainty, and as a result of this fear pawn our time for an eternity we cannot even imagine is most assuredly outside the confines of intelligence.

The habituated use of intelligence gives us a rational approach. What is *Rationalism*? Albert Schweitzer says that "Rationalism is more than a movement of thought which realized itself at the end of the eighteenth and the beginning of the nineteenth centuries. It is a necessary phenomenon in all normal spiritual life. All real progress in the world is in the last analysis produced by rationalism." [10] Rationalism is the process of reasoning. What is *Reason*? "Reason can be defined by three characteristics," writes Bertrand Russell. "In the first place, it relies upon persuasion rather than force; in the second place, it seeks to persuade by means of arguments which the man who uses them believes to be completely valid; and in the third place, in forming opinions, it uses observation and induction as much as possible and intuition as little as possible." [11]

Reason is our tool for constructive criticism. Socrates once said, "The life that has never criticized itself still lingers on the level of the brute." [12] John Dewey said, "Philosophy is a generalized theory of criticism." [13] Criticism is based on analysis, and "the true function of logic is analytic." [14] Analysis, if it is to be meaningful and verifiable, must be based on experience or a knowledge of things acquired out of experience; this is the *scientific method* and the approach to investigation, learning, to a verifiable philosophy of life. "What is the method of science?" states Dr. A. J. Carlson, President of the American Association for the Advancement of Science. "In essence it is this—the rejection *in toto* of all non-observational and non-experimental authority in the field of experience. . . . The principle of the scientific method, in fact, is only a refinement, by analysis and controls, of the universal process of learning by experience. This is usually called common sense." [15] Dr. John Dewey tells us that "scientific methods simply exhibit free intelligence operating in the best manner available." [16] William James explains further that "the only things that shall be debatable among philosophers shall be things definable in terms drawn from experience." [17] Thus "philosophy is a valuable, perhaps an indispensable auxiliary and guide in the co-ordination of the operations of living." [18] With our feet firmly fastened to the lighted path of reason we can proceed to observe the experiences of men and mankind, and of history, and allow both reason and knowledge to form for us, if possible, a practical, knowable, verifiable philosophy of Time.

Life is a process. Time is merely the measure by which the life is able to identify it, in degree and in kind, and intelligence is the means we process and possess to place a meaning and a value upon it.

I hope we have, in a very brief way, accomplished some understanding of the nature of Time. Now let us apply ourselves to try to discern the different meanings and values placed upon it in the lives of people, by organizations, nations, religions, social customs, cultures, and upon which they have built and now live in. And if we come, because of this inquiry, to understand ourselves and things about us a little better and strive to make our lives better, then it should be all worthwhile.

The greatest sacrifice is the sacrifice of Time.

ANTIPHON, *Plutarch's Lives*

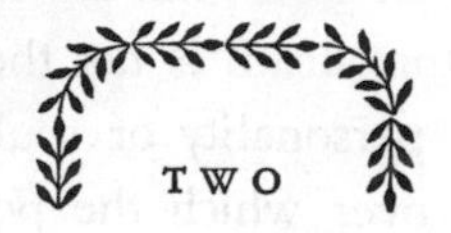

TWO

TIME AND THE *RELIGIOUS FACTOR*

Before we fall into any misunderstanding regarding the issue of this chapter, it is fitting that we make clear what is meant by the term *Religion.*

It is true that even an atheist can have a "religion" of his own, that is, a way of generalizing life into some sort of philosophy. John Dewey emphasized this point well when he wrote that "the positive lesson is that religious qualities and values if they are real at all are not bound up with any single item of intellectual assent, not even that of the existence of the God of theism." [1]

The scientist may also have a "religion" of his own and in his particular, specialized way. "It is precisely among the heretics of every age," wrote Einstein, "that we find men who were filled with the highest kind of religious feeling and were in many cases regarded by their contemporaries as Atheists." [2] Pope Pius V made the "infallible" statement of *"Ubi sunt tres medici ibi sunt duo athei"*—"Where there are three physicians there are two atheists." [3]

Ethics and specific cultural principles of peoples, nations, and groups, have gradually systematized philosophies which are designated in the broader sense as "religions" of a sort. The word might be applied, and rightly so, to a man's devotion to any kind of principle, good or bad, or both.

However, we are here concerned with *theological* religion, and the word *religion* thus applied, in its most recognized and accepted usage, is a system of *worship,* a relationship between man and the *supernatural,* his *faith in a god or gods,* to whom *obedience* and *reverence* are due, and the organized and institutionalized dogmas and cults by which this obedience and faith are carried on from generation to generation; the belief that death is not the end of personal life but that some kind of spirit, personality or soul, continues to live on in some form or other and over which the power or powers of supernaturalism has or have jurisdiction, or control the actions and destinies of man. It is this general recognition of religion as that department of a man's faith which deals with the acceptance of supernaturalism, personal immortality, and the ultimate realization of *eternal* security for one's soul, it is this accepted term of religion that we contend with here.

Dr. Alfred Weber, in his *History of Philosophy,* states: "We may define religion as follows: subjectively, it is the fear with which the givers of life and death, be they real or imaginary, inspire us; objectively, it is the sum of ideas, doctrines, and institutions resulting from this feeling." [4] Sir James Frazer defines it thusly: "Religion is a propitiation or conciliation of powers superior to man which are believed to direct and control the course of nature and of human life." [5] Dr. Theophile Meek gives his version: "Religion may be defined as man's belief that there is that in his environment which is greater than himself, and upon which he feels to some degree dependent, and with which he accordingly attempts to establish a relationship of mutual interest and good will. Just as soon as man developed to the point of reflective thought, just as soon as he became aware of his environment and his dependence upon it, he became religious. He sought a way of coming to terms with his environment in order to live." [6]

Life is indeed so fragile and delicate, so momentary and always moving, so soft, either growing or shrinking, so nebular that it slips from between the fingers or vanishes with a sigh, and yet it is strange that so many people think of it as a monument of glorified marble epitaphed by a hard chisel. The wonders of life, its joys and heartaches, are not things in themselves, but affected by things outside of themselves. Relationships identify the surfaces of all living things. The realities of life are so sober and shrewd and unforgiving that a sermon filled with abstract beauty, flowering phrases and celestial promises, is a base and unkind misdirection of life's great moral force.

It is needless to go into a long treatise on the origins and development of religion in primitive cultures, and the subsequent development of the doctrine of a future life from the savage cultures into the barbaric periods and on to the so-called higher or civilized religions of the later eras and up to the present. Though definitely related to a more thorough understanding of our subject, necessity of detail is not highly relevant. There are hundreds of fine volumes, if not thousands, by competent ethnographers, anthropologists, archaeologists, and other specialists of human history, which provide libraries of information for those who desire to proceed further in their search for knowledge instead of blind belief. Out of this vast tonnage of ethnographical and anthropological material that has been released in the investigation of primitive cultures of humanity there are many significant and salient revelations regarding our ancestors and their way of life.

In the first place it is understandable that the primitive savage, with hardly any knowledge of the physical world as we know it today, *feared* it because it was stronger than he and he knew that from morning to night each day he was subjected to its terrors, favors, changes, and uncertainties. "If the fear element were dropped out of any religion, that religion would not survive a year," wrote John B. Watson.[7] Everything about him was personified by him into creatures like himself, into spiritual, magical beings that he could possibly communicate with as he did with other savages and thus possibly control by the saying of certain words that indicated his secret power—the *origin of prayer*. In his dreams he saw his departed relatives as he saw them in life surrounded by the same environment in which he himself was living. He saw his enemies in his dreams, trying to attack or waylay him. He saw the animals, birds, the trees and rivers, the moon and stars, the sun, the other things in his day and night life, all these he saw in his dreams. He then knew that they existed as spirits, and he made spirits out of them and out of everything else including the pebbles and the dust at his feet. Theophile Meek wrote that "the belief in the reality of the other life arose not only from the impossibility of realizing the fact of death, but also from the experience of dreams. The dead appeared in dreams, and seemed just as real and alive as ever they had been. With this experience apparently came to man his first idea of spirit, which presently he applied to the phenomena about him; so naturism developed into animism or polydaemonism." [8] "The mind of the primitive," stated Lévy-Bruhl, "recognizes no fundamental difference between entities, even between those which we call animate and inaminate beings." [9] The primitive didn't understand death. "Primitive man cannot reconcile himself to the notion of death as a natural and necessary event; he persists in regarding it as an accidental and unnecessary disturbance of the proper order of nature." [10] Schopenhauer philosophized: "The greatest, and generally speaking, the worst misfortune that can befall any one is to die, and there is no fear equal to the fear of death." [11] The primitive couldn't stand the stench of the corpse and it was necessary to move it to the outskirts of the tribal area. He hardly understood the creation of life and that he was a partner in its procreation; he rather attributed magical powers to the women who attracted him so strangely and who gave birth to the "rebirth" of spirits that had lived previously—*the origin of man's fear of woman* and subsequently his condemnation of her as the accomplice of evil and sin. It wasn't until the 17th and 18th centuries A.D. that man definitely and clearly learned some of the truths about the processes of sex in the creation of life; monks and priests argued and fought over their council tables for almost two thousand years regarding the exact moment when a "soul" enters into the embryo or the fetus, yet knew practically nothing about the physiological processes of sex and embryology. Actually, it wasn't until 1827 when finally Dr. Karl Ernst von Baer brought out the proof that it takes two to make three.[12]

To the primitive there was one world—the world he lived in. Any lands or abodes of departed spirits were thus located here—were part of the natural world—whether above or below, near to or far from, the only world he knew about. "Even the savage cannot fail to perceive how intimately his own life is bound up with the life of nature, and how the same processes which freeze the stream and strip the earth of vegetation menace him with extinction. At a certain stage of development men seem to have imagined that the means of averting the threatened calamity were in their own hands, and that they could hasten or retard the flight of the seasons by magic art. Accordingly, they performed ceremonies and recited spells to make the rain fall, the sun to shine, animals to multiply, and the fruits of the earth to grow. In course of time the slow advance of knowledge, which has dispelled so many cherished illusions, convinced at least the more thoughtful portion of mankind that the alternations of summer and winter, of spring and autumn, were not merely the result of their own magical rites, but that some deeper cause, some mightier power, was at work behind the shifting scenes of nature. They now pictured to themselves the growth and decay of vegetation, the birth and death of living creatures, as effects of the waxing or waning strength of divine beings of gods and goddesses, who were born and died, who married and begot children, on the pattern of human life." [13] E. O. James, in his *Ancient Gods,* brings to light that "in the Paleolithic background of the ancient civilizations of the Near East the religious preoccupation was mainly with fertility, the mysterious phenomena connected with birth and generation, coupled with the ever-pressing problem of subsistence and the food supply, together with the final dissolution at death and its aftermath." [14]

Moreover, the living and the "living spirits" were constantly trafficking and intermingling with each other as one grand party. And to this were added the spirits of the trees, the rocks, the animals, bugs and birds of every description who not only had spirits of their own but who were often human spirits in disguise or condemned or rewarded to lower or higher forms of life. Today the supernatural world, the heavens and the hells of modern religions, while believed in, are actually not considered as part and parcel of the present life we live in. But the primitive had no dividing line; when he ate a cocoanut or a banana, when he killed an animal for food or sacrifice he had to apologize to its spirits or try to deceive its spirit to be fooled into being eaten or sacrificed. It was one world, both living and dead, and there was nothing not given a spirit or soul, because all these things were seen in their dreams, their imaginations, in the dark, or in their fantasies. There was no supernatural world apart from the natural world—*the origin of the idea of the endlessness of life*—which any savage wanted anyway. Even his enemies whom he killed could become his servants or slaves in the spirit world. No one actually died as we understand the term—*the origin of the idea of personal immortality.* In his *The "Soul" of the Primitive,* Lévy-Bruhl tells us: "To the primitive the dead man's existence does not cease. It is only his life on earth that has come to an end; his life continues elsewhere, and the barrier dividing the two conditions is not an insurmountable one. The living may be imagined as already dead, and the dead as alive again." [15] . . . "This dual mystery, wherein the timeless and the temporal are the same." [16]

The poor primitives, when death came, had to make their way as best they could in the land of spirits, no differently than the way they struggled in life. The chiefs, kings, nobles, high priests, etc., who had large armies, servants, harems, slaves, etc., when death came, had their wives, servants, slaves, soldiers, etc., killed so they could serve in the spirit world as they did in life. Cows for milk were killed to make spiritual cows and spiritual milk. Food was burned or embalmed to make spiritual food. The fighting weapons, their horses and chariots, the bows and arrows, pots and pans, jewelry, together with their wives, servants, etc., were dumped into a common grave to renew the same life and to serve their master without interruption—*the continuance theory in the origin of heaven.*

Ashley Montagu relates: "If Neanderthal men buried their dead (with implements to use in the spirit world), then it would be clear that they believed in the persistence of the soul after death, that they believed in immortality. By inference we know this to be almost certainly true of the men of Aurignacian times, who lived between 20,000 and 15,000 years ago. These people ceremonially interred their dead, furnishing them with implements and food with which to assist them to make the journey to the other world, and in many cases they placed red ochre within the grave, presumably as a symbol of the life-giving properties of the blood." [17] "The extension of the process of rebirth and regeneration to human beings when their alloted span has come to an end under temporal conditions is apparently almost as old as mankind, since in the archaeological record, as has been considered, indications occur of mortuary practices suggesting a cult of the dead going back to the beginning of the Stone Age." [18]

Edward B. Tylor tells us: "The Caribs, holding that after decease man's soul found its way to the land of the dead, sacrificed slaves on a chief's grave to serve him in the new life, and for the same purpose buried dogs with him, and also weapons. . . . When the New Zealand chief had slaves killed at his death for his service, and the mourning family gave his chief widow a rope to hang herself with in the wood and so rejoin her husband. . . . The Tunquz has buried with him his horse, his bow and arrows, his smoking apparatus and kettle. . . . the ancient Scythian chiefs, the contents of the burial mound: the strangled wife and household servants, the horses, the choice articles of property, the golden vessels . . . in old Europe, the warrior with his sword and spear, the horse in his saddle, the hunter's hound and hawk and his bow and arrow, the wife with her gay clothes and jewels, lie together in the burial mound. Their common purpose has become one of the most undisputed inferences of archaeology." [19] He continues: "Many Greenlanders thought that the kayak and arrows and tools laid by a man's grave, the knife and sewing implements laid by a woman's, would be used in the next world. . . . Among the old Peruvians, a dead prince's wives would hang themselves in order to continue in his service, and many of his attendants would be buried in his fields or places of favorite resort, in order that his soul, passing through those places, might take their souls along with him for future service. . . . Turanian tribes of North Asia avow that the motive of their funeral offerings of horses and sledges, clothes and axes and kettles, flint and steel and tinder, meat and butter, is to provide the dead for his journey to the land of souls, and for his life there. . . . The ancient Gauls were led, by their belief in another life, to burn and bury with the dead things suited to the living. . . . In modern centuries the Japanese would borrow money in this life, to be repaid with heavy interest in the next. . . . The souls of the Norse dead took with them from their earthly home servants and horses, boats and ferry-money, clothes and weapons; when King Harald was slain in the Battle of Bravalla, they drove his war-chariot, with the corpse upon it into the great burial-mound, and there they killed the horse, and King Hring gave his own saddle besides, that the fallen chief might ride to Valhalla, as it pleased him . . . in North America, the funeral sacrifice of the Winnebagos has come down to burying a pipe with tobacco with the dead . . . the rite of interring funeral offerings survived in Christian Europe. . . . The ancient Greek burial of the dead with the obolus in his mouth for Charon's toll. . . . The old Prussians furnished the dead with spending money to buy refreshment on his weary journey, so to this day German peasants bury a corpse with money in his mouth or hand." [20]

Regarding the Incas, Dr. Charles W. Mead informs us: "On the death of an Inca, or great chieftain, his wives and favorite women struggled for the privilege of burial with him, that they might accompany him into the other world and continue their services in the other life. Cieza de Leon says that more than 4,000 souls, women, pages, and other servants, together with immense riches were buried in the tomb with the Inca, Huayna Capac." [21]

Saint-Foix relates: "When Bertrand Duguesclin was buried at St. Denis, in 1389, several horses were sacrificed. The Bishop of Auxerre first blessed them, laying his hands on their heads and then they were killed," [22] just one of the continuations of paganism by Catholicism. "At Treves, in 1781, at the burial of General Frederick Casimir, his horse, according to the custom of the Teutonic Order, was led in front of the bier, and when the General had been laid in the tomb, the horse was killed and buried with him." [23] "At Reichenbach in Germany, a man's umbrella and goloshes are placed in his grave even today." [24] "The natives of Dutch Guinea have an interesting custom. They always place upon the grave, instead of a stone, a spade, so that if the corpse regains its soul and comes to life it can dig itself out." [25] "The Turanians of Northern Asia put food and napkins on the grave, saying, 'Rise at night and eat your fill. You have napkins to wipe your mouths,' " [26] Sir James G. Frazer, in an article in the *Journal of the Anthropological Institute,* reveals that the most probable origin of mourning clothes is that the primitives, fearing a return of the magic or spirit that caused the death, thus disguised themselves in the *darkest* colors to frustrate discovery of themselves by the evil spirit.

Not only people, animals and things, had spirits. The moon, the sun, the stars, the rain, the rivers, the good earth, the cooling winds, all the things that made life possible and comfortable, or retarded it, became personified spirits. The raging floods, the destructive storms, the blistering dry heat, the ravages of plague and insects, the wild beasts and reptiles that killed and ate them, the strange, mystic phenomenon of death that came upon them, these were personified and became evil spirits. Which spirit was evil or good obviously depended upon the locale and the needs of each tribe or race and the part of the world it lived in. The sun was a good god to the Peruvians but an evil beast to the Arabs. The wind was a good spirit to the Arab, who furnished his heaven with cool breezes, but to the Norseman it was a sign that Odin was very angry. "The meanings," said John Dewey in an address before the College of Physicans in St. Louis, "of such words as soul, mind, self, unity, even body, are hardly more than condensed epitomes of mankind's agelong efforts at interpretation of its experience. These efforts began when man first emerged from the state of the anthropoid age. The interpretations which are embodied in the words that have come down to us are the products of desire and hope, of chance circumstance and ignorance, of the authority exercised by medicine men and priests as well as of acute observation and sound judgment." [27]

Gradually, as man did emerge out of savagery and into the periods of barbarism, new spirits or gods emerged also such as Love, Wisdom, Strength, gods for Cookery, a god for the good luck of sailors and a goddess to appeal to in case of a bellyache. The Greeks not only personified into gods the various departments and subdepart--ments of nature beyond their control but had their gods mix freely with the mortals, in romance, intrigue, war, politics, jealousies and exhibiting all the emotions of human beings, no differently than the way the Greeks themselves lived. "There was no separation between the spheres of religion and of ordinary life. Every social act had a reference to the gods as well as to men, for the social body was not made up of men only, but of gods and men." [28] Between Olympus and Athens thousands of half-god half-mortal beings crowded the skies and shores. "The conception of a human soul, when once attained to by man, served as a type or model on which he framed not only his ideas of other souls of lower grade, but also his ideas of spiritual beings in general, from the tiniest elf that sports in the long grass up to the heavenly Creator and Ruler of the World, the Great Spirit, God." [29]

In Asia Minor similar things occurred but in a different way. There, the many good gods became one good god, Ahora-Mazdao, the light, from which word came Mazda or light bulb; and the evil things became one evil god, Ahrim, the darkness. One was sunlight, the other darkness—as man always feared the darkness and its invisibility, for animals and children fear the darkness for the very same reasons,[30] with its unseen spirits and ghosts and unexpectancies, and he tried to overcome the evil of the darkness with a symbol and a piece of the good god, the Sun—*the origin of the burning torches and tapirs in the temples, later the candles we burn in the churches, temples, and even on the Sabbath at home,* and the origin, among a million other things, of the burning of the Yule-log, celebrating Xmas, which is really the celebration of the sun-god, Mithra. It is also the origin of keeping candles lit for the departed so that the deceased may find their way through the labrynths of purgatory or the weird, darkened and twisting roads on the way to the Spirit-land, in their efforts to reach heaven.

An Assyrian war-god by the name of Yah-Weh [31] became the god of the Jews, later the god of the Christians, even though from what we read of Jesus from legendary sources (so far there is no historicity of Jesus), the Christ died as a religious Jew, never intended to create a new religion, was rabbinical in profession and approach, and lost his life because he, unfortunately, lived in a time when the priests had more power than the people.

So Christianity was born and went about substituting every pagan god and goddess with a disciple or a saint until there were so many saints and angels fluttering about that the sunlight couldn't get through, the Dark Ages began, and every free spirit shriveled and dried up.

The pagan gods-goddesses, which the pagans would not give up and worshipped secretly, became evil spirits, demons, devils, witches. Hel, the sweet pastoral mother-goddess of the ancient Germans, became an evil old hag riding a broomstick through the night—*the origin of the word Hell.* The cat, the well-known symbol of fertility and procreation, the sacred animal of the ancient pagan world and consort of most of the mother-goddesses (including the Catholic Madonna according to primitive paintings, etc.) became the witch's accomplice and was so sought out and destroyed that the rats multiplied in comfort and security and brought on, helpfully, the great bubonic plagues which almost wiped out Europe.

Mahomet, seeing the possibilities of good business, became a godly salesman, gave the Arab in heaven whatever he lacked or desired in life; he also sanctioned the Moslem to kill any human being in the world who doesn't believe in Allah and to confiscate his property. Allah showed his gratitude to Mahomet for starting the business by having him poisoned, and most of the Caliphs who followed him either were poisoned or killed off by each other.

As stated in the beginning, it is not my purpose or need to elongate a drawn-out and detailed review of the primitive cultures and the later religions of the world. The sketchy, elementary ladder here outlined is sufficient, I hope, to bring to light one salient point: The idea of a future life, the belief in personal immortality, the idea of a heaven and hell, the idea of many gods, the idea of a single god and a single devil, and the whole pattern of gods, goddesses, saints, etc., of all the religons of the world, came about by an understandable, historical process of development from primitive times. That billions of people suffered and died because of it is aside from the point; that phase is gone. The important thing is that *today* billions of people still, because of their belief in immortality, continue to consider *their Time* on this world as something insignificant, worth throwing away to gain the limitless, eternal bliss of the hereafter. "This great belief (the Future Life) may be traced from its crude and primitive manifestations among savage races to its establishment in the heart of modern religion, where the faith in a future existence forms at once an inducement to goodness, a sustaining hope through suffering and across the fear of death, and an answer to the perplexed problem of the allotment of happiness and misery in this present world, by the expectation of another world to set this right." [32]

When any religion fosters the pursuit of happiness, of racial integration and mutual cooperation and assistance, peace and friendship, when it helps the suffering poor and cares for the sick and the lonely, when it inspires in people the ethics of a fair deal, of honesty, of integrity, when it respects and holds sacred the rights of every human being to the same opportunities of the peaceful pursuit of a happy, loving and satisfied way of living, and all this for the benefit of people living *here* and *now,* then I would say that to this extent the religion is performing a goodly, practical and worthwhile service which is intelligent, wholesome and without regret and remorse, and which deserves, in turn, the appreciation of any good meaning and tolerant person.

A religion or any other organization is not restrained from doing good for people. The missionaries, in order to bring about a conciliatory attitude on the part of natives they are trying to convert, bring them all kinds of goodies, medicine, straw hats, parasols, seeds, nails, hammers, knives, night-gowns, bras, etc. The natives are supposed to think that the god who brings them these presents means them well and gradually they are bribed into becoming new protégés of Christ. I don't think that Christ would like this. Whether it is a Greek Orthodox monk on Mount Athos or a Club for the Comfort and Relief of Lonely Girls in Brooklyn, we are not critical of any really good work that its members might do for the unfortunate poor, the lonely and the needy. We are critical, however, of any group or organization which, in its efforts to sell the idea of unknowable immortality to its members, causes them to part with valuable and knowable mortality here on earth. And this is the crux of the matter. "It remains true," writes Ashley Montagu, "that the belief in immortality has often caused men to take too careless a view of life, of the lives of others as well as of their own. . . . The sacrifice of human beings to accompany the dead to the other world, of livestock, and the burial of valuable property with the dead, not to mention the vengeance killings of men in tribes in which the belief in natural death is wanting, can scarcely be considered as anything but evil by-products of the belief in immortality." [33] Llewelyn Powys adds: "Unfortunately, after 20,000 years it seems more than probable that all our sacrifices have been endured in vain, the hypothesis of the deity's existence being today as uncertain as at the beginning . . . we might as well try to snatch at a ghost as to try to understand God." [34]

Religion, in its primitive fashion, served the savage in a way that did not take him or his mind away from this earth; his rituals and his gods were mostly for the purpose of making his life here, his daily life, comfortable, healthy, secure. Even when he died his living relatives and friends felt he wasn't far away, perhaps in that large tree or in some animal or just a friendly spirit scalavamping around at night. "In Mexico, the Tlascalans thought that after death, the souls of nobles would animate beautiful singing birds, while plebians passed into weasels and beetles and such-like vile creatures. . . . In Africa the Maravi think that the souls of bad men become jackals and of good men snakes." [35] All the objects he saw and all the spirits he imagined were really one mass, like termites in a nest, all doing their parts more or less for the party. "Animism is, in fact, the groundwork of the philosophy of religion, from that of savages up to that of civilized man. . . . Thus Animism in its full development, includes the belief in souls and in a future state, in controlling deities and subordinate spirits, these doctrines practically resulting in some kind of active worship." [36]

But when the religion created a mythical heaven and hell far beyond the reaches of this life, points of no return, so to speak, and inoculated these fantastic ideas into the minds of children who grew up to believe them fanatically all through their lives, it created a Juggernaut that trampled down the hours, the days, years, of people into a worthless waste of Time. The valuable few years of people's lives became meaningless, frozen, of little interest compared to the ecstasy of a "forever" heaven or the terrifying, paralyzing horror of an "everlasting" fire of hell! Humanity is forever the dog with a bone in its mouth, forever dropping it into the reflecting brook. Aesop once said, "Beware lest you lose the substance by grasping at the shadow." [37] And Goethe so wisely rhymed:

"Always lightsome, never tiring,
Spell-impelled thy wings to ply,
Till at last the light desiring,
Thou art burned, butterfly."

Where in the world has religion a better hold on the masses of people than in India, and where in the world is life so much in transit *via karma,* so trivial and self-punishing as in India? "The old days of faith," writes Nehru, "do not appear to be adequate in a changing world; living should be a continuous adjustment to these changes and happenings. It is the lack of this adjustment that creates conflicts." [38]

More religion, less desire to live and less value to whatever living does go on. More religion, more desire to reach or "aspire" to a oneness with whatever anyone desires to call the ontological wholeness of the universal spirit, something which no one really could possibly understand at all. It is little man's self-inflicted but understandable privation to try to become a god and overcome the strife and mortality of nature by joining with or becoming part of, or simulating, the god itself and so be above nature and beyond the reach of its pressures and limitations.

Religion has been the greatest influence upon the life and mind, hopes and fears, of mankind. No other factor has been so powerful as the belief in a god or gods, in a supernatural world realm beyond death, and the continuity of the human personality in the form of a soul or spirit in the same manner, more or less, as it existed on earth.

As stated before, in primitive times the savage believed in his gods because they affected his daily life, death being some continued phase of the same existence. The idea of punishment or reward, heaven and hell, came much later. His belief in a "continuance" did not depreciate his daily life until the first priests took over and became the only agents or doormen to the gateway of the spirit-world. It was the savage-priest who, by taking over the management of the rituals, their many secret, mystic exaltations, magical recitations, and later prayers, became, by his own authority, the only means by which the common man could appeal to the gods for his earthly security as well as for some assurance of continuance after death. "The origin of priesthood is manifestly to be traced back to the earliest stage of social evolution and is doubtless to be found very close to the beginning of magical and religious practices. There was a time when each individual invoked the god for himself without the help of a mediator, but the idea early developed that certain individuals could get better, easier, and more intimate access to the spirit world than others. These were the first priests in religion. They were shamans, wonder-workers, medicine-men, individuals credited with the possession of mana, or spiritual power; or they could be men who lived near sacred places and so were supposedly on more intimate terms than ordinary folk with the spirits residing there." [39]

This is how religion came about, and it gradually grew into complex institutions or systems. The modern religions grew out of the older ones; they were not especially inspired or spontaneously created fresh and green from a new god. The new god was a composite of one or more of the older gods and the older gods evolved from the much earlier animisms of savagery and ancestor-worship. Different phases of this transition are still going on today in different parts of the world, and the various steps of religious development can be exhibited by people living today in many and far-off corners of the world. This has been clearly established, in essence and in substance, by the various sciences that deal with the history and development of human and related forms and cultures—anthropology, paleontology, archaeology, ecology, ethnology, and a hundred more specialized fields of investigation in the story of mankind.

If we are going to try to find the basic value of Time we must face up to realities and certainties. "We seek a method," states Dr. Joseph K. Hart, "a great dramatic highway of understanding through the fields and forests of experience, where we may find orientation without dogmatism, and illumination without loss of sight." [40] Barrows Dunham put it more on the line when he wrote: "Even if we cannot avoid the blows of circumstance, there is some consolation in knowing what hit us. . . . The performance of modern men proves beyond doubt that it is knowledge, not faith, which moves actual mountains." [41] Mark Graubard adds: "In primitive society, as today, knowledge did not come from heaven." [42] Benjamin Cardozo advises, "There is no safety in ignorance if proper inquiry would avail." [43] Anatole France, with his witty satire, wrote: "If you have a fresh view or an original idea you will surprise the reader. And the reader doesn't like being surprised. He never looks into history for anything but the stupidities he knows already. If you try to instruct him, you only humiliate him and make him angry. Do not try to enlighten him. He will only cry out that you insult his beliefs." [44] Alas, we will try to be optimistic rather than cynical if we are to attempt to accomplish anything, although the study of human nature can provide some very interesting material to support the feelings of Anatole France.

The belief in a god and in a hereafter only proves the reality and the certainty of the belief, not what is believed in. Were this merely a game of logic betwen philosophers and theologians, one would never come to a clear understanding of what might be true, because each side would depend on the facets and foibles of reason and faith to outwit each other. But this is a case of *history,* of *evidence,* of the gathering of material and materia from all the corners and crevices of the world. It is not a situation where doubt exists, where faith and hope are the predeterminators of the results desired, nor is it an *a priori* fantasy of any poetical philosopher or mystic who is sure the universe conforms to his ideas or dreams. This is history, and if we ever want in our lives to feel the full impact and pulse of Time, we must realize that eternity is just a way of taking our Time and, like a drop of ink, throwing it into the ocean as if we can turn the oceans into ink just as black. Try it. It will not work, regardless of your prayers, neither will your wishing bring you eternal life while you may go on wasting your lives dreaming about it. "In spite of all claims to the contrary there has not been a single authenticated case of communication between the dead and the living. . . . It has been argued by some that the proof of the existence of life after death does not depend upon the demonstration that those who have died can communicate with the living. That may be so, but if there is any other way of demonstrating the truth of this kind of immortality, I, at least, do not know of it." [45]

Normal people die involuntarily, which is logical proof that we desire to keep missing the bus in getting to heaven. "Survival after death," wrote Bronislaw Malinowski, "is probably one of the earliest of mythical hypotheses, related perhaps to some deep biological cravings of the organism." [46] "If we were not afraid of death," states Bertrand Russell, "I do not believe that the idea of immortality would ever have arisen. . . . Fear is the basis of religious dogma. . . . It is fear of nature that gives rise to religion." [47] Even Plato, the metaphysician, wrote that "the mortal nature is seeking as far as possible to be everlasting and immortal." [48] "Death is a reality which human beings pulsing with life cannot face with equanimity. Hence, what more simple than to deny it as a reality? Why should the life of man have definite limits in time and space? Why should there be a term to his existence?" [49] "Death, because it is not desired, is interpreted away, and life because it is desired is interpreted into eternal being, and hence, the belief in immortality . . . the belief in immortality is man's answer to the challenge of death; his answer to the insecurities and unhappiness of this life on earth. So believing, men can indeed find life more tolerable upon earth than they might otherwise find it, and this appears to be the principle function of the belief in immortality." [50]

Everyone seems to believe in God but no one seems to be in any particular hurry to meet him. Like running forward to meet the horizon, so is man's fantasy and dream to scale the heights of heaven, when it really lays at his very feet. The horizon is everywhere and so is heaven, if only he will understand and make it so. "There is no God in the sky," lectured A. Powell Davies, the brilliant liberal-minister, "God is in the heart that loves the sky's blueness. There is no army of angels, no hosts of seraphim, and no celestial hierarchy. All this is man's imagining. But there is angelic purity of motive, and seraphic joy, and a celestial heightening of the human spirit." [51]

The belief in immortality is the greatest waster of life. If people seriously realized that this life is the only one they really know of, and it is, they would be more careful in wisely spending it as long as possible. Why sacrifice for a future state of being dead when we are already making all sorts of sacrifices in order to stay alive? Isn't it strange that people will make a hard bargain over trivial purchases, and yet pay out every week of their lives for something they don't know anything about and which they only expect to receive after they die? All progress must be based on the ideal of constructing a heaven on earth and allowing the angels to shift for themselves. Many people are always wondering whether the world began last Saturday and whether it'll end next Sunday, forgetting in the meantime to take into account the rest of the week.

It is the stretching of life to limitless eternity that makes the hour so valueless. Please remember, the Sabbath day is every day and every day can be your only day, for yesterday is no more and tomorrow may never come.

Organized religion depends on blind belief and rigid conformance. Conformance occurs when a person does not allow himself the privilege of making up his own mind. Credulity is a state in which a person is afraid to think on the grounds that it might make him live better. To remain faithful to a belief just because it is a belief, without any attempt to analyze its just, equitable, and true nature, is not the mark of a sensible person. It is neither intelligent nor intellectual. Dogmatism and religious freedom do not go hand in hand for, truly, the freedom of religion means that one can deny, another affirm, what still another doesn't even want to be bothered about. J. B. S. Haldane wrote, "There is a worse evil than intellectual starvation, and that is the deliberate suppression of free thought and free speech." [52] It is said that there was a man who was so serious about his belief in hell that he started digging. He was the first to actually venture to go down to prove the existence of hell. He struck oil. Now he is a millionaire and he's forgotten all about hell!

What is human is not necessarily universal. Man is an imperfect animal and will remain so as long as nature exists, because nature is motile, and adaptation and variation go on constantly. Man evolved by the same process and principle as all else in nature. Those who think this universe was made *ex nihilo*—out of nothing—should try to quench their thirst by not drinking. Even the gods have been imperfect and have no free will but must repent upon the jests and sorrows they have "created," and they had to be simply satisfied with an imperfect world, regardless of their wishes. Sébastien Faure, the great entomologist, wisely wrote, "Perfection cannot determine imperfection." [53] . . . "The multiplicity of gods is proof that none exists." [54] "Just as our science of biology," wrote Joseph Campbell, "came to maturity only when it dared to reckon man among the animals, so will that of mythology only when God is reckoned among the gods." [55]

For many milleniums human savages throughout the world believed in *magic,* all kinds of it, sorcery and taboos, sympathetic or homeopathic, positive and negative, white and black magic, contagious magic, and many other classes of magic, all of which funneled itself into good and evil spirits, demi-gods, gods, and finally into a supreme being idea with his partner in co-existence, the devil. In savage times, souls rambled about as free spirits, especially when people were asleep and very often the medicine-man had to go out to recapture one, as a wayward child, and bring it back to its owner who, of course, paid for bringing it back. "It is a common rule," Frazer tells us, "with a primitive people not to waken a sleeper, because his soul is away and might not have time to get back." [56] "Lest the soul of a babe should escape and be lost as soon as it is born, the Alfoors of Celebes, when a birth is about to take place, are careful to close every chink and cranny in the house, even the keyhole; and they stop up every chink in the walls. Also they tie up all the mouths of all animals inside and outside the house, for fear one of them might swallow the child's soul." [57] "The people of Nias think that every man, before he is born, is asked how long or how heavy a soul he would like, and the soul of the desired weight or length is measured out to him. The heaviest soul ever given out weighs about ten grammes. The length of a man's life is proportioned to the length of his soul; children who die young had short souls." [58] "In Transylvania they say that you should not let a child sleep with its mouth open, or its soul will slip out in the shape of a mouse, and the child will never wake." [59] "When an Indian of Brazil or Guiana wakes up from a sound sleep, he is firmly convinced that his soul has really been away hunting, fishing, felling trees, or whatever else he has dreamed of doing, while all the time his body has been lying motionless in his hammock." [60] "Amongst the Alfoors one way of recovering a sick man's soul is to let down a bowl by a belt out of a window and fish for the soul till it is caught in the bowl and hauled up." [61] "The son of a Marquesan high priest has been seen to roll on the ground in an agony of rage and despair, begging for death, because some one had desecrated his head and deprived him of his divinity by sprinkling a few drops of water on his hair." [62]

Edward B. Tylor, considered by many scholars the father of anthropology, adds his comments about the soul: "Africans think that souls of the dead dwell in their midst, and eat with them at meal times." [63] "Men who perceive evidently that souls do talk when they present themselves in dreams or vision, naturally take for granted at once the objective reality of the ghostly voice, and of the ghostly form from which it proceeds." [64] "The notion of a vegetable soul, common to plants, was familiar to medieval philosophy." [65] "The conception of the human soul is, as to its most essential nature, continuous from the philosophy of the savage thinker to that of the modern professor of theology. Its definition has remained from the first that of an animating, separable, surviving entity, the vehicle of individual personal experience. The theory of the soul is one principal part of a system of religious philosophy which unites, in an unbroken line of mental connexion, the savage fetish worshipper and the civilized Christian." [66] "In the Hervey Islands," writes Dr. H. R. Hays, "fat men had fat souls, thin men had thin souls. Sorcerers used fiber loops to capture them, large for the fat souls, small for the thin ones. . . . The Andamanese thought of a reflection in a mirror as the soul, and feared cameras, lest the white man, by capturing the reflection, take their souls away in his black box." [67] "Cannibalism," states Dr. E. P. Evans, "originated in the belief that the soul resides in the blood, and that by drinking the blood of the bravest foeman their courage, cunning, and other distinctive and desirable traits may be acquired and thus serve to increase the fighting force and efficiency of the tribe." [68] "We know much more," satirizes Bertrand Russell, "about the education of salamanders than about that of human beings, chiefly because we do not imagine that salamanders have souls." [69]

The magic of the primitive savages predated the original

priests or medicine-men. Magic originally was the prerogative of every person and the principles of its operation were entirely related to natural things and events, not supernatural realms and spirits. Take for example the primitive way of getting rid of pimples. "If you are troubled with pimples, to watch for a falling star, and then instantly, while the star is still shooting from the sky, to wipe the pimples with a cloth or anything that comes to hand. Just as the star falls from the sky, so the pimples will fall from your body; only you must be very careful not to wipe them with your bare hand, or the pimples will be transferred to it." [70] These are the cures that superstition endowed man with. "Where the mind is filled with superstitions," wrote Lester Ward, "supernatural causes and spiritual beings, all of which are synonymous, there can be no science, no knowledge, no attempt to control phenomena, and we revert to savagery. Since the scientific era began, there has been no such faith in the supernatural as exists among savages. Science was made possible for the diminution of this kind of faith and the concomitant increase of faith in natural causes. The history of science shows that those who still possessed a large amount of the faith of primitive man, opposed science and stubbornly resisted its advance." [71]

Milleniums later, it seems, with the increase of man's mental capacity and a greater degree of self-consciousness as an individual, it dawned on him that magic somehow or other was not as successful or fruitful as it was cracked up to be. When the first priest became initiated, it started a class of leaders and kings who gradually came to be the only agents who knew magical words, who had the authority to pronounce them, and who could command nature and even the gods to do their bidding. To this day this concept of the priest's power still continues even in our own patterns and Western cultures and, of course, in most of the world. Whether it is the priest, the brahma, the rabbi, the minister, or the king and queen, the people of these religions and cultures throughout the world and through history, have looked up to these agents as the actually appointed and authorized representatives of the respective gods and only through whom was it possible to appeal to the gods or gain their favor or reward.

While magic still persists in all the cultures of the world in myriad customs and superstitions, the emergence of the spirit-world, and finally the concept of heaven and hell, created something the pre-religion savage never dreamed of. Here was a situation where a heaven awaited each person, a living "forever," an eternal stretch of perpetual happiness! While "science cannot admit the immortality of the conscious soul, for consciousness is a function of special elements in the body that certainly cannot live for ever," [72] yet the "rulers hope that the promise of 'pie in the sky' will make men willing to live on rice, turnips, and potatoes now." [73] To ensure conformance the fear of hell kept everyone in line to abide with the requisites of each religion and which, of course, was to perpetuate the power and the prestige of the priesthood or kingship. Originally, each human was his own magician, even his own priest, then it passed into the specialized services of the tribal medicine-man or head magician, who later evolved into the earliest chiefs or kings. During this period of slowly changing society, and this took thousands of years, the idea of gods, goddessess, heavens, hells, and all the smaller scales of saints, angels, demigods, etc., came into being. "The conception of gods as superhuman beings endowed with powers to which man possesses nothing comparable in degree and hardly even in kind, has been slowly evolved in the course of history." [74]

Why should all this detail concern us here in our search for a philosophy of Time? It is obvious. This, clearly, is of paramount importance because we can never have a just and honest evaluation of Time so long as we can never use it as a measure or as a screen of finity against which we can project our actions, our hopes, our plans, and the world we find ourselves in. The more conforming to his religion a person is, the more is he likely to lessen the value of this life for the life in some hereafter. To the same extent he is lessening the value of his time on earth. It goes even further than the field of religion, because this lessening of the value of time penetrates, as a result, into all the other activities of living. It becomes a life pattern, or we may call it a dying pattern, inoculated into the young, which, if it isn't thrown off at an early age by enlightenment, becomes a malignant cancer which replaces the normal "tissue" of Time for a mass of incomprehensibility. "The fear of ghosts has led race after race, generation after generation, to sacrifice the real wants of the living to the imaginary wants of the dead. The waste and destruction of life and property which this faith has entailed are enormous and incalculable." [75] "Obviously, the belief in a happier hereafter is a great compensation for the unhappiness of this life and a considerable support to the tottering buttresses of one's consciousness of insecurity. It is this aspect of religious belief which has been called 'an opiate of the masses,' tending to make them uninterested in improving their lot on earth." [76]

It has been said that if all the religions were obliterated in one sweep, man would be more cruel, more dishonest, less moral, more selfish, less tolerant, etc., than he is now. True, they say, evil cannot be extinguished but there would be far more evil and cruelty were it not for man's fear of his god and fear of the punishment resulting from his sins.

History does not bear this out. Professor Morris R. Cohen, the great liberal and a great teacher of philosophy with affection, in his *Meaning of History,* writes: "Faith in the supreme value of righteousness and the pursuit thereof does not depend upon any religious belief in a personal God or other supernatural influence in human affairs." [77] "Like grass," writes Dr. Samuel F. Dunlap, "we are born from a parentage, not from a philosophy or a theology." [78] John Dewey plainly states that "men have never fully used the powers they possess to advance the good in life, because they have waited upon some power external to themselves and to nature to do the work they are responsible for doing. Dependence upon an external power is the counterpart of surrender of human endeavor." [79]

The ages of savagery, during which time man tried to live by magic, were at least natural and empirical, that is, the radius of all his daily living and all his hopes and dreams were limited to this earth, this life, and if he extended a spirit beyond his death, it appeared to him to be a continued phase of his earthly life, often an intermixture, and the spirits, all around him and the spirits of his environment, were all in physical contact with him, dead or alive, and the powers of the living nature and the powers of the spirit nature were one homogeneous process of idea-associations. "Naturalism," wrote Professor John H. Randall, Jr., "has traditionally protested against several types of metaphysical dualism. It has denied the religious supernaturalism that has set up God as a supreme Reality or First Cause." [80] If the age of magic had skipped the age of religion and thus shortcutted to the age of science, the human world would not have been sunk in darkness for so many thousands of years. But this didn't happen. Somehow or other, out of the age of magic evolved the age of religion, which separated completely the mortal world and the immortal world and put the priests in charge of the bridge between. Thus started the greatest fraud in human history and also started the biggest business.

It is during this period of the age of religion, which is still continuing, that the greatest extent of cruelty, hate, massacre, selfishness, intolerance, immorality, slavery, war and human wastage, existed and still persists to exist wherever religion finds itself strong enough to supersede law and where it can overcome the democratic principles that protect the rights of minorities. It is impossible for me and for you to really feel the contemporary pulse of living in the Dark and Middle Ages of European history. But whatever historical evidence is offered to us is sufficient to make us shudder and to be happily grateful that we happen to live now instead of then.

Then the skies were filled with demons and angels, foraging constantly over the housetops, each sneeze was a discharge of six platoons of dangerous imps and ogres, vicious little devils tearing away and playing tug-o-war with the entrails of people. From awakening to bedtime each hour was controlled by some religious activity of a kind. The priest was a mortal diety who was capable of judging a person to eternal fire, even could make the god do his will. This life, this time on earth, was of no significance; it was merely a brief period of testing one's soul for the eternal, never-ending bliss of paradise or eternal damnation. This fear of the unknown, this fear of unpredictable judgments, this fear of the priest-deity was not only the greatest mass neurosis of humanity, it was also the greatest perfidy inflicted upon people by wasting their lives away upon the altars of ignorance and blind, slavish faith. Oscar Wilde pleaded the cause of the victims. "The terror of society," he said, "which is the basis of morals, and the terror of God, which is the secret of religion—these are the two things that govern us!" [81] "Religion is a disease born of fear and is a source of untold misery to the human race." [82] "I believe," wrote H. L. Mencken, "that religion, generally speaking, has been a curse to mankind—that its modest and greatly over-estimated services on the ethical side have been more than overborne by the damage it has done to clear and honest thinking." [83] Professor James Harvey Robinson, one of the great historians of our period, wrote: "To the candid historical student the evil workings of religion are, to say the least, far more conspicuous and far more readily demonstrated than its good results." [84] Dr. George A. Dorsey simply states, "Religion is a disease." [85] Sigmund Freud: "Religion is an obsessional neurosis of humanity." [86] Albert Schweitzer: "Christianity is dualistic and pessimistic. . . . Christianity abandons the natural world as evil." [87]

During these many centuries men's minds and hearts were not made more kind, more merciful, compassionate, sympathetic, neighborly and tolerant. It was just the opposite. They slaughtered and massacred each other, enslaved and became slaves. "At least 10,000,000 Jews, Hindus, Mohammedans, Armenians, etc., have been murdered by their religious rivals since 1900." [88] Each creed and sect murdered each other with cold indifference. Popes poisoned their enemies and kings built bastilles and dungeons. During this period every conceivable contrivance to torture and to kill was ingeniously created. People were slowly roasted alive, while around the stinking pyres stood priests, monks, nobles, and people behind them to watch these things. A person must be of the lowest stupidity not to see that such a religion could not be a godly one if we assume that God means peace, pity, mercy, and compassion.

During these many centuries every progressive move of man, whether it be in medicine, democracy, abolition of slavery, freedom of trade guilds, sciences, astronomy, and a hundred others, was thwarted and crushed down constantly by the religious forces in power. Andrew D. White, co-founder and first president of Cornell University, wrote that because of the precepts of St. Augustine "there were developed, in every field, theological views of science which have never led to a single truth—which, without exception, have forced mankind away from the truth, and have caused Christendom to stumble for centuries into abysses of error and sorrow." [89] Professor Burnham P. Beckwith states: "Perhaps the most unfortunate result of religious indoctrination is that it makes men authoritative and dogmatic. It teaches them to rely upon traditional doctrines and prejudices rather than upon their own observation and judgment or upon that of qualified scientists. This discouraged new scientific research on problems of personal and social conduct." [90] Professor H. A. Overstreet writes: "Most of the traditional thinking of men has suffered from the infantile fallacy of anthropomorphism, the fallacy, namely, of conceiving the forces of nature in the image of the human being." [91] Arthur C. Clarke writes in *Horizons,* regarding exploration of life on other planets, commenting: "When we make contact with superior extraterrestrial intelligences we shall find that belief in a supernatural order of things marks an early stage of development amongst most rational creatures and perishes with the rise of science. Most disconcerting of all would be the discovery that man alone is a myth-making animal, forever impelled to fill the gaps in his knowledge with fantasies." [92]

Volumes upon volumes have been written and could be written which would only nauseate us in reviewing the history of religions throughout the world. But this is not our purpose here. Here we are trying to find some realistic analysis of the nature, meaning, and value of Time. It is sufficient for us to realize that religion did not increase or project the true meaning of Time. It dissolved the "substance" of Time into a wish fantasia. It tapped the dog with the bone in its mouth and directed it to drop it into the mirage of a bigger one, only to result in losing all.

Religion divides the world into a thousand camps of intolerance, each camp intolerant of the other; and if for a moment there is peace, it is not because of the religions but because of the political and societal laws, police power, and fear of civic punishment, that keep the creeds from tearing away at each other as they have done so often before. So much kindness, morality, honesty and human sympathy were taught to people by religion that, if the law and police were removed, there would hardly be a store or home unlooted and hardly a person could walk the streets without being thugged and robbed, raped or murdered. A person who has little or no value for his own time on earth has little or no value for the time of other people. The meaning of peace is understood by those who value their lives so much that they get to understand that life is also valuable to their neighbors. True peace is the culmination of the understanding that Time is more enjoyed in love and friendship than in fear, fantasy, and feud. Sir Arthur Keith made this point well when he said, "If men believe, as I do, that this present earth is the only heaven, they will strive all the more to make heaven of it." Religion has been successful because it has built its towers upon the fears and fantasies of men and used the weapons of hate and hysteria to carry on crusades and atrocities upon the human race. Someone wrote: "Be wary of the theological man! He is either a fool or a faker. While he is mumbling prayers he may be planning to rob you, dead or alive!" Gaetano Mosca wisely observes that "in societies in which religious beliefs are strong and ministers of the faith form a special caste a priestly aristocracy almost always rises and gains possession of a more or less important share of the wealth and the political power." [93]

Strange as it is, when the power of the clergy and Church were stronger than all the kings in Europe and dominated the political and economic life of that Continent, in no instance did they raise their voices to free the serfs, to stop the slaughter, to affect public and private sanitation and education, to promote constitutional, democratic government for the people instead of standing behind the power of despots and despoilers. Strange it is that such reforms finally came about by people who fled from religious persecution and who fought its power and who created constitutions by which the peoples became free of the power of the clergy and free of the power of kings. This is history and not a sermon flavoring the froth of human credulity on Sunday morning by taking their hard-earned money in exchange for "something" neither priest nor pope nor parishioner knows anything about. "I cannot understand," writes John Dewey, "how any realization of the democratic ideal as a vital moral and spiritual ideal in human affairs is possible without surrender of the conception of the basic division to which supernatural Christianity is committed." [94]

The religious man will often say: "Well, no matter how true or false is the idea of god, religion, etc., look at the goodness and fine living it has accomplished. While things have always been bad, without religion it could have been infinitely worse. Religion has been like a dam, holding back the floods of human ferocity and selfishness from annihilating the race. Even upon the assumption of false grounds, it has built up a moral imperative of *oughtness,* a distinction between good and bad which the savage did not possess. . ."

"It has evolved," the theologian will continue, "a system of moral principles for the good of man, for peaceful living, and has ritualized a mode of life under the Golden Rule. Without it," he will contend, "the relationships of men would be as-catch-can and unorganized. Without it life would be a road without purpose or destination; a state of things without meaning. . .

"In brief," the theologian will go on, "without religion the world would sooner or later drift into a chaotic form of nihilism in which force would spend itself upon its own destruction and dispersion."

The man of theology may continue and say thusly: "Man is friend unto man because both fear God, and it is God's will that peace should prevail and good will flourish among the peoples and nations. If the fear of the Lord will accomplish peace and good living, then it is good for mankind, as the means justify the ends, and vice versa, and disregarding any physical or logical basis for the truth or falseness of the doctrine."

The man of theology may go on endlessly, claiming that a man without religion is like a ship without a compass or rudder, just drifting on a sea, without direction or purpose. He will claim that religion has been a comforter for the poor and miserable, a great hope for the aged who can look forward to something bright and happy after death, irrespective of the truth or falseness of the doctrine.

The man of theology will claim that it has brought individuals and groups more closely together, and centralized peoples into a common thought. Therefore, he will claim, religion has been a prime factor in civilizing humanity for the sake of peace, progress, and of still greater importance, it has instilled into the hearts of any religious group a common and agreeable relationship for their mutual guidance and comfort.

If these claims were not mere spouting of oratory, but were proven by a sincere bulwark of historical investigation, there might be applied some sort of pragmatic justification for the existence of religion. But it is not so. History reveals otherwise.

The sphere of influence that religion, superstition, metaphysics, dogmatic ritualism, mysticism and occultism, have spread over the story of man's descendancy from prehistoric times, is so tremendous and powerful and penetrating that it has affected in major ways the entire emotional and cultural patterns and circulation of human events. It plays a direct, indirect, and influential part during every day of our lives. Its radius of play covers the widest section of society, of individual relationships. It is a definite factor, hopes and resolutions notwithstanding, in the choice, association, and living philosophy of most human beings. It is for these reasons that we should inquire as to the origins of these philosophies and their effects upon the human mind in its search for the most perfect potential of living and association. We have a right to ask what they mean to the individual and to society. Winwood Reade, in his classic *The Martyrdom of Man,* stated: "Doubt is the offspring of knowledge; the savage never doubts at all." [95] Professor Friedrick Paulsen also aptly stated: "Philosophy is the product of the inquiring and thinking mind; the mythico-religious conception of the world is the product of poetic fancy." [96]

History disproves beyond any doubt that religion has been of any encouragement to live better, not only between man and man, but also between man and woman, and between them and their neighbors and the world. On the other hand, it has depressed and discouraged these natural relationships, constricted them, and blinded man's mind so fiercely that all the beautiful things in life, as well as living itself, have been thrown to the discard while the ego has been falsely inflated by the hopes of immortality and heavenly bliss. Dr. A. D. White relates that "out of the Orient had been poured into the thinking of western Europe the theological idea that the abasement of man adds to the glory of God; that indignity to the body may secure salvation to the soul; hence, that cleanliness betokens pride and filthiness humility. Living in filth was regarded by great numbers of holy men, who set an example to the Church and to society, as an evidence of sanctity. St. Jerome and the Breviary of the Roman Church dwell with unction on the fact that St. Hilarion lived his whole life long in utter physical uncleanliness; St. Athanasius glorifies St. Anthony because he had never washed his feet; St. Abraham's most striking evidence of holiness was that for fifty years he washed neither his hands nor his feet; St. Sylvia never washed any part of her body save her fingers; St. Euphraxia belonged to a convent in which the nuns religiously abstained from bathing; St. Mary of Egypt was eminent for filthiness; St. Simon Stylites was in this respect unspeakable—the least that can be said is, that he lived in ordure and stench intolerable to his visitors." [97] Alas, the animals in the woods know better!

We have noted that originally the primitive's purpose in the belief and appeasement of spirits was for his immediate comfort, for his earthly existence and not for any life completely isolated from this mortal life and beyond the grave. Any idea of life beyond the grave was something he didn't understand or didn't think about. The very spirits of his dead ancestors were part of *this* world, living among his tribe, helping or harming him according to their desires and his cunning. The dead ancestors fought with his tribe against the enemy and their ancestors. The gods of Greece went forth to battle with the Greeks. Yah-Weh, later Jehovah, was also a war-god and went to battle with his followers. The spirit or god of food helped the vestals in the technicalities of table delicacies. In general, these spirits or gods were not apart from this earthly existence but part and parcel of man's everyday existence. As yet, they were not elevated by any abortion of Aristotelian inference to a heavenly bench of super-judgment. So far the worship of the ancients, while steeped in superstition, ancestor-worship, mana and nature-worship, was understandable in the sense that, so long as they didn't possess the knowledge to properly understand these things, they personified them and worshipped them in sundry ways so that their *own* lives in *this* world may be bettered and more prosperous. The more they became steeped in their blind beliefs, the more tortures, sacrifices and crimes, were committed in their names. But when the mere brief moments and comforts of this life were brushed aside for eternal bliss in some hereafter, religion "became of age" and found the greatest key to the weakness of man, his docility and gullibility. As we described before, the dog dropped the bone to catch its reflection in the water, only to see them both disappear. "It is the people with their hands in the till and their eyes on heaven who ruin existence. There should be open-air temples in every town and village where philosophers can expound this soundest of doctrines. Why is half the population tormented with restraints, obedience to which in no way furthers the public good? Because the priests for generations have been confederate with the money-makers, and they both know very well that if natural happiness were allowed the generations would no longer accept their shrewd worldly maxims, no longer be docile, so easy to be exploited." [98] Theodore Dreiser once wrote that religion is "only an illusion of the rankest character, yet which for the many at least has served as a nervous or emotional escape from a condition too severe to be endured." [99]

The history of religion disproves the notion that it has been of any civilizing influence, if we mean by civilization the nurturing of a genuinely higher, finer, freer, and more peaceful, moral and equitable character of living and not merely the building of cities, walls, bell towers, etc. On the contrary, it has been a very great retarding agent. The historical development of democracy, of medicine, of physics, of astronomy, of the political and economic sciences, of any of the sciences, reveal that religion was their greatest enemy. Religion never reared its influence against the hatreds of centuries. On the other hand, the bacteria of religion fostered hatred, caused wars, massacres, tortures and misery, so much of it that the books of human history overrun with the innocent blood of millions of people. Religion has never been able to avert a single war. Professor Morris R. Cohen said, "The God which is on the side of the heaviest artillery is not necessarily the God of love and righteousness." [100] And J. H. Denison wrote, "War is never so sublime, never so heroic, and never so brutal, as when it is inspired by religion." [101]

Religion has never stopped a slave ship and made the captain return the captives to their homes. It has not influenced the kings and queens and tyrants of thousands of years of history to improve the social and economic status of the poor and the common man. On the other hand, it has taken away from the poor what little they had left. It has also blessed the slave ships to have successful trips. "The governing classes tend to be exceptionally religious, because they wish to regard the misfortunes of their victims as due to the will of God." [102] Briffault tells us that the "Chaldean priests helped themselves to first fruits and death duties, that is to say, they plundered the farmer and left the widow naked." [103]

History reveals that religion came up from the most brutal periods and was most influential and powerful when the age was most tyrannical and murderous. History reveals that as the spirit of social, political and individual freedom expressed itself more and more, so much did the influence of religion decline. The theologian cries aloud that wars appear when religion is discarded; this is pure nonsense and is intended for consumption by people who never read a book. Even the apologists wouldn't dare to put forward this false and entirely misleading statement. On the other hand, religion intensified struggle, gave rise to the bloodiest of religious wars, to poverty and unrest, and thus additional struggle and war. In many ways religion has been a boiling pot, always stirring this emotional, predatory and merciless cataclysm within the general history of mankind.

Let us examine the "civilizing" influence of religion further. We find that it has subordinated man and his life to a hope or faith in something which has never been demonstrated or substantiated, and which has not been proven by anything objective, scientific, or rational. This subordination of man's life to a higher power and an ethereal bliss, has promoted a constant devaluation of the only life we know—the one we live. Religious extremists and fanatics have withdrawn from normal life and its functions, have denied themselves even as much as a glance at the sunshine or the clear sky, and have subsisted as mere pawns of prayer and ritual. Millions have sacrificed themselves upon the altars of the gods, waves of ascetical depravity, waves of ritual suicides, and the destruction and misery of other millions who disagreed with them in the least degree. It cut man off from the rational processes of normal and reasonable living and desires, and built up, instead, a fantasia of dreams unreal as well as untrue.

Take the subject of friendship among people. Actually, friendship is a very irreligious thing. It has not seen the golden portals of heaven or smelled the frankincense of the bishop's pot or kissed the Blarney Stone. Were it part of the supernatural realm, the human race would have long ago annihilated itself at the hands of class and racial hatred, and yet it might succeed in doing so without it. In the furious torrents of our history, furious with the stupidities of exalted vagaries, blind ambitions, and cloud-straphangers, here and there friendship at least cooled the sweaty brow of man before he went on again, bent forever on dashing out his ill-used years against the beautiful walls of his own Valhalla. He appears dazed at his own shadow projected against the spectrum of nature. He stands before the aurora borealis of his all-too-hasty life; instead of evaluating himself by its definite measure, he tries to encompass the horizon with his arms, only to strangle himself.

Accordingly, and strangely and not too frequently, we do know that friendship exists with us, with the *knowable* world, although many have applied its name to countless misconceptions. It is purely a mortal thing and we must be honest and content ourselves, in our attempt to learn what it is, with relegating our search to mortal things. Heaven does not need it and hell will have none of it. It belongs solely to our own little world on which we find ourselves, willing or not willing but nevertheless, much alive. We seem to know, quite frankly, that we are the most intelligent beings on earth, if not the only intelligent ones. However, no matter what our conclusions might be, they arise from the ceaseless experience of *being*. And in order to *be* what we are today, for the better or for the worse, we know from acquired knowledge that we have evolved up from lower or down from higher forms of life, as either poet or cynic might decide, but evolve we did. If a god did create us, as the preacher would want us to believe, the divinity has surely shown us very little friendship since.

God is not only on the side that has the heaviest cannon but also with those who possess a greater cunning and indifference to the welfare of others. This might seem to many a violent accusation against the divine host, but it also seems strange that the victor often feels that God is on his side; the loser feels, in a whisper, that God has let him down; the cruel and tyrannical often feel that God must really favor their means simply because they have been so successful. "Pope Alexander VI was the wickedest man known in history, but he had great and unbroken prosperity in all his undertakings." [104] "The crusaders were greatly perplexed by the victories of the Mohammedans." [105]

Almost every tyrant in history has either appointed a god as his chief minister of affairs, or has adorned the robes of a god himself. However, so far as we have gone, we have not noticed any specifically religious ceremonies among the other animals, insects, and plant life. Besides, if anyone really thinks that no one but a god can make a tree, as a so-called "brilliant" poem tells us, it is only because we have to allocate a power which we do not possess to something which we believe in as closely related to us as possible. Incidentally, the devil is also a close relative of ours, but somehow or other we do not allocate to him the creation of things we like but things we do not like or which we are told we should not like but continue liking, despite the instructions. Whether it be a tree, a bug, or a fern, they are all wonderful so far as trying to duplicate them is concerned, good and bad things alike. It is sufficient for us to realize that a tree, a sunset, a snowflake and the ripple on the stream, are all relatives of ours and they are all part of our family tied by the umbilical string of Mother Nature.

The worship of a single god, instead of many, is purely a modern idea. If we are to be impartial and weigh the various religions of the world today on a balance of one god or many, we will find that most of them, if not all of them, worship many more than one. It is innate in human nature never to close the door completely on the expansion, change, or reinterpretation of worship. Gods have multiplied or vanished as the whim of human nature desired it.

We find that early primitive religions have associated the sex life with their religious ceremonies. Lee A. Stone writes regarding this factor: "What is unknown, what is mysterious, is the basis of the clustered sentiments and emotions that we call religion. . . . All the evidence collected from the beginnings of history, all the later investigations conducted into primitive customs and beliefs, have shown that the worship of the human generative organs, the protagonists in the life drama, is or has been universal." [106] The fact is that religion, as we know it today, owes much more than we think to the sexual act, and volumes upon volumes of data have been gathered and published in regard to the interrelation and synonymity of sex and divine worship. Every church door, every steeple, and countless other structures which are sacred today, are nothing but phallic symbols. The swastika, the mogden dovid, the cross, the Egyptian sistrum of eternal life, the Hindu yoni symbols, and a thousand more, are all ideas that came to the human mind to symbolize and consider sacred the creation, adoration and unity of life.

However, sex did not probably have as much to do with primitive religion as it had later on. Human nature is full of predatory instincts because man at bottom is a predatory animal. He was a brute in the raw, although primitively not warlike. Only the regulation of social relationships by law has harnessed to a limited extent the ever-present brutality of human instinct. The primitive feared *many* things, either because he realized by unfortunate experience that he was too weak or helpless to cope with them, or because the things he feared were beyond his reach or understanding. This lack of power plus the lack of understanding gave rise to the first principle of appeasement or compromise with nature. The things that man could grapple with, overpower, enslave or control, received very little worship if any at all. He treated these things with indifference, and he still does. Those things which he could not control but which controlled him and upon which his very existence depended, these things he treated with cunning, appeasement, trickery and compromise of pseudo-friendships. This basis of appeasement, which in biological terms might be deemed nothing but the adaptation of a growing mind and body to its environment, might have been the first signs or causes of institutionalized tribal custom. Man, as yet a social automation with no self-consciousness as an individual, fitted into these patterns upon which the existence of the family group, tribe, clan or community, depended. Even today, where primitive people still exist, one will find that tribal custom has little to do with esthetic or moral experiences but mainly, so far as religion is concerned, with the fear and appeasement of things which do harm or may do harm to their actual existence or upon which their very lives depend.

The arts of the early primitives show that they had more to do with the constant experiences of living and the things they lived with, rather than with the adoration of gods and goddesses. These divinities came milleniums later and only after the human mind had broken through the prime chrysalis of tribe submergence and into the first earliest stages of self-identifiable individualism; that is, when the individual recognized the value of his own life apart from the lives of others. From this beginning came the gradual attempt to spiritualize in order to extend his personality to eternity. He did not realize that, while his purpose was rational, his method and direction did the very reverse of his intentions. Instead of bettering the few years of his life, the "spiritual" extension of the personality to infinity diluted his existence and thinned the fullness of earthly joys into the bleak, blank spaces of a contorted mind that believed more in what he dreamed at night than in what he actually saw during the day.

When belief in spirits and supernatural beings was in the making, the group or social masses of people grew into it because this belief aided them in the constant struggle to exist. So long as tribal dogmas went along these lines the people also went along with it even to the extent of sacrificing their very life-blood, their own children, their most precious possessions, so long as it might appear to benefit the group of which they were a part. Individual sacrifice, as we use and consider the term, had not yet been fully felt or known; everything was still in terms of the group or tribe.

The idea of eternal life in heaven or hell is, as we have stated, comparatively a fairly modern idea and came as an offshoot of religion-evolution rather than one of its causes or prime purposes. So long as man was interested in his little communal affairs from sunrise to sunset, he worshipped and sacrificed so that he might live better. But the idea of eternal life or life of the ancestors in the spirit world, and then into two worlds of heaven and hell after death, gave rise to a devaluation of life and the search for how to die better. It can thus be seen that while the beginnings of superstitional belief and tradition were born of man's fear of and desire for immediate things, they demoralized into a wild dream and scramble for eternal life. Is there any objective, scientific basis to sustain this dream? None whatsoever. Dr. A. J. Carlson, President of the American Association of Science, states clearly and emphatically: "Has science anything to say on the theory of personal immortality? The idea of persistence of the individual after physical death came down to us from the ancients in most if not all races. What credibility are we going to give to the idea solely because of its venerable age? So far as I can see, we can give no greater credibility to the ancients' views on immortality than to their views on other things about which they knew nothing. Conscious phenomena and intelligence in man, that is, personality, appear to be just as much an evolution of the material world as is the rest of the body processes. We seem to be forced to this conclusion from the evidence of the intimate dependence of all phases of consciousness and memory and personality on the quantity and quality of the nervous system, and these, in turn, depend on all the rest of the body mechanisms. . . . On the basis of the known and the probable, immortality of the person is at present untenable." [107]

Let's consider another change that took place in those early times. In the pupa stages of religious belief, man's feelings, whims, the animals about him, plants, objects, elements, these were symbolized and later worshipped in the order of how man used or feared these things. As the discovery of the individual expressed itself, these symbolisms, originally mere words of designation, were translated and transformed into human actions and forms. The tree became a tree-god who looked like a great human with many branches, the symbol of creations of strength, of stability, of a great world-mother. The thunder became the thunder-god with fearful strength and a thunderbolt in his hand. The wind became the wind-god, blowing warm winds when appeased and cold winds when angry. The sex act became a sex-god with a phallic emblem that women would prize and adore. Wine became a wine-god, being an animal or man, or both, of good cheer, always drinking, laughing and drunk. The lower half of the goat, being an animal of unusual and inspirational virility, was joined with the upper and directional portion of man's body, and became a satyr, one who shows the women the most enjoyable paths through the pleasurable woods of existence. The ancient sea, with its constant and ceaseless tides, became a sea-god, a great old, scrawny man, with long wave-swept whiskers and seaweed in his hair. The desire for war became a war-god, a terrible, cruel and courageous man with plenty of power, armor and sword. We could go down the line and continue describing the simple and endless anthropomorphic concepts of ancient man. He felt that these gods, the compound of which he discovered in his own personality, were really physical beings or things in similar likeness to himself, but endowed with supernatural powers, the favor of which required worship, faith, or obedience in myriad ways and forms. Professor Overstreet wisely counsels: "We must never again commit those childish follies of translating the great forces of reality into human forms. We must not crowd the hills and valleys with spirits, make the moon into a pallid female, or the sun into a god of flaming beauty. We must not have deities descending unexpectedly out of the heavens, changing the course of things. We must have no Jehovah walking in the garden, or delivering tablets of stone on a mountain top." [108]

The Garden of Eden was not the only place where gods were born. They sprouted everywhere and anywhere where humans trod or settled. Let's take a couple of simple examples. For instance, *fire*. Among the Aztecs fire was a god called *Xiuhteuctli,* or Lord of the Fire. The Delaware Indians called him *Manitu.* In Polynesia it was *Mahuika.* In West African Dahome it was *Zo.* The Ainos of Yesso called him *Abe Kamui,* and way up in Mongolia it was just *Ut.* In India his name was *Agni,* from which we get the Latinized word of *ignis,* to ignite. Old Persia called it *Ized* and in Circassia it was *Tleps.* We can keep this up for another page but it isn't necessary.

Let's take another example, the *good earth.* Among the Algonquins *she* was called *Mesukkummik Okwi.* As the earth gives forth food and food gives life or sustains it, the earth was usually considered a mother. The Peruvians called her *Mamapacha,* the Mother Earth. In China she is *Tu* and in India *Prithivi.* The Aquapim in West Africa called her *Bosumbra* and the Bygah Tribes named her *Dhurteemah,* also Mother Earth. The Khonds of Orissa titled her *Tari Pennu,* the Earth Goddess. The Finns had their *Maan Emo;* the Greeks had *Demeter;* and the Romans, *Terra Mater,* and even the ancient Germans had their *Nerthum,* a buxom strong woman driving a chariot drawn by cows.

One more example, the *sea,* the giver of floods, of food, of means of transportation, etc. The Peruvian Incas called her *Mama-Cocha.* In eastern Siberia it was *Mitgk.* The Eskimos had their fish-goddess who brought them seals and fish. The Japanese have two,

Midsuno Kami and *Ebisu,* the latter usually depicted as a gay, fat, laughing fisherman with his fishing pole. In Bali the sea-god is a demon, the equal of the devil in whose domain live all the bad spirits.

Achor was the Cyrenean god of flies and *Hobnil* was the god over the beehives for the Mayas. The Finnish god *Agras* was the protector of turnips and the Peruvians had their *Axo-Mama* who protected potatoes; *Fejokoo* is a Nigerian god who supplies yams to his people, and *Pani* is a Maori god over sweet potatoes; *Xilonen,* the Aztecan goddess of the young corn demanded human sacrifices in order to persuade her to allow the corn to mature, and the Hawaiian shark-god *Ukupanipo* was the head man to appease for good fishing. The Romans had a goddess *Fama,* who promoted and specialized in gossip and rumors about town, while the Egyptians worshipped a god of silence, *Harpocrates.* The Romans had a goddess called *Cardea* to protect their door hinges, the Chinese had their *Shen-Shu* to protect their doors, and the Hindus have a god *Vattuma* to protect the doorstep, while the Romans worshipped *Laverna,* the goddess who protected and "mothered" the thieves. And to clean up the mess the Aztecs had their goddess *Tlazolteotl,* the Eater of Filth. On and on goes the parade of gods and goddesses for this and for that, here and there, almost everywhere.

The rest of the story is the unfortunate deterioration of these beautiful, esthetic and simple fairy tales and metaphorical allegories and the gradual abortion of primitive fantasies, reestablishing them into dogma, ritual, ceremony and institutionalized religion. From this pudding came the stew of a long line of *a priori* delusions—*metaphysics.* Its malignant roots crept through the centuries and gave birth to religions of one kind or another, depending on the time, place and circumstances.

The evidence, furnished us by almost every form and direction of investigation, is overwhelming in the conclusion that there is no basis for supernaturalism. The facts cannot establish any existence of any god or spirits, not only from the point of view of any lack of physical phenomenal proof, but also because we now know sufficiently of the history of the evolution of the various religions dominant in the world today. Any honest student of comparative religion and anthropology will sooner or later come to this conclusion.

Organized religion will restrict itself to the worship and explanation of their particular creeds, not because of the lack of curiosity, but because their material dependence on others for earthly existence and continued influence and power, is of more immediate importance and concern than the truth or falseness of their religious doctrine. "It would be expecting too much from human nature to imagine that pontiffs who derived large revenue from the sale of the Agnus Dei, or priests who derived both wealth and honours from cures wrought at shrines under their care, or lay dignitaries who had invested heavily in relics, should favour the development of any science which undermined their interests." [109]

If one is to realize the essence of Time, he must also realize, as a prime factor, the essence of his own reality, his own value, the extent of his own existence by every rational and empirical process known. He must realize his rights as well as the joys of living. He must realize what the rest of the world means to him so that he may obtain the logical and rational ends, not only security and peace, but also a culture of pleasure in and a devotion to things and people that would merge as greatly as possible into the highest possible pitch and extent of happiness for him. By pawning all these intentions for a ticket to a paradise built of legend and metaphorical myth, he loses his bearings as well as his desires for these rational and natural ends of a normal person; he ceases to be an individual, reverts to an item in a hive of religious frenzy. Thus he is in no condition to weigh the potential of Time, not only within himself but also within the general framework of his relationship with others.

The prime requirement of Time appreciation is that the individual should realize his own significance. In so doing he will not only be able to realize the significance of others, but of more importance, he will become a *free* individual. Only a free individual can adopt ethical principles as a guide for action, thought, and association. One who sees favor and partisanship in a button, badge, or other signs of special group or class association, is not a free individual. "Nationalisms" among individuals, groups, and countries, breed sectarianism, groupism, bias, prejudice—the bacterial ferment for inequity, blind judgment, often misery and war.

We come to another point. The greatest Utopian fantasy that religion has tried to build, especially in Christianity, is the supposed equality of individuals. There is no question as to the good or evil intent of such a venture, nor does the hierarchy of the Catholic Church with all its orders, superiors, and subordinates, indicate any sincere attempt at equality, but it lacks even a meagre knowledge of human nature and events, and replaces whatever knowledge one might possess with an illusion that all men are equal in life and therefore before the judgment of a god. "That fine-sounding phrase 'God made all men equal' is as yet in the prophetic stage, for there still remain a few who wish to differ from their fellow creatures," wrote Charles Duff.[110] Not only can one never find two things exact in form, whether it be man, bug or plant, but he cannot find two people with exact mental attitudes. This is not possible because the experience of each thing differs in some degree with everything else; therefore it is not possible for any two people to have exactly the same desires, attitudes, interpretations, etc. On the other hand, it is fortunate for man that he *is* what he is—an individual with a different or varied perspective and experience of things. It is this difference of opinion, this difference of experience, it is this difference among individuals that is the very life of progress for a higher culture and for intellectual advancement among men. It is this very difference among people that gives them the desire for association. If the clerics would only preach that all men should have the same right to equal *opportunity* to better themselves and that all people should share the reasonable burdens to insure their own comfort, then these very same clerics would have to cease begging for alms and instead they would be compelled to flex their muscles and minds to earn their own keep.

Organized religion has retarded individual and social progress because it has tried to sterilize and compound men into a common alloy to suit the dogma. It has never occurred to them to change their dogma to suit mankind. The acceptance of the myth of the equality of men is to automatically regiment them into one static degree and then to believe that all experience from that moment onward is alike in every way of expression, action, and reception, *for all of them!*

During many centuries religion has strived to fuse individuals into a purpose and belief securing the existence of the particular creed. It lacked any real genuine regard for the earthly existence of the individuals involved. When has organized religion come to the aid of the slave, the underdog, the exploited and the oppressed? True it mouths soft, flowery phrases and prayers and epistles intended to show concern over the plight of the unfortunate and the enslaved, but these things belong to the realm of the tyranny of words. Has religion ever gone forward on a *practical* and *active* basis to relieve the social and economic stress and strife? To do so would be to undermine their own position and security. To do so would be to enlighten people as to their economic rights, and when people can apply their reason to their economic rights, they might apply the same reason to their moral and intellectual rights as well. This moral, intellectual freedom is one of the principle reasons and essentials for the culture of social progress and genuine good will among individuals and groups. As Thomas Jefferson once said, "The earth belongs to the living, not to the dead." [111]

Asceticism is the very life-blood of religion. It subjects the individual to a supposed power about which he knows nothing. It overcomes any of his objections to misery and struggle, so long as that beautiful mirage of paradise is screened before his eyes. For this paradise he has not only killed his own kind but killed others because they didn't believe as he did. The existence of so many religions and the history of religious persecution are no arguments for the existence of a benevolent, merciful and all-wise god who hath created all these things by "creating" the means by which all these things have been carried on. Reciting Latin phrases before people who do not know a word of it cannot be the answer.

To religion the group or communal mind is of prime importance. In order that the particular creed should continue it is apparent that the religion ordains that this group mind does not change adversely. The Church may condemn so many individuals to hell, but it will not be so stupid as to condemn the entire human race. This would put the Church out of business.

In order to accomplish this group mind, it is necessary to submerge the individual mind to the group or "congregational" mind.

As soon as any individual begins to think for himself in matters of religion, he may or may not agree with it, and more often comes to conclusions which religion does not care to invite into the group mind. Individual mind, if free, will disturb the tranquility of the group mind; this divergence is the essence of all progress. The finest things in life, material, moral, political and economic, were achieved because individuals were brave enough to dare the traditional organizations of society, to disrupt the lethargy of the group mind, to awaken it to newer and better and finer ways of living.

"I spent twenty-one years in Catholic schools," writes Father Emmett McLoughlin, ex-priest, "But I did not learn to think. . . . The term 'brain-washed' is applied to Chinese Communism, but the practice is as old as the Catholic School system. . . . Of the world's great literature, I learned practically nothing. Its greatest lights were locked in the prison of the *Index of Forbidden Books*. . . . I had become an automaton, a priest of sacred, half-known rites as meaningless in their efficacy as the chants of the Puerto Rico voodoo priest or the blessings of Tibet's sacred Lama." [112] He states further: "One of the greatest paradoxes of American life is that the American Catholic child is brought up in the land of Thomas Jefferson, Benjamin Franklin and Abraham Lincoln (with their heritage of freedom of thought and complete mental liberty) still adhering to a school system and a rigidity of thought reminiscent of the Middle Ages and its mental tyranny." [113]

People get a better idea of their life rights when they begin to think. When people think more they become less religious. Knowledge dilutes superstition as a ray of light dilutes the darkness. The scientist need not say a prayer before his test-tube or throw incense upon his Bunsen burner. Man, too, in order to improve his road for happiness and comfort that should be his, must dispense with his gods, and begin to realize the importance of living. When he will realize his own importance, he will begin to respect the importance of his fellow men. Only by the intelligent and intellectual freedom of the individual can the peace of all the people be secured. Only by the realization of the value of his own life can man realize the importance of Time.

Any fair-minded man with even a moderate knowledge of social history feels that organized theology is slowly declining and that it persists only in ratio to the ignorance and traditionalism of any group or class. The worlds of science in general and the departments of neuro-physiology, biochemistry, microbiology, anthropology and comparative religion, in particular, have proven beyond any reasonable doubt that the existence of religion does not rest upon any rational or justifiable grounds, but solely because it still clings on by its historical strength and because it continues to fill the psychological need of the human being to believe in the continuity of life after death and the survival of his own ego.

We have noted how the origins of superstition and religion were born out of simple savage cravings and fears and that these took shape and color as the savage turned primitive, as the primitive turned barbarian, as the barbarian turned into the semi-civilized state. "What primitive man was first conscious of was not the universe as a whole, but his immediate environment, the things at hand among which he lived and moved and had his being. He could not possibly know or think of the universe as a whole, and he assuredly had no occasion and no urge to think back of that universe to the First Cause and Creator of all. Primitive man was practical, not speculative, and the category of cause and effect simply did not exist in his psychology, as anthropologists have long since recognized. What he had to do was to live and to live he had to come to terms with his environment, and in doing this he became religious. Primitive man had no conception of the regularity of nature (he had not lived long enough); he had no conception of forces and laws; the only activities of nature that he knew were those mysterious phenomena around about him which did things to him, and it was with these that he felt the necessity of establishing friendly relations. The very phenomena of nature were deity to him—the actual mountains, stones, springs, trees, animals, storms, and what not. They were greater than he; they controlled his destiny; upon them he was dependent; and their good-will was necessary to his well-being. He accordingly attempted to control them to his own advantage, and so gradually devised a system of control through the use of various rites and institutions. Thus is to be explained the origin of the rites that the Old Testament so frequently connects with sacred mountains, springs, trees, stones, and the like." [114]

Almost every religious festival, holiday and symbol, can be explained through historical, objective evidence without having to refer to supernatural revelation. The symbols and rituals that cannot be presently explained are no supports for supernaturalism but mean only that men are still searching for further material, artifacts and data. The investigational sciences show plainly and clearly that all the gods and all the religions did not descend from some heaven, that the structure of religiosity and of all theologies, regardless of creed, color, are built upon false but understandable premises. Ages ago, the translation or identification of primitive fears, emotions, desires, hopes and loves, was demonstrated by acts of appeasement, symbolism, tabooism, etc. One could reasonably understand why and how these things came about, were desired and justified.

Religion, by the nature of itself, is a rigid thing and therefore cannot be a help to us because it does not progress with the general nature of change and variation. The minds of individuals and groups can adapt themselves to changes, progress and broaden understanding, change objectives to meet new contingencies, new knowledge, etc., but the nature of religion forbids this luxury to itself. "It is true," Bertrand Russell satirizes, "that in heaven hymns are sung and harps are played, but they are the same hymns every day, and no improvement in the construction of harps is tolerated. Such an existence bores the modern man. One reason why theology has lost its hold is that it has failed to provide progressive machinery in heaven, though Milton provided it in hell." [115]

Any change in religion occurs only as a weakness from its former strength. Religion is, by its own commitment, a dogmatic status. Its very dignity and supposed perfection are built upon the idea of immovability. It deals with absolutes, with infinitudes, with perfections. These are natural conflictions with the normal processes of conditional living; they cannot be reconciled with life, time, and the growth of understanding among peoples. Sympathy, appreciation, and the dynamic potentials of affection and sentiment, activate themselves in a field wherein perfections are not known; these things throb within a finite field wherein the actual living experience of imperfect people are polarized to sustain a peaceful procedure in the social order.

Religions have always opposed each other for supremacy, for the greatest number of converts, and for the necessity of greater material power to sustain themselves on a permanent, long-range basis. This natural antipathy for each other exists despite common idealisms preached in favor of their common purpose in the more Westernized countries where democratic governments exist and where religions are free to worship but are not officially any part of the secular authority, and where one particular religion is not closely or exclusively associated with the government. Where one religion is in full power in a particular country, that creed will not be polite or tolerant toward the entry of any other creed; religious courtesies exist only in such countries where absolute control by one religion is not allowed. Even so, in a democracy like the United States, each religion strives to outdo the others. By this method, each creed acts as an invisible nation with invisible borders and barriers. The world is composed of national countries and religious countries. Religious countries are more numerous, and because their borders cannot be physically defined, their real influence upon human association is more difficult to control. The breaking up of the human race into hundreds of religious masses and each mass into still more numerous cults of smaller masses, and each smaller mass into still more numerous petty sects of one kind or another, are in actuality little nations and large nations, child countries and parent countries, smaller spheres of influence and more powerful entities of control. This division of mankind into all these parts, creating emotional barriers to retard mutual and natural sympathies, becomes a great dam holding back the normal flow of the commonality of mankind.

We have seen that man was, is and always will be, a unit of the natural and not of the supernatural; he is helplessly in constant flux within it and with it. This flux is called experience. Accumulated experience becomes knowledge. Man is also an animal and has evolved in the same manner and out of the same principles governing all life. Dr. Philip Eichler states plainly, "Man belongs to the animal kingdom and never can graduate from it." [116] Samuel Chugerman states further that "the close affinity of all organisms, vegetable, animal, and human, among themselves and to one another, is indisputable proof of their unity and common origin." [117] Dr. L. C. Dunn puts it on the line, "All living things are composed of the same sort of stuff." [118] And Elie Metchnikoff of the Pasteur Institute of Paris, wrote: "The extraordinary quantity of rudimentary organs in man furnishes another proof of his animal origin, and puts at the disposal of science information of great value for the philosophic conception of human nature." [119] Dr. F. H. Shoosmith enlightens us further: "Organic evolution is based on evidence so abundant and so varied as to be conclusive. . . . The rudimentary gills that appear in embryonic birds and mammals, including man himself, despite the fact that no bird or mammal ever uses gills at any stage of its existence, are explicable only on the evolution theory—according to which all lung-breathers have a direct ancestry reaching back to the animal forms that first invaded the land from the waters in which life first appeared." [120] To say that man is an animal is certainly not to cause him any alarm. On the contrary, the truth should awaken in him a pride which is full and wholesome and upon which he can lay a firmer, finer, and a real concept for action to achieve the end of living, happiness and security.

Life evolved out of instinctively group or communal protoplasmic cells which, in turn, in their struggle to exist, have evolved nerve cells for detection, protection, control, and preservation. Beadle writes that "the lowly sea anemone has a simple network of nerve cells—but has no suggestion whatever of a brain." [121] Roy Waldo Miner tells us that "in the early history of the earth, the lowly hydroid polyp and its relatives, laid the foundation of higher animal evolution." [122] And Dr. George C. MacCurdy tells us that "in their simplest forms plant life and animal life are so much alike that biologists find difficulty in determining whether a given form is animal or plant." [123] "Every kind of living thing," wrote Julian S. Huxley, "from a disease-germ to a turnip or an oak-tree, from a coral polyp to an elephant or an ant, reproduces itself in the same general way." [124] Donald Culross Peattie, the esthetic naturalist, states that "plants share sex with the animal kingdom is one more proof of the oneness of life." [125] And Ray and Ciampi, the brilliant ichthyologists, write: "We humans owe what few teeth we have left to their origin as placoid scales.[126] . . .

Most sharks and rays possess a hole just behind the eye. This is the spiracle and is exceedingly important to the bottom-living species for the purpose of breathing. . . . In man a vestigal spiracle is seen in the Eustachian tube, which connects the ear and throat cavities." [127] Dr. Henry E. Crampton states that "every animal and every plant gains its final form by a series of transformations from simple beginnings." [128] And Dr. A. I. Oparin, in his *Origin of Life,* summarizes thusly: "Life has neither arisen spontaneously nor has it existed eternally. It must have, therefore, resulted from a long evolution of matter, its origin being merely one step in the course of its historical development." [129]

"It is the biologist," writes Dr. A. T. W. Simeons, "who now expresses scientific contempt of the supernatural origin of the human body.[129a] . . . The highest mammalian order, the Primates, had its beginnings in the tree-tops of the early Eocene, about seventy million years ago. One mammalian species that had never left the trees began to specialize in a further evolution of its brain. It retained its small size and with its pointed snout closely resembled the tree-shrews that still live in Far Eastern jungles. The inch-long skulls of these early shrew-like mammals have been found in Mongolian deposits that are forty million years old. Yet even those fragile remains show anatomical traits that exist only in primates. This small shrew-like animal that lived so long ago is the grandsire of all the lemurs, tarsoids, monkeys, apes, and man." [129b]

The nerve cells, in turn, have evolved into reflex nerves evolving into the brain case and the gradual evolution of nerve control box or brain. It is very interesting to note, as Ivan P. Pavlov says, that "one can hardly deny that only a study of the physico-chemical processes taking place in the nerve tissue will give us a real theory of all nervous phenomena, and that the phases of this process will provide us with a full explanation of all the external manifestations of nervous activity, their consecutiveness and their interrelations.[130] . . . Our objective investigation of the complex nervous phenomena of the higher animals fills us with a reasonable hope that the fundamental laws underlying the fearful complexity in which the internal world of man is manifested to us can be discovered by physiology, and in the not far-distant future." [131]

During this process of cerebral evolution, the nature of the life-organism, and the processes of experience itself, have compounded man into an intricate collection of interrelated and interdependent emotions. From these emotions flowered the rational processes of the mind, or what is called the *reason.* As one German scientist once said, "*Ohne phospor keine gedanke,*"—without phosphorus, no thought. Today, man's emotions and his reason cannot be separated from each other; each is intertwined and constantly fusing, influencing and creating the other. Out of the first spark of man's reason, out of his primeval reflections of himself and his surroundings, were born the great grandfathers of modern religious faiths.

It was natural that man should try, with whatever crude and limited knowledge he had, coupled with an embryonic mind steeped in brutality, savagery, and almost pure animism, to attempt some understanding of his life and the powers that controlled and influenced his daily precarious existence. Out of this dim and foggy age were born magic, spirits, the personification of trees, animals, stones, places, elements, and what-not, as beings like himself and with whom he had to contend in order to preserve himself.

Later he personified his feelings, whims, desires, notions, hopes. Further he personified the most "miraculous" thing of all—the "creation" of life by two beings, and especially by woman. Communal, group and tribal life gradually evolved rituals and taboos and organized worship to revere the sex act. As the emotions and reason of man flowered still higher, and this took many thousands of years, his personifications became more numerous, more rituals accumulated, and various superstitions of one kind or another filtered into every tribe and race. And still, for many ages, man did not yet think of anything like a hereafter, immortality, heaven, hell, or of any gods. All his desires, his feelings, and his crude thoughts, were all centered on the security and extension of the only existence he knew or dreamed of. The rise from these beginnings to concepts of a soul, of one god, of paradise, of purgatory, is a long evolutional distance.

The greater the mind became through evolutionary processes, the greater and more numerous became the emotions and the desires of man. The revolt of man's reason against the tragedy of death, the incomprehensibility of complete annihilation, the natural refusal to even accept the reality of it, the dreams about departed relatives, friends, enemies, the natural desire to continue living in whatever form so long as self-identity is preserved, and many other contributive factors, gave rise to the notion of the soul. Finally, man imagined gods who were, of course, in the image of man, and who could dispense good things and bad things according to the ideas and dictates of man's own intentions, desires, and magical formulae.

Mythology is the poetry of the primitive heart and mind. It is the music he heard in the wind of the gentle breeze and the thundering storm, and the patterns of colors he seemed to see in the mirages about him. Mythology is the creative expression of wonder and excitement, of the joy of birth, the sense of being, and the sorrow of death, of the puzzles and bewilderment of both. Mythology is natural as the imagery of poetry, as natural as the widest and deepest ideation that could come to man in his attempt to explain himself and the world he finds himself in. Mythology is man's own childhood trying to reveal the concealed, to focus his mind on the colors in the dark, give voice to silence and description to what he actually could not see. But all this mythology is, by his own pulse, touched by his own emotions in traffic with time and the world. Mythology is the expression of man's art and the rainbow is his paradisical stairway to fulfill his being and metamorphose his finite, frail little frame into a universe of pretended perfections, just as a little child imagines herself to be the real mother of all the rag dolls strewn about her; she is queen of the day, and at night all her subjects turn into beautiful princesses, handsome princes, and wicked witches. Mythology is the instrumental symbol man uses to repeat the echo of his identity, to recreate from his own perishability and limitations the recurring endlessness of his personality, its needs, and his universal household. It is man's poetic chemistry to turn the substance of his egocentricity into hopes, the hopes into beliefs, and the beliefs into scriptures of imagined certainty and his dreams into sacred illusions. Mythology is the symphony of the life and death drama of the human being trying to fathom by sense and search, not only the beginnings and the ends of all things, but also the whys and wherefores of their common umbilicalized identity. It is the *Isis* of man's endless hope to find all the pieces that could fulfill the unfillable, the never-ending, never-ceasing passionata to find the key to the cosmic story. Can it be that *Isis* is hidden in the *Horus* of our own inner sanctum that *demands* the recurring cycle?—project its image against the worldly screen?—so we can see the world as if the world were a person? *Cogito ergo sum* identifies the self, not the world. Where Myth ends, Science begins. The pity it is that so often a man spends a lifetime vainly trying to find the unknowable and hardly lives a day to enjoy the knowable, as the flower opens itself to the sun and the baby smiles into your eyes, *just to live and enjoy the moment of living.* It is more wise, I think, to enjoy life than to seek the mirages that draw us away from life and which, after all, have only delayed us from the ethico-idealistic appreciation of the Time it is ours to live within. Pangloss had his myths and Candide was wise to counsel him to cultivate the garden.

The psychological metamorphosis of human nature in its evolutionary stages most often frustrates its own purposes of change. The individual becomes free only to build a colossus to enslave others as well as himself. People revolt to destroy a tyrannical individual or group and build instead a tyrannical state. One form of sovereignty evolves into another; until all forms of sovereignty are abolished or become unnecessary by the fusion of all forms into a world sovereignty; until this can be accomplished there can be expected to be more misery, poverty, strife, and the spectre of war.

Within the nature of growth and change lies the heritage of error. Progress is like a rubber-band; it grows by stretching, only to contract again with the first gust of leisure and complacency. Man seems to enjoy uncertainties, curiosities, new thrills and new hates, as well as new joys, even if it might let loose dragons of destruction upon himself and his kind. The essence of life being an uncertainty of direction and event, it would be foolhardy to lay down any definitive plan or philosophy whereby man can build a perfect society. Such a plan belongs to a god and the metaphysician. But it would not be too much to expect of a rational person to reflect upon the past and present experiences of his fellow men so that his own life can be much more endowed with the good things of life, peace, happiness and the ecstasy of pleasurable satisfaction that comes from a cordial, affectionate and understanding heart and mind. "We have to stir up the men of today to elementary meditation upon what man is in the world, and what he wants to make of his life. Only when they are impressed once more with the necessity of giving meaning and value to their existence, and thus come once more to hunger and thirst for a satisfying world-view, are the preliminaries given for a spiritual condition in which we again become capable of civilization." [132]

There must be some intelligible way to correlate and coordinate the natural fullness of human emotion with the scientific advancement of the race so that upon both premises may be built a political and social structure that would ensure the liberty of the individual, that would secure his existence as a harmonious unit of a society restrained from destroying itself by the rational application of these very two factors.

Millions in America and elsewhere still live, to a certain extent, in the Dark Ages. They still consider and judge people by the color of their eyes, their skin, according to a button, pin, badge, or password. Nowadays they are even talking about and taking to uniforms and shirts of various colors, as if it were possible to dress up a person's mind, his good intentions or cultural standing. These minds are as shallow as the depth of their own skins and they would become beasts for slaughter, hungrily foaming from their mouths in a crazed fit of hysteria, were it not for the law to prevent them. The most ferocious and heartless jungle is very often within the area they call "civilization." Therein are millions of beasts and millions of prey; some are dumb brutes and others accomplished stylists of manner, tact and cunning. Only the invisible cages of law and order and police power keep them within bounds and restrain these people from their full vent upon other people whom they little understand. Sectarianism is the internecine warfare of the human family. Little has been done to harness the human race in the middle of a tug-o-war between moon and sun. Human beings will continue to misunderstand, hate and kill each other so long as theologies remain to divide and mislead them into frenzied whirlpools of fantasy and emotional delirium. In all the thousands of years of religious and racial warfare and massacre, not *once* did a god come down, make himself known in some non-miraculous manner so that people might love each other instead of killing each other.

Many people, without thinking, often come to a hasty conclusion that one who attacks religion is not only a bad person but that he must also be a fascist, a communist, an anarchist, or a nihilist, or the likes of one. The truth is that belief or disbelief does not conclusively mean that a person is good or bad. It should be apparent, however, that a person who is usually cautious and rational about what he believes in, is usually cautious and rational about what he does. Reflection not only dilutes blind belief but it also sows the seed for a greater regard for oneself and others.

Communism and Fascism, while they may or may not seek the destruction of theology, are not, in themselves, constructive cultural and ethical philosophies. They seek to destroy the religious god by competing with him in building up another colossus-spirit, the *State.* Theology wants man to sacrifice himself for the sake of paradise before a god; totalitarianism wants man to sacrifice himself for the sake of the "future" before the State; both sacrifice man for something which is never present but always coming; both are built on emotion, submission, blind belief and absolute discipline to the priest or leader. Both resent every form of intellectual, cultural and educational freedom. Thinking and a critical attitude are just taboo in their tastes. Both thrive on propaganda and racial division. Both seek to capture the minds of children and mold them into befitting slaves and knaves. One says, "God wills it!" The other says, "Follow the leader!" Both lead in the same direction: the degradation, devaluation, and the misery of the human body and mind. Both change what could be a beautiful and joyful life into one of ignorance, blind obedience, hate and self-emasculation.

During recent years there have been upheavals and setbacks in the economic, political and social life of the world. The failure of man to adjust himself to a new emotional sphere upon the submergence of previous traditions and conventions, the terrible plight of the younger generations in their attempt to look forward and realize some fullness of life-happiness obstructed by a cruel, haphazard and catch-as-catch-can economy, the rise of communism, fascism, and other social and political diseases oozing out of the chaotic slime in which the world finds itself today—these and many more have given the theologian something to bite on. The minister will claim aloud from the pulpit that the reason why man and his world find themselves in this awful and miserable plight is because he has been overcome by materialism and paganism, forsaking his god and following the doctrines of science and evolution. He will claim further, that all the malignant and political diseases are due to the paganism and the denial of all that is godly and supernatural. No longer exists the fear of the Lord, and that is why people have no regard for life and property, for womankind and for the poor.

Offhand, the pulpit oratory may appear reasonable. But when one begins to think, he finds an entirely different picture. The decline of religious influence in the academic, student and professional fields, as well as among the masses of the people, came about because of freedom of thought, press and education which gradually became stronger and more extensive. Further decline in superstition came about by the increase of transportation and communication facilities and inventions, which brought the far corners of the world within a common and closer focus. Besides the absolutisms of theology, there also existed the tyrannies of state, royalty, and military oligarchies. As science and free expression grew and widened, the people not only revolted against the supernatural narcotics, but also against the other tyrannies affecting their economic, social and intellectual welfare. Many peoples and countries are today unfortunately very vulnerable to communism and police state politics because of the serfdom of the people, because of their poverty and miserable living conditions.

However, let us return a bit to our general premise of this chapter and delve further into some analysis of the religiosity of people. In order to understand the human being, his experience and his present status and institutions, it is very essential to obtain some understanding of his growth to the present day. It is essential to know his nature, his traits, his arts, etc. To attempt to eliminate his biological and psychological history is to eliminate any proper understanding of his social, political and esthetic nature. It is for these reasons that it is important to survey his natural and historical heritage, in order to understand more clearly the nature of his "religious" feelings.

During the many thousands of years of religious belief, one can readily admit that in spite of many evils and evil ones among the religious cults, the various creeds of man possessed really great men and women, people who have devoted their lives and still do for the material and "spiritual" welfare of others. Many of them are sincere and honest and lead good lives. The same admission can be applied to some kings, queens, military leaders, popes, men of letters, and all types of leadership. It isn't the life of a good person with whom we have to contend here. It is the realization that religious faith in the supernatural came about because of the tragedy of misconception in the primitive mind. In his groping search for an explanation of his life and in his attempt to secure himself, if he would have followed the line of expanding experience without jumping ahead of himself in thought, civilization would today probably wear a different costume and a much more sensible one.

But the mind of man evolved faster than the rest of the body; in fact, the body of man today is far more under-specialized than many animals, and the reason might be that the fast-growing brain of man rapidly overcame many disadvantages of his body and therefore retarded any natural adaptation and selection that would have become necessary with a slower and smaller thinking apparatus. Before man could possibly realize the fullness of his actual experience, his mind had already wandered off too far ahead of him, left him on earth and sprouted a halo and wings. The first fantasy of man was also the first product of his vanguard imagination that had gone too far ahead and couldn't find its way back to the feet.

When man was so much like a bug he was the more perfect social animal, perfect in the same sense that the state, group or colony, was the conscious factor and he was part of this consciousness. When man emerged from this condition and became *Homo sapiens,* his mind gave him freedom and with freedom came insecurity and uncertainty. Man was first unconsciously social, then he became consciously individualistic; if we can get him to become consciously social, the ideal society might be at hand. To become social within the fullness of one's rationality entails the solution of many different problems, difficult because of the natural obstacles in the heritage and traits of human nature.

Theology and the supernatural bloated the individual; it showed him the road to a greater paradise for his individual self, instead of a cowering fraction of a cowering group. It raised to a higher and wider level the selfish hopes of man. In doing so he lost sight of his mission on earth—to live. To obtain the mirage of heaven, not only would he forsake normal living but necessarily forsake normal and happy relationships with others. When it comes to heaven, the holy books plainly tell us that it's every man for himself and the devil get the hindmost!

Apart from evolving myriad classes of people all over the world with myriad faiths and cults, each opposed to each other in some form or degree; apart from the hate, massacre, degradation and misery it has caused from this division of thought and action; apart from endless other calamities which the spectre of heaven has heaped upon humans, supernaturalism has deteriorated and devalued the actual life of the individual. It has induced him to pawn his few years and stretch them into an invisible, elastic eternity, making them meaningless and turning them into concepts of fear and distortion.

It is this abortion of the rational processes into distorted emotions which has kept the human race from the social and individual happiness which by life-right it deserves. This doesn't mean that by eliminating blind faith and religion all the problems of the world would be summarily solved. The belief in supernaturalism is just one of the major diseases of mankind. It is only by investigating every contributive factor that makes for disharmony and misery that we can arrive at some reasonable basis for constructive action. The more dogmatic, tyrannical and superstitious a country's traits are, there you will usually find the people steeped in ignorance, in misery, in poverty, in social and economic stress and tension, and you will find them deeply religious and their faith strongly impregnated in the fabric of their daily life and actively expressed by them.

The process of disintegrating strong emotions does take long periods of time, and many people may be right in feeling that such strong emotions may never be completely eliminated. Liberalism, free thought, and a free-flowing intercourse between various peoples can help to bring about a better understanding of things and closer ties. Whether this can come about is very difficult to calculate as the ways of people are strange and unpredictable, and the traits of the human being difficult to change, even for his own good. It is true that superstition and organized religion are becoming very slowly less and less engaged with the actual daily life and thought of people. This may continue to increase with each new generation. The trend is fairly definite and clear. Economic stress will help deliver us from the religious stupor of believing that the best meal will yet be eaten after we die; it will gradually dawn upon us that we have to eat now to keep ourselves alive. When the people are delivered from the bondage of economic peonage, it may occur to them that there are other pleasures besides just eating and sleeping.

It is not easy to tell people that there is no god, no soul, no eternal life, no personal immortality. "In India the wisest have never talked of good and evil," [133] for the truly wise do not really know, neither do they know of the gods or of heaven. George Bernard Shaw sums it up thusly: "The idea of God, which is the first effort of civilized mankind to account for the existence and origin and purpose of as much of the universe as we are conscious of, develops from a childish idolatry of a thundering, earth-quaking, famine striking, pestilence launching, blinding, deafening, killing, destructively omnipotent Bogey Man . . . evolving finally into the incorporeal word that never becomes flesh, at which point modern science and philosophy takes up the problem with its *Vis Naturae,* its *Elan Vital*, its Life Force, its Evolutionary Appetite, its still more abstract Categorical Imperative, and what not!" [134]

Yet, if it were known that heaven is a fiction and that hell is the only awaiting reality, most everyone would undertake to look into the claims of religion and see if it were really so. The same religious discrimination between good and bad (hoping that we know the difference), in accordance with our wishes, must continue after death if any belief in such things is to survive. It is natural that people should want to live forever. It is the will to live, the *ding-in-sich,* the ego, call it what you will, but it wants to continue. It is in the order of things. Even plants and animals, without either consciousness or reflective brain processes, struggle on in every possible direction to survive and lengthen their stay in their present form or in the forms made possible because of internal and external forces playing tag constantly in the realm of existence. "The cult of the dead," writes Felice Belloti, "is based on a conviction that something of the dead man must survive and continue to live in the place where he led his earthly existence. The *Manes* continue to take an interest in what happens to their family, to their village and, if they were powerful during their lifetime, to their tribe. Therefore it is well to appease these invisible guardians. Since the living have not the faintest idea what the spirit of a dead man might require, each man makes arrangements to honour him as he thinks best or following what he has been told by his father. Today there are still some who want to bury a pair of young slaves alive, bound to the corpse of the dead father, simply because they will make him feel happy in the beyond; others raise little altars to their *Manes,* on which they place victuals which they constantly renew—until they begin to forget. Others still bring daily to the grave of the dead man those things which he held most dear in life." [135]

I, too, would like to live forever, if I knew of any such possibility. But I know of no such possibility. Sir John Bland Sutton, one of England's most famous surgeons, said, "Death is the end of all." [136] The subject about which Frazer remarked "that great mystery of which fools profess their knowledge and wise men confess their ignorance" [137] only reveals to us that "the belief in immortality is ultimately an act of faith, whatever its origins, and whatever its end, and faith is untouchable. It cannot be demonstrated as a fact." [138] Mencken put it more plainly: "I believe that the evidence for immortality is no better than the evidence for witches, and deserves no more respect." [139] Technically, "the belief in survival of the body or soul," writes Chugerman, "is wholly unnecessary to support the fundamental postulate that the law of causality outlives all the transient things of the earth." [140] H. G. Wells confessed that "I do not believe I have any personal immortality." [141] And Bertrand Russell expostulates that "immortality appears exceedingly improbable. Memory is clearly associated with the brain, and there is nothing to suggest that memory can survive after the brain has disintegrated." [142] Ashley Montagu also states very clearly that "unless this world and our experience in it be but a dream—which may be doubted—and we are destined to wake up in some other world of reality, I cannot but see that the notion of a physical soul surviving after death is forever shattered. For if body and soul are one then death of the body means death of the soul. We can be reasonably certain that if there are other habitable worlds human souls do not journey to them." [143] Elie Metchnikoff adds: "The idea of a future life is supported by not a single fact, while there is much evidence against it. The phenomenon of intercommunication across a distance, sometimes called telepathy, may be actual, but affords no support to the conception of the existence of souls apart from bodies. . . . It is easy to see why the advance of knowledge has diminished the number of believers in the persistence of consciousness after death, and that complete annihilation at death is the conception accepted by the vast majority of enlightened persons." [144] "Man has the seeds of immortality in him," says Sir Arthur Keith, "but the gift is for the race, not for the individual." [145] Confucius, the amiable philosopher, asked how can a person understand immortality or the life after death when he doesn't understand life before death? But the best punch line of all is that which Fred Hoyle makes in his book, *The Nature of the Universe.* He says: "It strikes me as very curious that the Christians have so little to say about how they propose eternity should be spent." [146] Or as the witty Bernard Shaw had written: "If some devil were to convince us that our dream of perpetual immortality is no dream but a hard fact, such a shriek of despair would go up from the human race as no other conceivable horror could provoke. . . . What man is capable of the insane self-conceit of believing that an eternity of man would be tolerable even to himself?"

The world of reality is not so generous to us as little particulars in the big pot of nature. Experience proves often otherwise and contrary to our wishes. It tells us plainly and bluntly, in spite of our fantasies, that we know of one thing only: of that which lives. It doesn't tell us anything about the dead except that which we see—

a process of decay and disappearance. It's a bitter pill to swallow, the most bitter, especially when our loved ones go. Alas, no one can bring the dead back to us, but in our hearts and minds they need not leave; memory stops them at the bridge and sends them back to us. This is also part of reality.

If the nature of experience and of experiencing is so constantly fertile and full to the brim with the realization, sad as it may be, that death is the end of individual consciousness, wouldn't it be logical to assume, if immortality were a fact, that nature would also be abundant with things of certainty that would be ample proof of this immortality? Then faith would not be necessary. We would *know,* as we know many other things. This may seem to many that it is bombastic and conceited for men to think that they can comprehend more or as much as a god can. In reply we can say that man has already comprehended many things in almost every direction far greater, more wonderful and more powerful than that very god. We can open the windows of great telescopes and look into the fiesta of stars millions of miles away. Even on earth we have fairly conquered, through science and not through prayer, various diseases, overcome plagues, lengthened life, and many other countless things just as wonderful, if not more so, than ourselves. Further, if I am honest enough to admit that I cannot comprehend a god, being just a man and not a god, how is it that other fellow creatures of mine have been so naive as not to inform the rest of mankind as to their divine discoveries, instead of asking us just to believe them. The truth is that man cannot compete with a god because the reality of a god has not been established, but man can compete with the theologian who claims there is one. God and theology are really one, the child and the parent together. Only a short time ago this very same theology, claiming it speaks for its god, also claimed that the world is flat and burned many who differed with them. *"E pur si muove!"* whispered Galileo, as he was forced to submit to a "flat" world.

Many people think that the idea of a god is intuitively native or born into the mind of man. This has not been established except by grown-ups. Gould brings out that until deaf mutes are actually educated or taught about religion they do not conceive or perceive any idea regarding a god or religion.[147] Frazer says, "If we could trace the whole course of religious development, we might find that the chain which links our idea of the Godhead with that of the savage is one and unbroken." [148] "Man has, properly speaking, no immediate or *intuitive* knowledge of God." [149] Tylor adds: "As prayer is a request made to a deity as if he were a man, so sacrifice is a gift made to a deity as if he were a man. The human types of both may be studied unchanged in social life to this day. The suppliant who bows before his chief, laying a gift at his feet and making his humble petition, displays the anthropomorphic model and origin at once of sacrifice and prayer." [150] Albert Einstein states the scientific impossibility of gods: "The man who is thoroughly convinced of the universal operation of the law of causation cannot for a moment entertain the idea of a being who interferes in the course of events—that is, if he takes the hypothesis of causality really seriously. He has no use for the religion of fear and equally little for social or moral religion. A God who rewards and punishes is inconceivable to him for the reason that a man's actions are determined by necessity, external and internal, so that in God's eyes he cannot be responsible, any more than an inanimate object is responsible for the motions it goes through." [151] Morris R. Cohen, in his book *The Meaning of History,* states: "In practice, religious teachers recognize and even stress the evil in the world, but they have not offered a real explanation why a benevolent and omnipotent Father tolerates it." [152] . . . "The fact is, however, that the great interest of religion is not to explain the evil of human history but to strengthen us to bear it by offering us faith and hope in a better and higher world." [153] Another beloved teacher, esthete and philosopher, Irwin Edman, said: "The hope of the world certainly lies in intelligence. Certainly, there is no hope anywhere else. I cannot look to anything so remotely definable as God for aid, nor do I regret not being able to do so." Auguste Forel, the great entomologist, stated: "As for a God incomprehensible to man, whom some believe they can know or feel, whom many adore and others fear or even curse, and whom many people bedeck with their human attributes—we may leave everyone free to adorn this idea with whatever metaphysics he pleases. The religion of the social well-being of man has no concern with Him; neither have I seen His temple among the ants." [154] "There is nothing in nature to justify a scientific inference for the existence of the God of our infantile conception. If men could once get it clearly in their heads that the affairs of the earth depend entirely upon their own efforts and that the last hope of outside intervention must be abandoned, our situation would be on the way to improvement." [155] Alfred Weber, in his *History of Philosophy,* clearly states that "St. Augustine's chief aim is to elevate God by debasing man; to represent the latter as a wholly passive being who owes nothing to himself and everything to God." [156] John Stuart Mill recalls with wit and humor his childhood days. "My father taught me," he wrote, "that the question 'Who made me?' cannot be answered, since it immediately suggests the further question 'Who made God?' " [157] But coming down to earth, as George Jean Nathan put it, "One's faith in God increases as one's faith in the world decreases. The happier the man, the farther he is from God." [158] And the gods change as men and their cultures and societies change; no god's life is exactly static or unchangeable. As Elie Faure, in his *History of Art,* points out: "A god only becomes a god at the moment when he assumes form. This is true. But it is also true that at the moment when he assumes form he begins to die." [159] It was Schopenhauer who said, "If a god has made this world, I would not like to be that god; the misery of the world would break my heart!"

Thus man must recapitulate his life and the Time he lives within it. We begin to see how priceless are the very moments we breathe. We cherish the ability to see and feel and hear and do so many things we know as commonplace but which are the very essence of living, enjoyment and sober appreciation. Once we evaluate life in terms of living, we are grateful for what we have and what we could still achieve. We begin to have good reasons to take courage to better these conditions so that the universe may be unfolded in its possible fullness before us and within us. In our own perception of life, we come to realize the same relative values for other things, including our fellow-beings. *Life becomes too sacred.* Hate becomes obnoxious and a waste of precious time. Man sees himself, not as a member of a puny little set of beliefs, but as a living part of a wondrous universe of living things, all fused into and out of one great principle of nature. *Only the sacredness of life can stand as a principle for its own preservation.* Time has meaning in terms of the hour, not in terms of eternity. Albert Schweitzer feels this pulse when he writes: "Day by day, hour by hour, I live and move in it. At every moment of reflection it stands fresh before me." [160]

Voltaire, after jostling Candide about a world of terror and cruelty, had him conclude, "Let us cultivate our garden." We have just begun to do this by uncovering an opiatic weed. When we shall have cleared our worldly garden of weeds, we will be able to evaluate much better what remains, and cultivate that which is *really* true and possibly good. When man rises above all the ignorance and silliness which theology has filled him with all these thousands of years, he rises above the bog of hate and creedism. He becomes a man, in the true sense of the word. He identifies himself as a living thing and is willing and humble to go through life on a basis of rational principle. When a man can find himself through and by reason he will have sufficient reason to find others. Thus he can also find the meaning of Time.

The clock clicks alike for all, for the one who is grateful and happy to live another day and for the one who is willing to throw the day away for the blinding mirage of eternity. The clock clicks alike for the fisherman smelling the fresh, fragrant air of the woods by a stream and alike for the monk sitting on straw in his windowless cell repeating prayers monotonously hour after hour with a sombre, tragic, cheerless face. The clock clicks alike for the mother radiantly happy with her baby, digesting its every whimper, cry and smile, with delight and deep exhilarating satisfaction, and it clicks alike for the nun who has forfeited motherhood to marry a myth and resign herself to ascetic slavery. It clicks alike for the Indian boy riding his buffalo home with a song and a lust for life, and alike for the old Brahman sitting on nails and who has not spoken a word for years as his way of trying to bribe his god for an eternal ride on Shiva's bandwagon. On and on we can go, parading before you the people who want to live in the sun, who want to be happy, experience normal living in its fullness of flavor, to live conscious of its brevity, grateful for its continuance, peaceful because life, to be enjoyed, must be peaceful. And on the other side, the people who grow old when young, who shy away from life and from the living and await, in fear and uncertainty, the death for which they have been constantly thinking of and preparing for. For all these the clock clicks off its precious moments of Time.

It is highly doubtful whether we will see in our time the elimination of religion from the life of mankind. Whether it will ever be eliminated is also problematical and even dubious. To be honest about it, whether I am right or wrong doesn't matter but I feel a doubt very seriously whether it will ever be *completely* placed behind the history of our world. My reason for thinking so is because I feel that man's ego, which is a natural and constantly functioning part of man's physiology, simply doesn't want to die or to end itself. This is understandable. It will always, it seems, refuse to acknowledge, even in the face of realities or in the conscious sense of self-deception, the naturalness and the eventuality of personal oblivion. There are some philosophers and scientists, who in order to palliate the religious need for personal immortality, will try to explain it in what is understood as *biological immortality* or the continuity of cellular life, the constant rebirth of living organisms through what is considered an almost "never-ending" process of life and death, growth and decay, reproduction and disintegration. There is another class of thinkers who like to soothe this compelling desire of man by saying that the accomplishments, the personality of his deeds and his influence upon others continue as testaments of his life and persist, in a way, after his death. While all these things and claims may have psychological value to people and may be pragmatically justifiable, they do not have anything whatever to do with the specific idea that a person wants his personality or soul to continue after death on a similar basis as he is conscious of during his life.

To indicate the powerful influence of the idea of personal immortality affecting people who will accept it in the face of knowledge and reality, this is what Dr. Frederick Barnard (1809-1889), once President of Columbia University (1864-1889), had to say: "Much as I love truth in the abstract I love my sense of immortality still more; and if the final outcome of all the boasted discoveries of modern science is to disclose to men that they are more evanescent than the shadow of the swallow's wing upon the lake . . . if this, after all, is the best that science can give one, give me, I pray, no more science. I will live on in my simple ignorance, as my father did before me; and when I shall at length be sent to my final repose, let me . . . lie down to pleasant, even though they may be deceitful, dreams." [161] Even they who know better refuse to part with it, and considering the human constitution, it is understandable, expedient, consoling, and forgivable.

Miguel de Unamuno speaks out for the consoling need for God's existence: "I am presented with arguments," he writes, "designed to eliminate it, arguments demonstrating the absurdity of the belief in the immortality of the soul; but these arguments fail to make any impression upon me, for they are reasons and nothing more than reasons, and it is not with reasons that the heart is appeased. I do not want to die—no; I neither want to die nor do I want to want to die; I want to live for ever and ever and ever. I want this 'I' to live—this poor 'I' that I am and that I feel myself to be here and now, and therefore the problem of the duration of my soul, of my own soul, tortures me." [162] . . . "We aim at being all because in that we see the only means of escaping from being nothing. We wish to save our memory—at any rate, our memory. How long will it last? At most as long as the human race lasts." [163] Thus, Unamuno in his *Tragic Sense of Life* makes the same tragic mistake as Kant; first he acknowledges that God is not verifiable to the human mind and then with the same human mind proceeds to rationalize God into existence out of necessity and desperation.

Nevertheless, if man must live as a plural identity of body and soul and must have deception, either in ignorance or in spite of knowledge, at least let them live together for the purpose of living, not of dying. If the soul is to be the acknowledged master, which is, of course, infantile to any psychiatrist, at least let us educate ourselves to the point of influencing this "master-soul" to direct or help direct the body to live better, to a possibly higher and finer and healthier sense of being and to achieve happiness on earth, during our earthly lives, and not to draw us away from life but towards it and within it. Let us follow, at least, the last words of Socrates who said to his friends that whether there is a god or not doesn't matter but what does matter is that a man should live a good life and no harm can come to a good person now or after death, and in this way we can have the sustained satisfaction that we at least have lived a good and happy life when we were able to and fortunate enough to accomplish it. "There is such a thing as faith in intelligence becoming religious in quality—a fact that perhaps explains the efforts of some religionists to disparage the possibilities of intelligence as a force. They properly feel such faith to be a dangerous rival." [164] "If thought is to set out on its journey unhampered, it must be prepared for anything, even for arrival at intellectual agnosticism. . . . Previous thought imagined that it could deduce the meaning of life from its interpretation of the universe. It may be that we shall be obliged to resign ourselves to abandon the problem of interpretation of the universe and to find the meaning of our life in the will-to-live as this exists in ourselves." [165]

Let us lean, then, upon the value of Time and all the gods will be with us. Because, once you have realized the importance, the irreplaceability of this very passing moment, you will be placing a high value on living as the prime motive in life, of itself, for itself, and by itself, and this is exactly the original idea for which the first god was created by man. Make Time your god, if you must have one, and you will be just as strong as your god and your god will never leave you because it will exist by your side so long as you *know* you are alive. And if you must have an altar to worship at, go into a garden. A garden is the only temple worthwhile to go into. There you will find gods and goddesses ever ready to compose you with peace of mind. There no one will threaten you with hell or opiate you with dreams of ethereal bliss. Each little garden is a speck of heaven in which one can enthrone himself in the fine, free, fragrant air of fresh and joyful philosophy.

True love, sympathy, friendship, appreciation, goodness and character, cannot be achieved or attained by the adoration or worship of images, amulets, books, pendants, rings, special clothes, the following of rules, the conformance to holidays, or the mere repeating of language. It is not what our beliefs are, nor whatever we worship or glorify; it is what we ourselves have done to make this world a better and happier place—this is a reasonable aim and this constitutes our morality.

These are the wise words of the poet, George Eliot:

"So to live is heaven."

Nothing is ours except Time.

SENECA, *Epistulae ad Lucilium*

THREE

TIME AND THE *POWER FACTOR*

Hegel once stated, "To judge what has content and solidity is easiest; to comprehend it is more difficult; but what is most difficult is to combine both functions and produce an account of it." We shall try here to apply this effort to the nature of *Power* and its relationships to a philosophy of Time.

Power is the natural heritage of man and it cultivates the desire and inclination to express itself by building a stronger force to ingest lesser forces or by destroying them, or by trying to overwhelm less resistant forces than its own, only, in turn, to be overwhelmed sooner or later by still stronger forces. It seems to be a manifestation not only of man but of all else and it identifies itself by the cyclic processes of nature, by the continuing regeneration of young or renewed forces and the disintegration of old or weaker, deteriorating, degenerative forces. "All life is action and the tension of force. All the faculties, sensation, imagination, understanding, reason, will, are of the same nature; they differ only in degree." [1] "Conflict is the basis of life." [2]

In man we discern this manifestation as the subconscious desire or *will-to-overwhelm*. It is the generating energy of life to climb or fall, extend or contract, to spread or shrink, to absorb or to be absorbed. It is the chemistry of life to live by growing or dying. Growth and decay are the fused, collated processes of one process.

Very often we see men of eighty, even ninety and more, still actively engaged in building power, in trading, banking, politics, controlling, striving to gain more properties, more assets, more power, cumulating more money constantly even though, if they lived another five hundred years, they could not possibly spend their present fortunes. When Queen Elizabeth was dying she exclaimed, "All my possessions for one moment of time!" She was very willing to give up everything for just a little more life, even though she was a shrewd and ambitious, property-coveting queen all her life. One obsessed with power wears dark glasses. What is it that makes people behave in this way when they have so little time left to live? Why do they enjoy so passionately and helplessly the struggle for power and possession until they are almost on their death beds? Why do they fall sick and wither when they are taken from their race and forced to "catch their breath" on a doctor-compelled vacation? Why does it seem, this frenzy for power, like an opiate which they must constantly feed themselves with in order to sustain themselves from day to day? It couldn't be based on ignorance or a lack of intelligence, of reasoning ability, as many of them are brilliant, well-educated, and in the social sense could be considered highly intelligent, so-called mature, shrewd, sharp, smart, and calculating. I heard about a millionaire businessman who dropped dead at fifty-five. This man raced through life, intoxicated himself with power, authority, and money. He was a multi-millionaire when he was thirty. He never had a real friend because his ambitions were so strong that he stepped on many people in order to leave behind millions of dollars as an epitaph to his own empty existence. He had no time for marriage and children and was unkind and unsympathetic to his relatives. He never knew why he lived and what for. Aesop again points his finger and tells us that "wealth unused might as well not exist." [3] "The source of all wealth is the peasantry of grass." [4] Lewis Mumford wisely says: "The person who fancies he has made his own career, or the inventor who fancies he has the sole right to his invention, or the philosopher who announces a completely new system of thought, is merely ignorant of his sources." [5]

We have seen in the previous chapter, concerning the religious factor, that the biological and psychological analyses of man's acceptance of the belief in personal immortality indicated that man, regardless of the truth or falseness of the doctrine, firmly believed in it to the extent of sacrificing even his own beloved children, parting with his most prized possessions, even killing himself to "realize" it.

To resign oneself to the idea that death is the final end of the person is unthinkable and unacceptable to the modern, Westernized primitive, and to the primitive of twenty thousand years ago quite unnatural and impossible. The idea of life continuity after death was an uninterrupted manifestation and relationship of the entire personality in a more or less identifiable form as it existed during life.

The idea of the *Power* factor concerns the *nucleus* of this personality form, the *Ego,* or the *Super-Ego,* or the *Id,* or the subconscious drive of people who, lacking a clear, discernible, meaningful value of Time plus a quantum of ignorance regarding its essence and measure in relation to the other factors in their lives, are either helplessly victimized by the norms of society or self-victimized by their own neuroses or by their unwise philosophy of values and meanings. Looking back over the centuries, Frazier Hunt tries to analyze the behavior of peoples in their struggle to achieve some kind of life satisfaction. "I cannot stress too often," he says, "this strange and incredible release that comes to a lowly people through revolution. Longings for beauty and happiness and leisure and security are not limited to those to whom the chance of birth has given wealth and place. The human heart varies only in the tiniest degree. Universal is the dream for the good things of life. For ten thousand years and more the struggle for mere food and shelter and clothing used up most of the energies of man. Chiefs and priests and fighting men combined and took tribute from the common herd. They looked down upon them and thought of them as slaves ordained by God to their inferior status. But no one generation ever passed that did not see these lowly and abused lift their eyes to the stars and pray that soon justice would come for them. Inarticulate, submerged and chained to a slave's psychology of failure, yet the fire in their hearts never went out." [6] No wonder the wine of power and its soothing flavor are so enticing and so wanting to those who were denied it so much.

In the religious factor man threw away his life *here* for a belief in a life *after*. In the power factor man also throws away his life, not for a life after death, but a phantasmagoria *during* his life, like a disease which can only envelop a person during his life and which relaxes only when death relaxes all. Both factors are biological and psychological. The former is sure he is going to live *forever;* the latter is sure he is going to live *tomorrow.* One believes in infinity, the other assumes it as an immediate appendage of the present; both deny themselves a rational, realistic, equitable approach to the value of the Time factor as a concomitant relative of their lives. Like the other animals, the humans, too, rarely consider themselves perishable until the undertaker proves it. No one desires death, not because it is lack of existence, for a stone exists also, but because it is not sufficiently troublesome. It is the striving against things, the struggle on virgin paths, the pioneering, aggressive spirit against opposing things, the breaking through of walls that confront and halt us—this is part of the interest that makes life what it is—the *will to overcome.* This is the life spontaneity of each thing. The strange as well as the naive plan of nature has it that by doing so, many are overcome by that which is least suspected. Vigilance, but not too much vigilance, is the peculiar measure of keeping the worlds in balance.

What is *Power?* It is the expression of life, the means by which life manifests itself. What is *life?* It is the expression of death, the means by which death manifests itself, and vice versa. It's all a carousel on which we think we are going forward when we are only returning to where we started. Life is always in the experimental stage, but an experiment that is based upon the same thesis and brings the same synthesis. Creation is not a thing that has been done; it is rather a thing which is being done and will probably never be done doing. The spark of life doesn't come with any special decorations. It comes very much unadorned and goes out the same way. Simplicity is the torch it carries and sincerity is the light it bears. Matter identifies itself through mind; thought through action; philosophy through practice. It builds out of one cyclic recurring principle. Life and death are words merely signifying states of being of the same substances. Death is a capsule that comes with every seed. Life is the growing, flowering, fruitful plant, but death within it is the potential soil for its growth and existence. Death is the window-dresser of the universal store; it takes the old and changes it continually into the new. Life is the continual effort to support life, and death is the vehicle of this effort. "It is questionable whether man is born for any other reason but that an egg has been fertilized. If it is true that while there is life there is hope, it is truer that while there is life there must be death." [7]

I think it is wrong to consider death as being outside of life. Biologically it is always within life. It is the stream in which is carried on the substance and the manifestation of living things. What is the stage called on which all of this play takes place? It is called *Nature,* the blind mother of the Siamese twin, life and death, with interchangeable personalities. To understand the meaning of power and to be able to value it more properly we should understand a bit about the *nature* of Nature, through the arteries and veins of which the ertia of power flows and recedes.

There are many people who profess to believe, religiously, metaphysically, or philosophically, that the whole of Nature, of which we are a part, is the *body* or entity of God. Further, some claim that this Nature is a necessary and unconscious harmony, geometrical and physical; others claim that the entirety of Nature is not only a harmony, because all its related parts and attributes fit into a general pattern of things, but that it is a purposeful design that indicates cosmic, universal intelligence; that the over-all governing principle is of a moral nature or essence. Harmony indicates design, they say, design indicates cosmic intelligence—cosmic because this design must include everything to be universal and harmonious. It therefore follows, they claim, that there is an universal intelligence.

Let us see what all this brings us. If there is universal intelligence, then the intelligence cannot be a harmony, because harmony is perfection while intelligence is perspective and the power to vary; therefore, there is neither intelligence nor harmony. Also, it would reasonably follow, as Nature is supposed to be a harmony, that the destructiveness of Nature in the form of volcano, plague, violence, cruelty, and all the other destroyers of mankind, are part of this harmony, which we cannot reasonably call good, moral or just, and which *we* cannot reasonably and consistently love by intellect or emotion because our intellect and emotion cannot see either the justice or the acceptability of such harmony or design. When one is about to die in battle, should he say that he must do this to perpetuate or not to upset this harmony? When a mother sees her dead child, broken and bleeding on the pavement, killed by a hit-and-run driver who got away, should she say that "All's well, this is part of the great harmony, the moral force of the universal intelligence!" When the hunter aims his rifle at a curious, staring deer, who is to accept harmony and who is to deny it? If a steer could walk into the butcher shop and see himself all chopped up into little pieces with price-tags, could the steer acclaim that this is the great harmony and that he has at last found paradise on the meat counter? What is good for one may be bad for another. What is fine for the victor is death to the loser. Who can proclaim the universality of the harmony of Nature?

This harmony design of Nature is such a lazy proposition. If one had a bed-bug crawling down upon his nose, he would, to be true to this idea, have to say, and with dignity, "I suppose, noble creature, that God has sent you down to me, and as everything has its place and purpose in creation, it must simply be that your place was destined for my nose!"

If Nature is harmony why should the English little owl have to kill a pheasant, chicken, rat or rabbit, for the sole purpose of burying the body so that the corpse may attract beetles, the food which the owl seeks? [8] The idea of God in Nature is critically analyzed well by John Stuart Mill; he says: "A single hurricane destroys the hopes of a season; a flight of locusts, or an inundation, desolates a district; a trifling chemical change in an edible root starves a million of people. The waves of the sea, like banditti, seize and appropriate the wealth of the rich and the little all of the poor with the same accompaniments of stripping, wounding, and killing as their human antetypes. Everything, in short, which the worst men commit either against life or property is perpetrated on a larger scale by natural agents." [9] Speaking of design and purpose, Haldane writes, "The majority of species have degenerated and become extinct, or, what is perhaps worse, gradually lost many of their functions. The ancestors of oysters and barnacles had heads. Snakes have lost their limbs and ostriches and penguins their power of flight. Man may just as easily lose his intelligence." [10]

Life confesses its own disharmonies. "Our strong will to live is opposed to the infirmities of age and the shortness of life. Here lies the greatest disharmony of the constitution of life." [11] Disharmonies are abundant throughout nature; trial and error are its heritage. "Rudimentary and useless organs are widely distributed, and we find them in many places. Familiar instances are the atrophied eyes of animals that live in the dark, and the sometimes rudimentary sexual organs of many plants and animals." [12] "The tiny protozoa multiply in such a way that if all the descendants of a single individual should live throughout a period of five years, their protoplasmic mass would be thousands of times greater than the earth in bulk." [13] Forest Ray Moulton states that "if the sun were created expressly to light and heat the earth, what a waste of energy!" [14]—because the earth uses far less than one percent of the light and heat from the sun.

Is there a moral purpose in this grand harmony of the cosmic intelligence? Fabre plainly says: "Pity is unknown among insects." [15] Donald Culross Peattie describes the affection of the sea anemone: "Elemental greed in a sea anemone closes upon a fish. This flowerlike animal, so low in the evolutionary scale that it has no consciousness with which to enjoy its meal, reacts by mere reflex to the presence of food. The fish is seized, enveloped, digested and the tentacles wave again in thoughtless readiness." [16] W. C. Allee states that "one species of animals may destroy another and individuals may kill other individuals, but *group* struggles to the death between numbers of the same species, such as occur in human warfare, can hardly be found among non-human animals," [17] which accounts for the excellent harmony of the human race. Alas, Nature, of which we are a part, has no way of showing its kindness, morality, purpose, or inclination. Its entirety, as a whole, is totally blind, and this might be a way of kindness, perhaps. But "What we plainly see is birth and death—the result of chemical and electrophysical processes . . . and beyond that—murder, the chase, life living on life, the individual sustaining himself at the expense of every other, and wishing not to die." [18]

One needn't be a scowling cynic, a bawling pessimist, or a worshipping, dreaming optimist, in order to observe things of certainty. Seeing is believing, and anyone observing Nature with a prudent analytical lens of fair judgment can detect the constant curricula of error by which Nature learns to continue to exist. Irving Adler in his little book *How Life Began* finds that "Nature chooses automatically and without purpose, and as a result of the *struggle for existence.*" [19] Tolstoi confessed that "Life is a blind alley." [20]

We observe that Nature is a constant flux of relationships of attack and defense. Experience is almost definable as either attacking (absorbing) something else, or defending itself (resisting) against something else. Those who attribute a sort of intelligence in the pattern of Nature fail to properly understand the meaning of intelligence or of any true, discernible sense of moral equity.

Let's try to face reality instead of trying to twist the universe (as if we could) to suit ourselves. Were mankind extinguished completely today, Nature wouldn't shed a single tear tomorrow. "Nature certainly has never wept in sympathy, nor stretched a hand protectively over even the most beautiful or innocent of her creatures." [21] "The sun will shine whether or not there is a human eye millions of miles away waiting to intercept an infinitesimal amount of its radiation." [22]

Before Nature nothing has preference; it treats all with its own justice, silently, both minute and huge, near and far, material and immaterial. Joseph Needham philosophizes thusly: "Existing infinite ages before man appeared, destined to exist ages after his disappearance, the universe moved in its cycles treating his whole realm of interests with complete indifference." [23] Nature forgives no error, praises no discovery, shields no one, fears no one, loves no one. "The grand, hard truth is that nothing in Nature happens in order that some thing else shall happen, but only as it must." [24] The constant rebirth of necessities is the generating power of Nature. "God helps those who help themselves because nothing exists which cannot in some way help itself." [25] Nature never gives; it always sells. There is always a price. The question is whether something is worthwhile buying. Therein is the test of wisdom.

Life, a manifestation of Nature, is a *nerve,* not a soul. It is a physical *mind,* not a fantasy of ectoplasmic magic. It is a *knowable.* It is a *Self.* It is *pain* as well as *joy.* Deep emotion, unrestrained sympathy and pity fathom in the unfortunate crevices of its own natural pains, but if pain is to be the dues of the considerate, of sympathy, and of understanding, it is also the first cry of life, the contant identity of living, and the unwilling appeal of the dying not to die. Eugene N.

Marais, author of the excellent book, *The Soul of the White Ant,* says, "If pain were to disappear from this earth, life would soon cease" [26] . . . "where pain is negligible, mother love and care are feeble. Where pain is absent, there is absolutely no mother love." [27] Pain and joy, peace and conflict, contain life, and life is confined within them. John Dewey wrote that "it is of the nature of society as of life to contain a balance of opposed forces. Actions and reactions are ultimately equal and counterpart." [28]

Life is worry, fear, anxiety, greed and hypocrisy. It is also love, devotion, sacrifice, charity, and appreciation. These are the opposing forces which make up life and give cause because of the effect of the other. Nature is a very blind mother, stone blind, is deaf and mute, and if she could speak she might say: "I am in all things of the mind and all minds enjoy and suffer; I am in all things of the mindless and they, too, enjoy, suffer, or just exist and change. I am pain because I am born of pain. You cannot take me in vain, for I am equity. To me you are sincere, though you may wear a mask before the world. I am judgment, and I am never too late to exact justice. I am accident, and Time is my field of play. The more I am known the greater the knowledge; the greater the knowledge the better the justice. I deny no special form, for I am always forming.[29] I am always born anew to suffer and to enjoy anew. Motion is my circumference; I am forever moving in circles without end. My worth to you is how you understand me and how you travel with me. Know, to know me, that heaven and hell are within me, and I am the *Moment.* It is not what you pray for, but how you live. It is not what you wish for, but what you do. My rewards and my punishments are sufficient." This is what Nature could or might tell us if she could only talk. But she can't. And we have to learn, as usual, the hard way.

Power is the *motion* of Nature. "Experience is *of* as well as *in* nature.[30] . . . The very existence of science is evidence that experience is such an occurrence that it penetrates into nature and expands without limit through it." [31] It is the test of wisdom to use this Power to bring us the fullest possible return. We can use it to blind us, enrage us, to soothe us, to give us further, clearer sight. We can use it to make money or we can use it for money to make us. We can use it to make happiness or we can use it to injure or destroy. We can use it to see the world or the world to see us. We can use it as a weapon or as a pen. We can use it as a means of giving or we can use it as a means of taking. We can use it for love or for hate, for shaking a friend's hand or punching someone on the nose. We can use it for charity, mercy, and compassion, or we can use it for theft, dishonesty, deception, cruelty or a confidence game. We can use it for peace or for war, as both contain, as life does, a measure of power. Perfect peace is the horizon that life cannot reach. Life always contains a measure of power, and power cannot be perfectly peaceful. But it can be applied in myriad directions, for both good and bad. There is no psychosis in the nature of power; the psychosis may occur in the application of power and it may occur in the mind of man in the use of this power.

It is this very power that the primitives sought to master, control and use by sorcery and magic. It is this very power that man has tried to extend to eternity by trying to fit the robes of gods upon himself. It is this very power that later philosophy tried to rationalize into principles for living. A man has a right to call the universe his own for the better or for the worse; but only his mind, his heritage and and experience, will help him decide for which. After all, the wise man realizes that possession is mostly a mental state, for the truly happy man possesses the world, and the unhappy one, though he hold title to the world, does not possess anything. Though a man own all the wealth on earth, he does not own more than his own life. And what good is it that a man own a million acres of land if he doesn't know how to live on one?

We see, therefore, that *power is a vehicle* that can be used by anyone, more or less, for good or bad as each interprets good or bad. Professor William H. Kilpatrick elaborates this point: "A goal cannot be intelligently set forth apart from the path which leads to it. Ends cannot be conceived as operative ends, as directors of action, apart from consideration of conditions which obstruct and means which promote them. If stated at large, apart from means, ends are empty." [32] It is for us to rationalize, if we can, the relationships of power to the Time factor in our lives so that, perhaps, with a clearer understanding of its certainties we can gain a fuller and a finer sense of value for its use. To understand that power is not an end in itself but only a means of a process and of processes is the beginning of revealing to us a possible ethic in its expression.

For Power is the expression of life and identifies the natural processes of existence. *It should not be used as a substitute for life.* Power should serve life, not life the slave of power. "Every impulse is," writes John Dewey, "as far as it goes, force, urgency. It must either be used in some function, direct or sublimated, or be driven into a concealed, hidden activity. It has long been asserted on empirical grounds that repression and enslavement result in corruption and perversion." [33] When life, and with it the co-existent quantum of Time, is enslaved to power, then an illusion is set up that blinds us from truer values by which we might reach a higher and finer realization of affirmative and positive factors in the enjoyment of life in general, and the happiness and satisfaction we seek every day in particular.

Take for instance, the power of money. Just a simple, small thing as a coin, shaped and stamped by man himself, a little rounded piece of metal one can flip up in the air so casually and freely, and let it fall upon the pavement to see whether it falls heads or tails up. Yet this really innocent man-made piece of metal has been the cause of untold miseries, misconcepts, injustice, anxieties, feuds, fears, murders, even wars between nations. When money rules, the rule is war to preserve it.

Now why are we so concerned with money in relation to the Time factor? Money is a form of power besides being a means of security and exchange. It is needless to state that the economics of life is a most necessary and prime constant in our efforts to sustain ourselves, to provide food and shelter for ourselves and our loved ones, to feel a sense of security and peace of mind for today and for tomorrow, to even give us the possibilities of luxury, easy living, travel, prestige, and afford us with expressions of charity, benevolence, gifts and all sorts of pleasures which money makes possible.

But it is axiomatic that money is not everything, as any sensible person well knows and will testify to. The problem we have is that we all too often give to money a higher meaning and value than it deserves and, as a result, it covers us with false and deceiving values, sometimes to an extent that it becomes a neurosis with us and blinds us to higher and stronger values. I have met rich people who had no money and I have met very wretched people who had lots of it. People share their crumbs, not their diamonds. People are more apt to share the little they have, rarely when they have much. Money cannot think. Money can buy anything as good as money, but not better. Money is truly important, very important, but when it becomes all important, then something has gone wrong with our sense of value. The greatest and most tragic mistake many rich people make is that they so often confuse the possession of money and the possession of brains. A man may have a million dollars, have cancer of the throat, lose his only child, and the wife he loved may have run away with the butler whom he always hated, but so long as he has a million dollars he is considered a *wealthy* man!

It is not correct to say that a rich man just died and left behind a million dollars. It is better to say he left it *upstairs*. Both time and money are spent by man; man can replace money, but Time will always replace man. Money itself is meaningless unless it is attached to something more valuable. What represents value is the test of a man's philosophy. After all, it isn't what a man possesses; it's what he contains that is really important and meaningful. Money, like all things necessary to let us live in security and comfort, is a subdivision of Time. He who can fully understand this makes money to live instead of living to make money. I would rather be on a horse and buggy on a winding dirt road overhung with birch and maple than an ingenious robot in a financial mausoleum. The man without a plan, fishing by a brook, is far wiser than the money king who spends his life storing a cache of dollar bills for the Surrogate Court.

Very often money becomes a burden to the rich and they spend so much time of their lives worrying about it, and trying to figure out how and to whom to leave it when they die, that they lose the time to spend it on themselves. There are two kinds of lamentors when one dies; one kind laments that the deceased should have left them more money; the other kind laments that it's a pity the deceased didn't have a chance to spend it on himself.

Often, too often, money brings so many possessions of one

kind or another that they become like anchors to hold people down like slaves, entanglements of property that frustrate the natural and free sense of movement and unburdened thought. If only they would consider things outside of themselves as perishable and as transient as they are, instead of considering themselves as permanent as things outside of themselves, they would know better how to spend their time, which is infinitely more precious than material things, and receive the greater wealth of more happiness through greater freedom. Lest we forget in our rush for power, possession, property, and authority—only a short time ago we were babies, helpless, dependent and broke, and they loved us for it!

Greed is like a stone rolling down hill. It picks up speed as it goes on and finally ends up in a ditch. The happiest man is the richest man and all his riches are within him. We often make the mistake of not reducing the complexities that pile up to simple elements so we can so much better take inventory of ourselves. Simplicity means that one can have all the conveniences, comforts, pleasures and luxuries of life and remain the master of these things, that he is above them and not below them, that they exist because of him and for him, that he can do without them when he should be called upon to do so and still remain the same person.

Contrary to many hard-boiled economists, statisticians, and money-men, a better economy necessitates and implies a rise in ethical culture as anterior to any such progress. Education will do more to enhance a more equitable economy than laws. Laws are over-ridden, misinterpreted, misapplied and abused through countless loopholes. A person with a more equitable attitude possesses a social behavior which recognizes the firmer and more substantial energy of fairness and reasonableness.

The most successful banker is the one who knows when to stop counting his money so he can go out for lunch. While a business is important to make money it should never be used or considered as the making of a man. Man should be more important and the time he spends doing business should be valued in the same way as the primitive who went out to hunt for food. If one spends all his time hunting for food, when will he have time to stop to eat? A business is a contraption that takes a man's mind off living, and when we consider that a day is what we pay for living twenty-four hours less, we should not make any business into a place of worship wherein we go about like austere high priests preaching the end of the world unless that order does come in or a shipment goes out, and where the monthly bank statement becomes the Ten Commandments. Captains of industry are very often really just inflated buck privates who believe they are generals, and if one of their employees doesn't do his daily dozen to the brim according to the expectations of the Univac, they would have him shot if it weren't for the law to cool their ambitious tempers. The feudal lords actually did, because they were the law. Silly little people who waste so much Time, precious Time. Decorum, system, effort, conscientiousness, are all important if a business is to succeed, but we still should keep in mind that people are born to both serve and be served; none are born purely to do one or the other. I wonder what would really happen if a departed spirit would return to talk to us. Would he say that if he had another chance to live, he would never complain about the weather, or worry about debts, or give a damn about how much of anything he has to leave behind? Or, knowing human nature as we do, would he?

There is no doubt that a major factor in the history of human association is *Power*. No one will question the fact that it has affected the lives of most people living today. Likewise it has affected the human race for thousands of years, intensifying or relaxing conditional to time, place, and event. The world seems to rest on this volcano of agitation and it never grows completely cold. Not many have taken the effort to find out whether there is not some other vent for this tremendous innate urge other than directly up into the groins of living people.

No one can deny that man possesses definite predatory instincts which are not in any way humorous. Relating the history of the later Toltec-Mayans, Gann and Thompson state: "The Nacons were special priests, whose sole duty was to tear out the hearts of sacrificial victims. They wore their hair long and matted with the blood of the victims they sacrificed, and were clad in long white cotton dresses that reached to the ground. Below them were a grade of lay priests, called Chacs, whose chief duty was to hold down the arms and legs of the sacrificial victims." [34] And Denison tells us about the guillotine in Paris during the revolution. He says, "A hole measuring about six cubic feet was dug underneath the platform to take the blood and also the water with which the engine was frequently washed. But this trench was quickly filled and began to poison the surrounding air to such an extent that the authorities, in the interests of public health, decided to fill it up and make another deep enough to absorb the blood." [35] And Bertrand Russell satirizes again for us: "We develop wonderful skill in manufacture, part of which we devote to making ships, automobiles, telephones, and other means of living luxuriously at high pressure, while another part is devoted to making guns, poison gases, and aeroplanes for the purpose of killing each other wholesale." [36] Man is a beast, and no other beast is capable of the bestiality of man. "To my mind there was very little difference between those jungle primitives and these Kultur maniacs with fanaticism written on their faces." [37] Man's diary has maroon covers and its pages are bound together with the pangs and shrieks of torture and death. "In spite of our modern surroundings, most of our instincts are still those of our savage ancestors." [38] And Fabre, the greatest entomologist of France, bitterly remarks: "To what ideal height will the process of evolution lead mankind? To no very magnificent height, it is to be feared. . . . We are made after a certain pattern and we can do nothing to change ourselves. We are marked with the mark of the beast, the taint of the belly, the inexhaustible source of bestiality. . . . Hence the endless butchery by which man nourishes himself, no less than beetles and other creatures." [39] The strangest thing about it is that all the murders, wars, and billions of crimes committed to satisfy the lust for power, all of it has not made the world one wee bit larger or smaller, nor has it extended the longevity of life. Carlyle said that history is the biography of great men; obviously the rest were the victims.

It is needless to elongate our survey by reciting all the bloody wars, feuds, massacres, and slaughters, in which man killed man. No other animal has been capable of such continuous and wholesale destruction, while maintaining a supremacy in numbers, intelligence, and material welfare. It will, nevertheless, be interesting to occupy ourselves with some possible causes for this strange suicidal trait of men.

In the beginning, man, either as the slave or master of woman, went out to bring back food to his little family and protect it from its environmental enemies, just in the same manner as the birds and other animals do. When the community group or tribe came into existence, he still served in the same way except that he united with similar members for the common purpose and his behavior and habit were integrated into the colony and cultivated into a homogeneous process. Still, as we have learned previously, he had not realized self-identity as a unit or an individual. His entire existence seemed to be social and purely communal. In this state of his evolution, his traits were as yet untainted with subtlety, conspiracy, lust for power, self-aggrandizement, sex-perversion, and greed. Quincy Wright, in his *Study of War,* states: "Wars of independence are unknown among the most primitive people because slavery, subjection, and class stratification are unknown. Slavery, social classes, empires, and minorities are phenomena of civilization and of the most highly developed of the primitive people." [40]

When the priesthoods and early kingships ascended, and with them the leadership of the few over the many, man slowly realized that it was possible for one to be more powerful than another and to profit by it. The medicine-man and warrior-leader were born of the same stem; both called upon the gods to assist them in their "good" purposes. The people listened, obeyed and were enslaved mentally, morally, and physically. The instinct to protect and sustain was channeled to destroy and exploit. The naturalistic trait to provide security, even blindly, was turned into another trait to imperialize, even intelligently.

In the same manner as different theologies, priests, ascetics and prophets, decentralized and broke up the natural evolutional homogeneity of peoples, the warrior-leaders and the magicians, priests, etc., became kings and found that a king must have been accorded "divine" privileges and wisdom for the benefit of his subjects. This "benefit" had to be expanded, and conquests were usual, bringing about reprisal, conflict and war. We find wars among the insects but they are not carried on for the benefit of the king or any individual of the colony or fornicary except in protecting the royalty for the purpose of propagation and in the preservation of the colony through egg-laying. Wright continues: "Ants and termites maintained highly complex societies in the Oligocene fifty million years before the origin of man. These societies have, therefore, had time to develop specializations and modes of maintaining social equilibrium superior, in some respects, to those of human society." [41]

Man, in becoming an individual, found others to fight for him. Military paranoia was born and history is filled to the brim with it. Megalomania knows no bounds until it is exterminated; it is sublimated to some degree in all of us, and when it reaches to such proportions that it transforms itself into action, it merely takes the proper bacterial event and circumstance to cause conflagration and social chaos. Dictatorship and tyranny are born of it. Greed is born of it. The lust for power is born of it.

Intelligence and emotion serve each other's ends by least resistance. When the emotional impulse is stronger, the intelligence will serve the emotions. If the intelligence is sufficiently submerged to the emotions, a self-styled prophet is born. When the emotions are suppressed by intelligence, a distorted snob is born. When the intelligence is properly cultivated to realize the importance, value and satisfaction that intelligently-channeled emotions can bring, an intellectual coordination takes place within the experience of a person. The functional and rational processes of the human body becomes homogenous. A waxened mannequin has very good manners and a crazed pervert hasn't any at all, but both are of no meaning or value either to themselves or to others. Civilization should be the process of rationalizing emotions to the point where such a social and political functioning can possibly bring about a greater potential of happiness for the individual without lessening the same potential for others. Instead, it has grown into an art by which the "civilizing" process is too often a pattern, religious, social, economic or political as the case might be, of exploiting one person by another. Herbert S. Dickey in his *My Jungle Book* relates that "Cinchona bark, from which the alkaloid quinine is derived, was one of the principal products of Peru. It was worked by

Indians who received as much as one pair of trousers a year for their labor." [42]

Within the arena of intelligence and emotion plays a historical cast of known and unknown leaders of men and women in all the fields of human thought and experience. He who is cunning enough to understand the audience of the masses dramatizes that which the people have not but would like to get. They applaud him because he promises to fulfill their dreams. Those who are truthful and sincere, who can promise little but try much that could be done without dreaming, it is thumbs down for these people and they are sacrificed in a million tragic ways and sometimes glorified later. The animal in the woods has to fight for his food; man would rather dream his nourishment into his stomach when his intelligence has made it possible for him not to fight for it, at least not like most other animals.

The nature of Power is not antithetic to life. It is the generating process of life. Power and growth are synonymous. Power implies the faculty of successful accretion or absorption, regardless how limited. The entire universe seems to be a dynamo of power and every living thing is merely a minute part of and within its constant change of variation and within the nuclei of every bit of existence. The most primitive forms of life had to fight for their existence, and because of this struggle these forms evolved changes of structure, size and function, and as these structures and functions changed, the quantum of power also varied. It doesn't matter whether an animal fights with its brain or with its claw, or with both; it merely does so because it has evolved its peculiar propensities and prehensilities for such behaviors and over which it has no control; this is the skeletal, basic function of life. Voltaire in *Candide* makes Cacambo say: "Indeed, the law of nature teaches us to kill our neighbor, and such is the practice all over the world. If we do not accustom ourselves to eating them, it is because we have better fare." Life asks no questions; it weighs no value nor gives its interpretations as to why the essence of life should carry on so. Man can interpret, but his functional potential to do this is controlled by this very desire to obtain the results of power, that is, to live more. Whether he succeeds in so doing or not doesn't matter to the nature of power. The failure is submerged and the successful continues. Nature is blindly kind in this respect: while it is a pity for a great good man or woman to die, it is true nevertheless that all tyrants must die sooner or later. Man can conquer man, but he cannot conquer Time. Nothing can conquer Time because *Time,* while it is a fourth dimension in measurement and relativity, is a relative factor in itself and the other three dimensions vary within it so far as they are related to it according to their finite extensions. It contains no power because the necessity for power does not exist.

What does all this mean to man? It means that if we are to be successful in obtaining a reasonable time-valuation it is essential for us to understand the nature of power, how it grows and declines, its purposes and how man can rationalize the direction of its growth and extension into the channels which use the plow instead of the sword and into other channels where men can equitably and wisely live by the fruits of the plow so that the need for the sword may not rise again.

It is futile, I believe, to idealize about the eventual elimination of man's instinct to fight. James Harvey Robinson, the historian, looks at this problem very realistically: "Man is always a child and a savage. He is the victim of conflicting desires and hidden yearnings. He may talk like a sentimental idealist and act like a brute. The same person will devote anxious years to the invention of high explosives and then give his fortune to the promotion of peace. We devise the most exquisite machinery for blowing our neighbors to pieces and then display our highest skill and organization in trying to patch together such as offer hopes of being mended. Our nature forbids us to make a definite choice between the machine-gun and the Red Cross nurse. Se we use the one to keep the other busy. Human thought and conduct can only be treated broadly and truly in a mood of tolerant irony. It belies the logical precision of the long-faced humorless writer on politics and ethics, whose works rarely deal with man at all, but are a stupid form of metaphysics." [43] It is the nature of life to fight, to be predatory, or to defend oneself, in some form or manner. Professor F. Max Müller gives a bit of ancient history: "If one reads the description of Babylonian and Egyptian campaigns, as recorded on cuneiform cylinders and on the walls of ancient Egyptian temples, the number of people slaughtered seems immense, the issues overwhelming; and yet what has become of it all? The inroads of the Huns, the expeditions of Genghis Khan and Timur, so fully described by historians, shook the whole world to its foundations, and now the sand of the desert disturbed by their armies lies as smooth as ever." [44]

It is so with man and he is no different from the rest of nature. However, it is quite another story to determine what is worth fighting for and why. The instinct of power in man brought about by necessity and carried on by each cell within his body and by every cell in all other forms of existence is an elemental trait of life and without which life cannot be what it is. Death, or the absence of life, has no need of it because *nothing within Time is of the same essence as Time itself*. Buddha was a wise man when he said that Nirvana is obtained when all struggle and the need for struggle no longer exists. He identified struggle with life. So did Schopenhauer, a disciple of Buddha. Both were right but only right as to their perspectives of life's ways. Here, too, it is quite another story to direct this struggle constructively to better and greater forms of expression and action so that the blessings of Nirvana can at least be partially extended to include the few brief years of a man's existence. This brings us to our first point in question.

It is natural and reasonable to understand that power should exert itself to the greatest potential as appears the urge for its necessity, no matter whether it is the power of the poor to endure or the power of the rich to enjoy. Brain and muscle combine in every animal, including man, to fulfill its needs to preserve itself and extend its area of aggression in ratio to its preservation and the possibility of growth extension relative to the line of resistance of external factors. However, man aborts this natural tendency by over-extending the area of aggression beyond necessity and against greater and more powerful external factors; this he contrives because his brain capacity is greater and the disharmonies of his body more numerous than these same things in other animals. We use our brain so much in selecting our foods that we eat many things which are not good for us, simply because in giving the brain a greater area of activity the body of man has lost out in other sections of his body, such as the loss in detection of danger, of poisonous foods, the loss of sensitive hearing, taste, smell and movement such as animals possess. So rapid has been the increasingly miraculous expansion of man's cerebral function that it has far exceeded its sphere of necessity. This is a strange assertion and it would be promptly taken that it is not good for man to rise in intelligence, that to be a dumb animal is far better, and the road to a better life is back to the savage cave. Far be it, nor is it possible. What we mean here is that man's capacity to expand his intelligence has exceeded his physical capacity to absorb the results of this intelligence, and this lack of absorption has resulted in disharmonies. In physiological terms it is the struggle of the diencephalon of the human brain to retain its original evolved purpose to protect the body and its clash with the cortical growth of the brain with its non-instinctual but calculating, reasoning potential. Dr. Joost A. M. Meerloo brings this out plainly: "Man is born like a monkey foetus, naked and unprotected, with a freakish brain, an overgrown computer, far too advanced for its body. . . . As yet, he has been unable to come to biological terms with his own overgrown brain. When this internal conflict remains unchecked it may even lead man to psychosomatic suicide: ultimate surrender to fear that devours him. Man's brazen and presumptuous brain has grasped too much in too short a time. His brain cortex, as a tool with which to master the world around him intelligently, even becomes an obstacle in the comprehensive understanding of himself." [44a] Dr. A. T. W. Simeons analyzes the situation thusly: "As far as one can now forsee the only enemy that seriously threatens man's continued existence is his own brain, with which he has so far been unable to come to biological terms. There seems to be ways of doing this, but it will be a hard struggle. Much of what he now cherishes will have to be dropped by the wayside, and it will be painful to tear himself away from many of those concepts to which he owes his spectacular rise. If he wants to free himself of the threat of psychosomatic suicide, he will sooner or later have to perform the painful operation of allowing new insights to criticize what he has hitherto considered the acme of his wisdom."[44b] Let us clarify this point further:

With few exceptions it is generally observed in the habits of most insects and animals that nature provides a social or group instinct for the purpose of protecting the species, such as in the ant colony or elephant herd. Primitive man lived by these instincts also and still does to a great extent today except that in primitive times his social conduct and relationships were activated by instincts, while today his line of social relationships follows a line of least resistance and necessity and therefore should be considered as resting upon a somewhat superficial basis. As previously noted, when man realized that he could do things alone and even get others to work or fight for him, he became an individual in the sense of identifying himself as an independent but related unit. His march through the centuries is the march of power racing downhill, gathering acceleration with increased and widened experience. Every faculty and agency within his grasp have been exploited and extended so that this power of independence and sense of security might be strengthened. It is a pity and a paradox, that the vast accumulation of knowledge, cultural, scientific and industrial, has so bloated man into a superficial and artificial robot in a whirl of dazzling achievement that man, in his very search for knowledge, self-identity and self-security, has relinquished his very hold on himself, subjecting himself as a child of Carthage, into a sacrifice for the new and modern Baal-Astarte of Power. He was so intense at trying to magnify himself, because of his subconscious fears of insecurity and the sense of futility in coping with the struggle to live now and to live forever after death, that he forgot to look at the clock and when he finally did hear it tick he was just too late to appreciate it. It was gone, and so was he.

For further clarification, let's go back a bit for a glance at the evolution of kingships. Very ancient kings were really the religious heads of small communities and tribes. When the warrior took over the kingship he had to obtain through the priestcraft the official recognition of the gods or god so that his authority could be established by divine right. There appears a two-fold purpose here: to obtain the blessing and assistance of divine guidance and protection; and of more importance, to psychologize the people into fearing their king as much as they feared their gods. This divine right of kings gave birth to the principle of eminent domain. During the revolutionary eras, when kings fell on block and spike, this right of eminent domain passed into the hands of new political and military forms of government, and continues with us to the present day. The eventual fall of monarchy, emperors, and dictators, was an evolutional phase of social and political change because "the entire history of the human race has shown that dictatorships of single individuals do not permanently endure." [45] The evolution of government follows a line from the power of a *real* person, in the form of chieftain, king, emperor, etc., to the *ideal* person in the form of government of a set of political principles by which communities of individuals are willing to be governed. Incidentally, the attempt of certain persons to stabilize themselves as dictators and fuehrers is merely a most perishable effort to revert the masses of humanity to primitive forms of political and social evolution.

Human nature, on the other hand, remains the same, often for the worst. Even though a modern government may be carried on by a set of more or less equitable principles, certain individuals carry on the eternal fight for power until the very essence and purpose of equitable government has almost vanished and what remains is a nation filled with confusing laws, vague institutions, propaganda, graft-ridden petty power-drunk tyrants of one kind or another in the civil and military departments, intricate and far-reaching monopolies for myriad forms of greed and growth, political platforms and party hatreds, all permeated with the purpose of self-aggrandizement and continual disregard for the common welfare. Lord Acton said that "power corrupts and absolute power corrupts absolutely." Edmond N. Cahn, in his excellent book, *The Sense of Injustice,* emphasizes that "power turns destructively on itself whenever it chokes inquiry and the free play of thought." [46] Robert Briffault remarks that "power results in injustice not because men are wicked, but because power corrupts judgment." [47] Carl L. Becker, late Professor Emeritus of History of Cornell University, puts it more strongly: "The simple fact is that politics is inseparable from power." [48]

The Founding Fathers of our Republic knew and understood the nature of power corruption as part of the political arena, and realized the risks involved in the great experiment of Democracy. Thomas Jefferson, in a letter to a Dr. Priestley, gave his version of this experiment: "In the great work which has been effected in America no individual has a right to take any great share to himself. Our people in a body are wise, because they are under the unrestrained and unperverted operation of their own understanding. . . . A nation composed of such materials, and free in all its members from distressing wants, furnishes hopeful implements for the interesting experiment of self-government; and we feel that we are acting under obligations not confined to the limits of our own society." [49] A lot of water has passed through since then and many have lost sight of this original purpose and the values originally intended. Unless Democracy fulfills its original birthright by cleaning house, its very foundation might be in dire peril and catastrophe might follow, not only to the structure itself but to the termites feeding upon it. History reveals the constant fight against the bulwarks of power and whatever culture we possess today and which is of any truly civilizing influence, has been attained by liberated educational, legal and equitable, processes by which the freedom of expression and a sincere and wholesome search for certainty and justice have become terms of meaning and action. "The hostile forces of nature have been largely overcome, but a new menace has arisen in the form of man's inhumanity to man, a force which has exacted more victims than the hostility of nature. The question therefore arises: Man has mastered so many antagonistic powers in nature, can he control his own social conduct and improve the relations of man to man?" [50]

The lust for power is a matter of ignorance and irrationality. One may possess knowledge but if this knowledge is not rationally applied to his life and his relationships with others, this knowledge is either constricted into a narrow channel of ineffectiveness or else is misled by nefarious and unjust purposes. A man who is rational and just but who possesses little knowledge becomes suitable for being victimized, duped, or betrayed by those who possess it and who appeal to the victim by various means to contort his perspective and stir his emotions to promote their evil plans. "In each generation the human race has been tortured that their children might profit by their woes." [51] A truly wise man is usually a good and just man, and the greedy and lustful merely clever. Wisdom must be associated with good meaning, honesty, justice and ethical intention if it is to have any place within the search for truth and its proper value and expression. The world suffers today because the scientists and sociologists are too often secluded and restricted but for their particular line of knowledge. "I am compelled to fear that science will be used to promote the power of dominant groups, rather than to make men happy." [52] The means of interpretation and application of the achievements of science are in the hands of those who are not scientific, ones who possess little knowledge, pawns of their own partisan minds, often neurotics and paranoics who consider themselves the saviors of the world but who really can lead it often into war, suffering and destruction. It is for this reason that in no matter what field of human work and association, the policy of a people should not allow the growth of any one person or group of persons who are not willing to allow themselves to be judged, controlled and examined by the free choice and mind of the people. Any such growth of power, regardless in which field it may be manifested, is a lessening of democratic principle. A man who cannot stand criticism it not worthy of it. He who would not permit it seeks no need of it and is necessarily dangerous to himself and to the community. A good leader never has to force people to follow him; if force is necessary then it will follow that his leadership is also unnecessary and misleading. Friendship, itself, is a process of sincerity for self and mutual betterment and happiness, and it can only be attained among people who are true unto themselves. The same must apply to a political and social structure if it is to endure and be of substance to its members. The pity of human nature is that it often is so similar to the life of the caterpillar. The history books are full of evidence, and this evidence is no compliment to mankind nor to the gods it worships. "The fact stands out before our eyes, grim and inescapable, as the controlling fact of our time—the fact that the present war [World War II] is a manifestation of power politics on the grandest scale ever seen." [53]

The lust for power, therefore, is innate in human nature and takes the path of least resistance to internal and external relationships which may either hinder it or give it vent for growth. As no creature is completely isolated from constant influences and relationships from external factors, so long as there is existence and life, it follows that no one person is entirely responsible for his own lust for power, success or failure nothwithstanding. Success often comes about because of the stupidity, weakness, appeasing and credulous tendencies of those very people against whom this power is directed. The pity of it all is that the nature of *power is an illusion* of relativity of the parts or units of existence which are able to perceive its rainbows, rainbows of force and pressure that are mere shadows passing between light and darkness, making up the continuous variabilities between the smile and the tear, tyrant and oppressed, the greedy and the impoverished. The oceans, with all their almost ceaseless force and beating upon the earth, and the mountains on whose sides little man daringly creeps, must still stay at one place. The nature of power is its own entanglement, because power, like everything else in existence, continues to vary and sooner or later reaches a state of self-expiation or recedes before still greater powers, *ad infinitum*. Thus power, in itself, as a separate and glorious end for either individual, nation, sect, or people, is an illusion; it is a temporary grief for victim and a temporary glory for the victor. Victory itself is a view out of focus which can only be corrected with time.

The philosopher, in his interpretation of these forces of oppression, is considered by many as an Utopian dreamer, a fool chasing fantasies of hope and ideals beyond the realisms of this world. The supposed "realists" of power are usually those who have aborted principles and betrayed peoples needlessly and stupidly, chasing their own fantasies of world power, conquest, or "realistic" statesmanship, in its many forms of economic, religious, militaristic, imperialistic and political aspirations. The urge for power and domination has caused the emergence of different status levels for people all over the world and has divided the masses of humanity into camps of *those who have and get* and *those who have not but only beget*. Thorstein Veblen stresses this particular point well. He says: "Chief among the honorable employments in any feudal community is warfare; and priestly service is commonly second to warfare. If the barbarian community is not notably warlike, the priestly office may take the precedence, with that of the warrior second. But the rule holds with but slight exceptions that, whether warriors or priests, the upper classes are exempt from industrial employments, and this exemption is the economic expression of their superior rank." [54] These upper classes made it their business to "educate" the lower masses to appreciate their good fortune in sustaining the power and security of the upper classes. Pierre Bovet, in explaining *The Fighting Instinct,* says: "Since ever schools came into being—and before—it has been traditional to rock the child in his cradle to the tune of battle songs. Stories about national wars and the sword play of mighty champions—these were the chief material for the epic tales which, as long ago as the Greeks, formed the starting point for education." [55]

The true dreamers are those who are so self-emulating that they cannot see beyond their own noses; they inflate themselves by false hopes of "perpetual" power. They do not know, in their mad race, that they cannot possess more power than they can possess themselves. Like life itself, it slips from between the fingers; the last dying moments of the cruelest tyrant were of meekness and helplessness even though he may have owned half the world. Like mercury, power may be cupped in the palm of ambition; it cannot be handled to our completely isolated purposes. The intelligent person realizes this and knows further that by chasing power he is trying to overcome his own shadow; this is often the sad and turbulent cycle of a man's existence. The aged man, like the placid lagoon, looks back and ponders why he was in such haste to overcome the distances of his own years. He who intoxicates himself with power is like the wave that lifts itself through the reef and over the bank to conquer the land only to spend itself through the sands of time and return to the sea, perhaps wiser but cleaner with the realization within experience that power is not a one-way tower piercing the skies but a cycle of rise and fall, like life itself and of which it is merely a manifestation.

The phantom of power, like the phantom of contentment, is itself the victim of a unit or a group of units of power. Great empires and great men, both good and bad, rose and fell like waves upon the sea of time, and those who can visualize that eventually the finite field reveals the predominating and prevailing time as a barometer of real meaning itself, move with Time but never try to outrun it. The pattern and style of power are not woven into the fabric of experience but they are rather a pattern of development and envelopment. It consumes, like fire, the potent elements of life and turns them into ashen ruins of devastation, misery, disillusion and often final tragedy. Whether it be the life of a person, of a nation or of an ideology, when its experience has been consumed by the avid processes of power, the final ruin of its cycle is inevitable. "The heroic hours of life do not announce their presence by drum and trumpet." [56]

You see people using the precious years of their lives for the major purpose of accumulation, for expansion of their material gain, for the promulgation of sinister ideas aimed at harming the peace and the integrity of other lives, for military might and the use of its weapons to destroy rational and democratic processes, and the question rises: Is this the art of civilization? If civilization has been meant to be a process of servility to golems, threats, monsters of economic and military pressures, then such civilization must eventually become perishable. A glutton either dies of obesity or a shortage of food; one or the other must come sooner or later. Because power is the struggle to live in its most elemental and primordial process, it follows the dictates, not of acquired knowledge and rational faculty evolved through milleniums of time, but the simple driving urge of animal matter in a blind race for its survival disregarding all else. This paralogistic attitude is the mental trend of all those who think only of themselves and have no urge or intelligence to think of the rights of others. It is pitiful but true that the human animal has achieved the supreme recognition of being the most selfish and cruel thing on earth. "Whatever we think of war, I do not see how we can possibly get away from the fundamental historical fact that we are all descended from a long line of savage ancestors who fought well and liked to fight." [57] With all its growth and power the human being is merely perfecting its own misery and annihilation because it cannot find a comparable adversary; it is a vulture feeding upon itself. The struggle for power is the katabolism of man; a wasting away of the precious moments of life and the curse upon the freedom and security of society. It merely brings to light by the terrible experiences of human misery, prejudice and war, that the brain incites the body to increase its desires so rapidly in the direction of self-aggrandizement that the body becomes a helpless mule wandering about in a golden lode of greater and still greater greed and power until it has lost its way from the light above and becomes entombed with its treasures that have always remained and will always be part of our world's crust.

We have seen in previous pages that the struggle to exist among the plants, insects and animals, is so similar to the struggle of man for existence that any marked difference shows merely variations of degree. The genus of struggle remains the same. However, while the ertia of man's struggle and his animal and plant neighbors remains a parallax of design, the motives show a parting of the ways on the part of man to an abortive degree. The animal or plant takes the path of least resistance in its effort to exist: its curriculum of ambition does not submerge itself above its wants. There are many freakish exceptions, that is, predatory animals, insects and fishes that enjoy or are uncontrollably impelled to butchery beyond their own wants of subsistence; in this instance, these freaks have attained, without the aid of intelligence, what man has accomplished with it. Man, too, used to take the path of least resistance; he lived from day to day never thinking of accumulation or *capital* for himself beyond his immediate needs, or deriving pleasure from the needless spoilation of others. The pity of it, here again, is that the path of civilization has been also the hard and cruel roadway of aggrandizement, self-interest, self-emulation, and the lust for power and the prizes of power. The brain of man led him to the land of plenty but it has not yet taught him how to eat properly. There are always the same mistakes of indigestion from over-eating or under-eating, depending on who is upper or lower, and within the irritated groins the lives of millions have been wasted away. Man's lust for power is the tape-worm in his stomach; it knows no satiety or reasonableness; it devours all at the expense of the body, and in the end, malignant, it runs a hating, furious race with death. "A tendency to self-destruction seems to be inherent in the over-developed human brain. It is a situation similar to that in which a parasite thrives so exuberantly that it destroys its host, thereby bringing about its own undoing. As the host can survive if the parasite's rapacity is kept within tolerable limits, so man will survive longer if he can release his body and his diencephalon from the cruel cortical grip to which civilization is increasingly subjecting him." [57a]

In the same manner as the growth of superstition, that is,

the belief in personal immortality and the worship of the supernatural, has been an evolutive process of the individual to *preserve* his ego and solve the problem of his security and comfort indefinitely in spite of all the apparent evidence around him to the contrary, so has the growth of power been an evolutive process of the individual to *expand* his ego on earth in spite of all the apparent evidence of the intangibility, briefness, and cyclic graphs to the contrary. Unfortunately, therefore, the growth and extension of *self-realism* and *rationalism* have been overshadowed by the paradoxes of self-unrealism and neurotic, subconscious megalomania. It seems that intelligence has paid a terrible toll and man has not always been at the receiving end. It also seems strange that an intelligent person should so misguide his life as to dilute it by infinite expansion beyond and by over-burdening himself with the weight and pressure of power that he is so willing to carry needlessly and ruthlessly, to his own harm and to the detriment of others. I assure you that the story of Atlas holding the world on his shoulders is merely a fairy-tale. Progress is one thing and exploitation of oneself and others is quite another thing. Any individual who cannot see the difference should either be sent back to school or else watched very carefully.

We have noted that man was originally an unconscious part of a group operating as a whole and to which he was so instinctively attached that his life was completely submerged in the group or colony. The rise of civilization is the slow and gradual release from this communal grip and the increasing attempt of the individual to express and live his life in so far and wide as the growth of his intelligence gave him the conviction, courage and knowledge to do. However, the individual, freed as a unit, has learned how to enslave other individuals in more ways than one. Extended to its potential limits, there is no question in my mind that the unfortunately less powerful groups will attempt to revert to primitive behavior, sacrifice their individualism to achieve the mirage of self-preservation, and cause a cataclysm of chaos which can be destructive and tragic. This occurred in Germany when Hitler came to power; also, in the case of Mussolini, and of course, it expresses itself in the rise of Communism in many poverty-stricken and have-not countries.

Heretofore it was the individual who sought release and self-preservation. Now it may be the group or submerged masses who seek release and preservation against the blind and irrational ambitions of individuals. However, the chaos arising from this social and political and economic confusion will be due to the fact that man *cannot* revert to primitive group mentality just as one cannot shrink a grown person back into the form of a baby. The experience and continuums of emergent evolutionary stages of man's knowledge and historical development cannot be eradicated with a wish, theory, or with a cannon. Man is no longer an unconscious unit of a primitive group and he will never be such again. Fascism, Nazism, and Communism have attempted and will continue to attempt to regimentate masses of people into emotional thinking but all this will be in vain, utter vain, because: firstly, the advocacy of these doctrines is not due to their success but to the failure of democratic principle to protect itself against internal abuse; secondly, the adherence of people to these doctrines is not because they are sure of gaining their objectives with them but because the driving, blind race for individual power within democracy has impelled them to shelter, and when man is in distress, especially economic distress, an emotional hysteria results and he will take the path of least resistance to find some outlet or haven for self-preservation; thirdly, because the masses of people are not interested in these new doctrines except in so far as they may correct the abuse of other types of government without loss of individualism; fourthly, they do not realize during the emergency of forced transition, that this entails the loss of individualism and its submergence to a new god, the State; fifthly, the fact remains that these doctrines are also heralded by *individuals* ambitious for power and because of this fact their doctrines will only prevail so long as they have the power to enforce them and the means of propagandizing the people into a temporary stupor and which, in turn, cannot be possibly a stabilizing form of social and political adjustment. "No lie is too great in time of war; and no measure too savage." [58] "When the political state is constituted, one of its main tasks is to make use of the social sentiments of the individual so as to canalize his fighting instinct." [59] In Japan during the last war the oligarchy "educated" the people that to die fighting is to bestow honor upon the dying and a higher place among the heroes of the spirit world. "Militarists are wise in encouraging the belief in immortality." [60] It remains that because of these factors, man will have to pay dearly for his ambitions, and sooner or later, after much tragedy, hate and drenchings of blood, will he possibly realize that the finest power in life is when a man knows enough to rationalize his actions so that his freedom of expression and movement can be, not submerged, but even liberated more than ever before and this can be accomplished only through equity and not by domination. Man can move mountains with his brain by thinking, not by dashing it defiantly against the rock.

The moral of all this is that no one unit or group within a nation's economy or political structure can be allowed to grow to the detriment of the economic or political life of the people. Such colossi eventually fall of their own weight but it falls upon the lives of millions and the debris is the beginning of catastrophe, revolution and the entry of varied politico-parasitic diseases such as existed in Europe under Nazism and still continues to exist under Communism. In the last war, "in Vienna an enterprising firm supplied atrocity photographs with blanks for the headings, so that they might be used for propaganda purposes by either side." [61]

Let us return to the life of an individual and his money, now a form of power. The same principle of the power growth of ideologies can be applied to individuals. One should evaluate the meaning of money and what it represents so that his or her life can be guided by a rational growth and pursuit of happiness and the really worthwhile things that make a life wholesome, with less regret and enjoyed to the utmost and so far as the intelligence and good fortune can extend it. This is no plea for spendthrifts; rather, it is a plea for a rational economy for individual and group. The elimination or lessening of power will increase the potential for world friendship through a more realistic appreciation of the value of Time to each individual. Hungry men unite to revolt, not to fraternize. A man worried sick to pay his rent seeks a benefactor, not a friend, as friendship implies the moral and unobligatory independence of an individual. The meaning of money is like the song on a musical sheet; it means something only when it is played and heard, money when it is used. A man can own millions of acres, have millions in the bank, a dozen palaces, and still remain *practically* and *realistically* a very poor man. This is again no plea against personal initiative and progress; on the contrary, it is a plea for a rationalization of a person's ambitions and *the rational person realizes that he certainly does not live forever.*

The inherent mistakes of human nature are the misinterpretations of meanings and the subsequent misdirection of life forces. If only man would realize that the true meaning of property and power is in its value toward the security of peace and happiness and not for the sake of physical accumulation which builds itself into an idol to worship. The principle involved here does not necessarily apply to individuals alone; it applies to groups, to entities, to races, to nations and to certain groups of high finance, industry, commerce. Any student with a fair knowledge of history knows that war and friction are the results of one power trying to supersede another power, one business trying to get above its competitor, one group trying to get the balance of power to scale in its favor, one race trying to hold down other races for its own supremacy. Within this constant struggle of various industrial, financial, social and political forces, the common man representing the main body of humanity and the main body of customers, has been overlooked, abused, misled, and mishandled in a million different ways. Propaganda of one kind or another has been used to abuse him or else make him fight those who stand in his way. "War is the most hopeless of all absurdities." [62] "There is no more argument in favor of war than in favor of going to bed with a rattlesnake or a plague-infested rat." [63] Thus man's emotions have been twisted, contorted, degenerated, to suit the needs of the forces in power.

To contemplate any beneficence or tolerance on the part of any such group is to invite illusions. It is only all-too-human to be greedy past the cancer stage and the misfortune of nations to be tolerant only after they have lost in war. The human being, with all his intelligence and material progress, cannot be depended upon to follow through consistently any such moral principle of mutual welfare and regard. Democracy is the attempt of individuals to restrain each other from harm and insecurity; unfortunately, it has not been sufficiently successful. "Democracy cannot obtain adequate recognition of its own meaning or coherent practical realization as long as antinaturalism operates to delay and frustrate the use of the methods by which alone understanding and consequent ability to guide social relationships are attained."—the words of John Dewey, one of America's greatest exponents of Democracy.[64] Nevertheless, it is remarkable, *considering the nature of the human being,* that it has accomplished so much as it has. Moreover, Democracy appears to be a definite step ahead in the political evolution of humanity, and if it has been partly successful it is not because it is too good for us but because we have not evolved to that state of being sufficiently worthy of it. Democracy was born of idealism, a courageous and brilliantly intelligent idealism. Only *necessity* will keep its fire alive, its future securely ahead, and its political and social nuclei as the fundamental ideal of the ideologies of the future. Or it may be that "the historians of the remote future will, I imagine, see Democracy merely as one of the early experiments tried in that age of repeated upheavals—our own—in which mankind was still groping its way to a rational mode of life." [65]

Most people will agree that war is not a salubrious dish for anyone except for those who profit by it. It is also common knowledge that wars are brought about by the financial and political ambitions of certain groups and their subversive allies. Peoples are stirred to the point of murder by fanaticisms of hate, fear, and super-nationalistic ambitions; peoples become scapegoats for massacre and persecution; diplomats begin their grind of dignified lying, connivery and deception; industry and commerce begin counting their profits on futures; a match is dropped and another war is on. Knowing all this, it is rational to expect any person to defend himself and fall victim in the whirlpool of circumstance and event. Often events bring out, even with intelligent and peaceful people, a contempt of passive resistance to a growing danger and the spirit of war is raised high even with those of us who are normally opposed to any kind of warfare. When a people are attacked, the people must defend themselves, even offensively until the aggressor has been destroyed. We are not concerned here with this problem. Our problem is to find out how many of us made profit by supplying the enemy with resources of power prior to the attack, and to prevent such recurrence for our own peace and safety. During our own wars with the Indians, many a soldier's scalp was traded for a furskin, with a rifle thrown in to bind the bargain—except that the scalp was delivered later!

Power will only decline as the need for it declines. This can come about when the opposing forces in commerce, politics, society and education, cease exploiting and fighting each other, and instead begin uniting for the common purpose of bettering the common welfare, in general, and their separate existences, in particular. The surging forces of material and scientific progress, the increase of populations, the rise of living standards and the gradual educating of the peoples, these and other coordinative factors should inevitably force the decline of the older and abusive form of capitalism, and evolve a new form of socialized, democratized, managerial capitalism. This, in turn, we fervently hope, may lead to *social individualism,* the possible harmony-partnership as much as is humanly possible of people and their societies. So far as fascism and communism are concerned, "neither theory nor practical experience has yet shown that state socialism will be essentially different from state capitalism." [66]

Labor, as an organized group-force, is much younger and at present it may be reaching or passing its apex of power. Abuses, within and without, make up the whirling tail of the comet of power. When the abuses of labor against society become sufficiently abundant and troublesome to necessitate their removal, the necessity for the existence of organized labor will have passed.

Previously we have made it clear that power is in the very nature of life force. We cannot eliminate it from living things. Nor would we want to even if we could. The initiative of the individual and the unity of a nation's people both emanate from it as do the abuses of individuals, groups, and governments. The essence of the problem is to *channel* this ever-surging force away from its present hit-and-run least-resistance meanderings, and into regulated and equitable canals of various levels, with ample strong dams and controls, so that its progress can be more evenly distributed through the free world and its power used as a *process and method,* not a goal or a prize, to make more secure and peaceful the homes of the world without diminishing the natural and unquenchable spirit of pioneering and search of personal and group initiative into new fields and actions. Then you might ask: Who will be at the controls? Who will lay out the plans for the gradual elimination of the old and the building of the new?

The elimination of the old will be achieved, as it usually is, by those who were and still are in control of the old ways of things. Their greed and lust for gain and power will be the instruments of their own elimination and gradual transition into the new ways. It may be hastened by their last minute ditch-fight to hold on, and in their eagerness to hold on they may become more arrogant, more lustful, the reality of their selfish purposes revealed more openly and bluntly to the people. They will eliminate themselves, as gangsters usually do, by fighting among themselves or fathoming in their own made pools of self-strangulating competition. The leftists and the rightists will swing their clubs of power in both extremes and by their weakened strength may allow a more equitable, a more rational and a more flexible process to take root and form out of necessity. This more flexible and less regimented process has no publicized power now. It is growing, however, in all the intricate mobilities of social and economic strata today as necessity urges in a process that must be more soft and adaptable to necessary change and modification to meet the needs of the world. It doesn't want power because power will be its process, not its wand or destination. One will not be able to say that *he* is this power or *it* is this power or *they* are this power. It will be, I hope, an energy-quantum that people *use* to live by, not to struggle for. It should be so diffused within the movements of human effort that it can be unidentifiable by pointing a finger. It can be mankind's adjustment by the principle of *necessity* to save itself and to preserve itself. It will not be a question of ideologies or idealisms; it will be forced upon us, I trust, by the sequence of events. This force of social adjustment and adaptation to pressures may make people unite their individual initiatives to the common purpose of preserving their individualism and individualities, their liberties, their comforts and the joys of peace that make life worth living for.

The reflection of history shows that the white race has

emerged in its gradual climb and extension of the intelligence processes more so than other races, if we were to gauge such progress by their material, social, political and economic advancement. This does not in any way appear derogatory to other peoples, but it merely indicates that the white race has been, historically, able to apply itself more extensively and profitably and practicably, and has accrued because of it not only most of the material progress of the world but also the power to dominate the world. Japan had been the only recent exception and she was the loser by it. In joining the race for imperial domination she was also to join in the final tragedy of this great error.

The point I desire to stress here is that the application of our intelligence and knowledge to gain power and which application meant the enslavement and exploitation of other peoples including our own, may be gradually spending itself into utter disillusionment. It means that we are observing now the gradual decadence of Europe as a criterion of world civilization. It means that the process of civilization may shift to the Far East and the Americas, where there are no pure "aryan" races and so cannot be "cleansed" by the "sympathetic" methods of European imperialism, pseudo-intellectualisms and moss-covered cultures. The next drama of the human race will principally emanate from resurrected peoples, peoples held in visible and invisible chains for centuries from within and from without.

The world is shrinking in distances. The dormant wisdom of these awakened countries, once used mostly for proverbs and soothsaying, will be revitalized into rational, practical and objective action. The impoverished peon of Mexico, the emasculated Hindu, the sharecropper of the Carolinas and the wandering Oakie of the West, the slum-infested Coolie, the Congo miner—from the ranks of these people will come the real new order—not the kind that Hitler had insanely dreamed in his horrible fits nor that of the Russian oligarchists who are trying to deceive the poor of the world to submit to their own sacrifice before a state god of power, but one based on the sheer attempt of the peoples to obtain for themselves some surcease from slavery without loss of freedom, a chance to unite for the common good without loss of equity, a chance to work for security without the loss of justice, and to reach a proper heritage of life itself and through it a simplicity of purpose and understanding upon which there may be a possibility to bring about for all the peoples some enduring peace. I have great faith in this new emergence of the Americas, Asia and Africa, for between and among them the foundation of what real civilization should be, might be laid. Perhaps from this new wisdom and equity the new European might yet reconstruct his part of the world and bury the Machiavellian intellectualisms which have buried his predecessors. Whether this may come about depends on the possibility of a new direction and mutation in the course of human nature and world polity. If necessity can bring this about, the human race will be most fortunate. If human nature continues to exist as it always has done, and this is a greater probability, then history will keep repeating itself with the rise and fall of nations, cultures, and races.

The essence of power often becomes its own psychosis. The tiny grain of matter contains within itself great potency. It is only for science to achieve the possibility for the release and utility of its power. The process of transformation, explosion, dispersion and absorption, is a psychological as well as a physical process. The kaleidoscopic panorama of experience and life is the constant attraction, fusion, dispersion and absorption of ever-varying continuums of experience into experience. Take any criterion of experience. Let it be the life of the bee, of a person, the history of a nation or a religion, or even a philosophy. Project these things against the screen-graph of Time-Experience and you will find that all these things, regardless how they may vary in structure or policy, reflect the same story—a growth upward or downward, a distension or extension to its greatest possible potential of self-identification and expression, then a dispersion of its elements and gradual absorption by greater, newer, stronger potentials. Universally, the criteria of power remain the same. Only the entirety of experience, that is, the universe or everything there is, remains the same. Particularly, each little power or segment of experience inevitably rises and falls according to its finite limitations and eventualities. Individuals, as self-identities, are constantly forming, growing, shrinking, and dying.

With this in mind it matters not how great a man hopes to be. It does matter how great has been his influence upon newer and stronger forces, not the mark of power upon himself. Such influence upon newer and stronger forces must necessarily be for rational and progressive action; an unwanted influence built for self-glorification will be absorbed by weaker potentials, never by stronger ones. One is never fused to many because he wants to be a bigger *one.* To be rational one must necessarily be reasonable and conscious of others. To be progressive one must necessarily be constructive. To be constructive one must necessarily be coordinative. Reasonable coordination requires a respect of interrelative and interdependent forces. Thus only can any unit of power reflect any intelligent justification for its influence upon the common experience of people.

The importance of a public official is no greater than and is equal to the importance of the aggregate which he is supposed to represent. The prestige involved is solely a public one, not a private one. His prominence and power emanate and exhibit the very life and power of the polis of which he is a segment as well as its symbol and leader. Such leadership does not mean a physical separation from the polis but an umbilicalized relationship in which whatever power the leader is endowed with is really the nourishment of political force and decision not unlike the brain of a person, which, while directing the course of the individual, is still part and parcel of its body, is affluenced and influenced by its parts and nature, and only functions by the welfare and harmony of the entire body or polis itself, or misfunctions for any lack of these things.

Mass or group opinion is not of itself an indication of good or evil; likewise, the freedom of expression and intellectual liberty for the individual is no assurance that he is right or wrong because of these inalienable privileges. These are questions of judgments for group or individual to weigh, evaluate, or verify. This means that a leader or public official does not necessarily have to behave or think as the masses do, or control or withdraw his individual freedom of expression, or conform to the group steric clingings without rational and ethical purpose or justification. Where there is integrity and the respect of persons, which in a way Schweitzer calls his "reverence for life," whether individuals or groups, there is little or no submission by either one. Like the body to the brain, it depends upon its leadership that moves the entire entity into whatever harmonious and coordinative action both are capable of according to its nature for its own possible good.

However, when an individual or group acts as an entity by itself, for its own possessive growth of power, prestige, or approval, and this detachment, though it be sympathetic and even perhaps idealistic, is real and separate, the nature of this process can only induce, as it so often has and does, a wider separation between the power of authority and the polis over which this power is expressed or carried on. For any such separation, whether the leader is conscious or unconscious of it, is of itself a misdirection for both the leader and the polis and can only lead to a more fixed dogmatism or unchangeability on the part of the leader and an increasing submissiveness on the part of the polis; as a result, both the leader and the polis are headed for trouble or tragedy. Human nature has rarely been able to overcome an increasing intoxication of power which, like a disease, becomes stronger as the body it feeds upon becomes weaker, until both disintegrate and vanish.

It is therefore important for the general protection of the polis, as individuals of an aggregate and as an aggregate for the protection of its individuals, to promote political, social, scientific and economic education and a free field of open and decorous discussion and constructive criticism for individual and group so that the leaders and other authorities will be always exposed to this process, and thus understand better and more clearly the purpose, extent, potential and limitations of their powers which are not separated from the body entity of the society but emanate from it and within it. This field of liberty and volitional direction and judgment within ethical restraint is the protective shield for the polis and a constant therapeutic preventative for the leaders from victimizing themselves into power intoxicants, which is the insurance as well as the sustenance of democratic procedures and practices for themselves and for the polis.

To sum up: It is not impossible for war to disappear from the peoples of the world. Such a trend depends also upon *the disappearance of power as an organized force* in the hands of any individual, group, class or nationality. Such a trend depends upon a more earthly, basic and discernible interpretation of the meaning and value of Time to each individual and from the individual into the group experience. When power will be a force activated by *all* the peoples instead of by a part of them, the abuse or mischanneling of this power may probably end. This entails the fusion of all power and sovereignty to a world sovereignty governed and governing by law, science, ethics and equity, and the natural implementation of rational procedures based on realities and the nature of man and his world. Professor Carl L. Becker says: "It is futile to base any plans for a new and better world in the immediate future on the assumption that the sentiment of nationalism will be replaced by the love of mankind, or that political power as now organized in sovereign independent states can be transferred, by pledges signed or treaties agreed upon, to a European or world federation of states . . . we can never be sure that political power, whatever form it takes or whoever has it, will always be used with wisdom and restraint. Where political power is, there danger lurks, and the greater the power, other things being equal, the greater the danger." [67] And Philip C. Jessup, in his *A Modern Law of Nations* resolves it to a simple reality: "Sovereignty, in its meaning of an absolute, uncontrolled state will, ultimately free to resort to the final arbitration of war, is the quicksand upon which the foundations of traditional international law are built. Until the world achieves some form of international government in which a collective will takes precedence over the individual will of the sovereign state, the ultimate function of law, which is the elimination of force for the solution of human conflicts, will not be fulfilled." [68]

The origins of war do not emanate from the prime instincts of man; its origins are found in the failure of men and peoples to satisfy their prime instincts. Happy people do not fight; a satisfied people without fears of hunger and security do not fight. When a culture loses its grip because of its failure to satisfy the prime instincts of its people, war is generated within the bowels of that people. The problem of mankind is to solve the problems of economic, social, and psychological famines—these are the horsemen who ride upon the miseries of millions, who set fire to peoples and countries. Their horses will never tire until men tire of their own power psychoses and resign themselves, instead, by sheer necessity of avoiding self-destruction, to build a genuinely equitable, healthful and economically sound culture. This, in turn, may sow the seeds of real affection between men. It could be a world friendship built upon natural factors, not ideological illusions. Unless it is built upon the fullness of equity, it can only fall again as so many others have fallen before. Perhaps this might come. It all depends on the evolutional course of human nature in adapting itself to new necessities and pressures. So long as the realization of this vision is even a faint possibility, so long as such a hope is maintained by even a few of us, so long will our dreams for a more perfect and just society hold any philosophic value.

Time travels with divers paces with divers persons. I'll tell you who Time ambles withal, who Time trots withal, who Time gallops withal, and who he stands still withal.

SHAKESPEARE, from *As You Like It*

FOUR

TIME AND THE *SOCIAL FACTOR*

IN our quest for a philosophy of Time, we have seen in the chapter on the Religious Factor that man tried to *overcome* the finity of Time by attempting to become so much like a god and even to become a god and thus bring "eternal" existence for his body and personality, if not for both at least for one, and to gain control and supremacy over the powers of nature. In the chapter dealing with Power we have seen man trying to express this ontological aspiration by manifesting it on a finite premise. Both seem fallacious and self-punishing, though very much human-like and expected, especially when emotivism rules and rationalism is silenced or suppressed. Now, in this chapter we have to bend our efforts to investigate what Time means to man and woman in their relationships to the entity which is the creation of a sum or aggregate of many men and women, and which is called *Society*.

What is *Society*? To define it is very difficult indeed, because it cannot be taken hold of, like an object, and examined easily. Its supplements and interdependencies and its fluidities are so varied and interminably interlaced in cultures and social states constantly being more or less mixed, brewed with additives influencing the general patterns. "Society is a growth, not a structure." [1] "Society is either an abstract or a collective noun." [2] It cannot be exactly isolated for certainty. Besides, each "identifies" it differently. The Englishman's idea of society is identifiable with English norms, traditions and cultures. Even within a city like London an Englishman of the East End thinks of society not like the Englishman of the West End. Even within a district like the East End the landlord of some tenement doesn't imagine society to be the same thing as the blighter who exists in some alley cellar. In the West End a high-class "chippie" doesn't appraise society in the same light as Lord Something-or-other who reads the small print of the *Times* worrying about the next decision of Ten Downing Street. Take, for instance, a small tribal colony deep in the forests of Africa. The tribal chief, with his hundred or more wives around him, bedecked with ornament, jewels, embroideries, fattened by a lazy, effortless life, eating all he wants whenever he wants, picking his favorite wife whenever it moves his spirit, accumulating a wealth of gifts from his own people and from other tribes, his absolute power over the life and death of any member of his tribes, even being worshipped as a divine person, a god himself on whose life and strength depends the general welfare of the tribal body or society—such a man's concept of society cannot possibly be the same as that of the poor soul who is just a nobody around this village. Or the concept of society held by the witch-doctor who sees every one as a double or triple and has to watch not only each person but also his shadow, spirit, and his ancestor spirits, a very complicated and cosmopolitan procedure.

Although in Europe it is familiar history for neighboring countries and neighboring peoples to kill each other wholesale and regularly, the European would hardly accept, at least in principle, a society in which it is not only customary but very delightful to eat your neighbor, especially the breasts if it is a woman, and the heart, symbol of courage and strength, if it is a man. But the Mangbetu savage of the African veldt thinks it is a great idea. The Hindu widow who threw herself upon the burning funereal pyre of her husband so she could continue to serve him in the next world, lived in a society which is definitely different indeed from the one in Reno where some blonde bombshell is arguing over the telephone about the size of her prospective alimony.

Luther's concept of society, if he had one, could possibly have been one of a skyful of demons attacking humanity from around every corner and from under every table. He himself fought the devil and hit him with a bottle of ink, so they say. Certainly his concept of society was not like that of Robert G. Ingersoll or Thomas Paine. Even Lincoln's idea of society conflicted with that of Douglas. The slaveholders of the South certainly felt that their conceptual pattern of society was just, and that the reformers of Chestnut Hill were just off their rockers. The medical profession in America surely thinks that society is orderly and proper so long as no one gets the bright idea of socializing their practice for the common good. The ravaged poor remnants of the American Indian cannot think of American society in the same way as the Board Chairman of the large lumber, oil, and mining company. Surely the Maori maiden who is respected and desired so much more because of her premarital sexual promiscuities, doesn't see society in the same looking glass as the frustrated prudes of Salem who threw stinky things at the poor girl locked in the stocks because she committed the heinous crime of winking at the schoolteacher. Syphilis did not come from the South Seas to Europe and America; it came from Europe to the South Seas.

And so it goes, within each country, within the various individuals and groups within each country, all over the world, from the Great Powers to the little nomadic Bushmen of the African deserts, from the tall Patagonians to the tiny Pygmies; each, in their own helpless way, contain different views of what society *is* and what it *means* to them.

From simple observations it appears at least that Society is a field of relationships, a means by which individuals *process* their experiences and which, in turn, influence the continuing nature of these relationships and which, in turn, influences the experiences of the individuals. An individual can, and many do, exist without society, but a society only appears to exist by the individuals who identify, contain, and carry on its field and process.

Then how can we ever get a fair, even a general, if we must, idea of what society actually is? And if we could possibly describe it, how can we affect a genus that could be a certain identifiable and verifiable containment of society that could possibly apply to a general analysis of social growth common to mankind, and its meaning and value common to man?

To attempt to do this we may have to go back a bit, say, fifteen to twenty thousand years at least and try to discern how it all started. Life is a photograph blurred at both ends; let's try to pierce the blur at the beginning, if we can, so that the other end may become a bit brighter, happier, clearer, for us.

To do this we must hold in mind as a constant that we are not seeking perfections or the keys to heaven. "Nature sets no standards, man does; and hence perfection is impossible. Thus, we must eat and breathe, and act; but what, how, and when, and why are matters of opinion." [3] Humans, like we are today, were always imperfect creatures groping for security, presumably peace, and happiness whenever they could get it. "If we consider man's course of life from birth to death, we see that it is, so to speak, founded on functions which he has in common with lower beings. Man, endowed with instinct and capable of learning by experience, drawn by pleasure and driven by pain, must like a beast maintain his life by food and sleep, must save himself by flight, or fight it out with his foes, must propagate his species and care for the next generation. Upon this lower framework of animal life is raised the wondrous edifice of human language, science, art, and law." [4] If they took the wrong or right road, as is all too human, on the many crossroads in history, at least let us try, too, to recognize them as wrong or right if we can, so that the roads now and ahead of us may be so much better planned. One would be all too foolish to assume that man must know everything, as if he could, to find the key to the panacea of human contentment. "What is most needed is more knowledge as to the true nature and origin of society. Unless it is known how society has been developed and of what elements it is composed, it is impossible to adopt measures that will help on its natural evolution." [5]

Obviously we are not gods. We are animals, similar to, but also different from, the other animals around us. The further we go back we find that we lived more like other animals do and less like we do today. Every child reminds us so much more of our prehistoric animal nature and each child fights back, like an animal, in its usually losing battle with society as it keeps growing up. "Children use the fist until they are of age to use the brain," commented Elizabeth Barrett Browning. Visiting a hospital baby ward, a nurse remarked to me, "The baby fights for its life because it does not know why." Grownups fight more, especially among themselves, and know less. "Man's hereditary nature, deep-rooted in the brain," writes Auguste Forel, "makes him an egoistic, individualistic, fierce, domineering, tyrannical, jealous, passionate and revengeful being, who wishes to enjoy liberty by the abuse of his neighbor's toil. For the slightest social defects possessed by this neighbor he is argus-eyed, but he unconsciously misinterprets or extenuates his own faults." [6]

The world was assuredly not created *ex nihilo*—out of nothing—all complete with triple tiaras and sistrums, saints and seers, laws and lanterns, kings and ministers, marriage bureaus and divorce courts, military manuals, rich and poor, prophets and prospectors, turkish baths and barber shops. Man and all his accessories of living evolved like everything else. "History in being a process of change generates change not only in details but also in the method of directing social change." [7] The same process, but not necesssarily the same circumstances and details, applied to all life in general. *Life is want* and it is this *want* that is at the root of the generating processes of evolution. And wants come with life, are life, and *life is more what it makes of you rather than what you can make of it,* even though we have anointed ourselves with "moral" illusions and categorical imperatives that some god gave us all the same sum of experience, heritage and wisdom, to pick right from wrong.

To understand our animal nature is to understand the basis of our social appetites, the origins of individualism, the evolution of our fears and fantasies, the submergence of the individual into the social state; the nature and history of those institutions that have germinated, festered and prospered upon this submergence through thousands of years of human slavery, misery, and misconceptions; the spasmodic attempts, most often out of sheer desperation, of man to liberate himself from this serfdom and from the Juggernaut that had swallowed him as hard clay had swallowed the termite; his attempts to reconstruct, reconcile, compromise and control, by peaceful or violent means, these institutions in newer or different forms; his attempts to find some way out of this social morass; his submergence today and the possibilities between individual and state. Unfortunately, we have given animation to the guillotine only to find ourselves holding our heads in our hands.

Many thousands of years ago our ancestors were roaming savages of the jungle. They wandered wherever there was food to eat, water to drink, and a cave to shelter them. Where there were no caves they sheltered themselves in trees, and when it rained, snowed, or blew hard upon them, they raised their arms to cover their heads. Do it yourself and look how the hair on your forearms runs down and away, just one lingering adaptation to the primeval environment. The savages were in small family or clan groups, later combined with other family clans or groups in order to kill the fierce and mammoth animals that existed at that time, just as lions, in a group, plan to stalk a herd today; just as a pack of gorillas or monkeys defend themselves against a common attack. Gradually, after thousands of years, certain human groups or kins took to agriculture and kept cattle, and thus stabilized themselves in particular places suitable for this great advance in human history and the beginning of private domain and property. The slow, creeping advance went on in different speeds in different parts of the world, depending on the geographic and alluvial circumstances of each location. Today there are nooks and corners, from the Australian bush to the Amazonian jungle and the bamboo thicknesses of the African wilderness, where humans are still savages, still living in the most primitive conditions, without clothes, utensils or the simplest tools, living like animals of the woods, in the same manner as tens of thousands of years ago. When ships had compasses and Paris had fancy carriages, there were great empires of peoples in the New World where stone was the only "metal" and where the wheel was unknown!

Here, too, we must reluctantly restrain ourselves from getting too deep into this most fascinating story of our ancestors, in order to keep our focus around our central theme—Time. For those who care to go further into the study of primitive man, there are libraries full of wonderful, exciting and enlightening voyages into the past. However, for us, presently, we have to be content with a sketchy, skeletal, introductory or *apéritif* outline of what society is, how it came about. Let us return to our savages.

In those pre-primitive times the savage recognized one law, one custom, one governor—the law, the custom, the government of Nature—the constant, hourly and daily struggle to exist. Eat or be eaten. Fight or be beaten. If man had no beast to eat, he ate man, but he had to keep eating, same as we do today although we can eat on tablecloths and order almost anything we please so long as the money is in our pockets to pay for it. We eat with forks although "in the England of Elizabeth it was declared from the pulpit that the introduction of forks would demoralize the people and provoke the divine wrath." [8] The savage had to stone his beast, rip its body open with his bare hands or a piece of sharp stone and eat it raw, just as the other beasts do, just as many Eskimos still do to this day. He recognized no adversary as his master except the pressing jungle around him and his fight to keep alive from day to day.

As time went on, larger groups settled and built small villages. Milleniums passed before the first town appeared. More milleniums passed before the first "city" and the first empire. It was a gradual evolution from the cave to the castle, from the firepot to keep warm to the furnaces of Moloch. This evolution of society went through a number of phases or metamorphoses. After all, the savage originally only answered to himself. To fight for his life and his livelihood, he joined with others into a group for the common purpose. The original social contract assuredly was not created for the "glory" of the group, as a new personal or self-identifiable entity, but for the welfare of individuals dedicated to a simple idea that by uniting forces a greater degree of security may be attained. This was the intent of the primitive, to create a human phalanx of effort and strength for the common good and thus for the individual good in the struggle to exist, *a group survival* of the fittest so that the most fitted individuals could keep surviving to continue the group survival. It was simple and elemental and only the simple and elemental could be possibly understood by the primitive.

The primeval human animal had no sense of consciousness of individuality. He "colonized" with other humans because nature automatically directed him to do this in order to survive. It was the unconscious but physically alive, instinctual drive to "group" over which the individual had no control or discretion, just in the same manner as "we regard the termite colony as one individual." [9] "The hive," writes Peattie, "is a unit to which all individuals are subordinated. It is itself a law, a psyche, a state so perfectly organized as almost to constitute a single organism." [10] We should take into account, as Dr. Otto Klineberg states in his *Social Psychology:* "There is no sharp dividing line between the social behavior of animals and the culture of human societies." [11] Even "Marriage in primitive or semi-civilized tribes is not an individual affair as it has become with us. It is an affair of the community." [12] The same primitive social trait of the human animal was common among other related animals. "There is a definite social order in the porpoise school. . . . Any novel disturbance in the tank causes all of the animals to band together." [13] Even the origins of the dance rise from the jumping and clapping of anthropoids and primates who are our relatives. "In many of the ape's rhythmic movements," says E. G. Boulenger, "are also to be traced the crude forshadowings of the human dance." [14] The expressions of affection, kissing, hugging, friendship, love and romance, are common among many animals besides man. "Highly emotional at all times, the chimpanzee is the only primate which habitually and spontaneously expresses affection by the very human medium of kissing and hugging." [15] Kissing and hugging are manifestations of animal affections and practiced by many wild and domesticated animals for the same reasons that humans do. Seals are real technicians in the art of kissing and would put many a Romeo to shame. Apes are not only good kissers but know how to neck in a very affirmative way. Dogs, cats, birds, and countless other animal kinds, show many outward ways of making known their romantic desires and their emotional feelings.

Tens of thousands of years later the first "leader"-man exposed and established himself over others, and slowly individualism arose out of the human constitution. However, in the beginning, it is doubtful whether man "knew" he was an individual or as something to worry about himself apart from the mass-worry of the group. "The individual (in primitive society) is of importance only in so far as he is a member of a group. It is the destiny, the needs, the safety of the living unit which constitutes the group, that determines the lot of any given individual member." [16] . . . "That the social group, clan or *sib* is the real unit of which the individual is a mere element is proved in many cases not only by the familial structure; the daily life of the primitive is another testimony to it." [17]

With the rise of leaders, the priesthoods, kings, the division of labor into specializations from soldier to statesman, from potter to priest, from servant to banker, the individual found himself willingly or unwillingly entrapped, depending upon his status of being a *taker* or a *giver*. In this human cage he either accustomed himself to it or rebelled. The complete freedom of the savage, as part of a group of more or less equal values, was understandably exchanged for the serfdom or lordship of the barbarian. Whenever the squeeze of this compromise brought too much pain, out of sheer necessity people revolted. Some of them won, most of them lost. Some of them lost only to win centuries later, others won only to lose centuries later. But the battle was never ended, and is not ended so long as there are individuals who keep burning the torch of freedom and the right to think, and keep high the principles of justice and fairness. And so to this day the battle is still on, between the ones and the many, the individuals and the societies, the thinkers and the believers, the leaders and the followers, the units and the masses, both good and bad on both sides.

Most of humanity, voluntarily or involuntarily, live more in filth and squalor than in cleanliness; more in poverty and want than in comfort and security. Most of humanity, therefore, remains basically and inwardly more savage, more resentful and more animalistic than the supposedly civilized and intelligent few who are more apt to become their exploiters and masters rather than their benefactors. This *is* human nature and while civilization, education and general ethical progress, have brought about great advancement over the ancients and have made possible open and constructive criticism as part of the public and private domain, the superficial cover which this has also wrought conceals the constant law of the jungle by which people live and the precarious thin veil of acceptance and practice by which ethics rule.

Originally, a human being, ignorant of any rule or contract, found himself being grouped in order to better himself. Why shouldn't he have joined with others so that each could live better, have a better chance to survive? While actually the course of nature directed him to join with others because at such time all the members of the group worked unconsciously together for the common and individual good in order to preserve themselves and their species, as other kinds of life behave likewise. But with the approach of leadership and the rise of individualism and individual identity, he would be foolish to join only to lose his own identity and become a dot of blackness on a bigger massive dot of blackness—Society.

Originally, people created money to facilitate the exchange of goods and services; it was not the idea to create money as a barometer of wealth. But money became so powerful that today it rules the world by its own impersonal power, a gilded god, and every bank a temple to its worship, although it does not bear pain, cannot love, doesn't really do anything, isn't born and doesn't die in the same manner as humans do. It has no personality, hopes or ambitions of itself and for itself, yet it has become a power, a gigantic robot of power, and in its grasp lie the destinies of the peoples of the world. Not that a dollar in the pocket is bad, or a building where we can safely keep our money or draw checks. What is bad is when money becomes a goal to live by instead of a means to live with. *The same sad equation seems to have gone on in the individual's attempt to socialize the struggle to live as a means of bettering his individual life.* Instead, the society which he created originally for this very purpose, like the creation of money, has become *the* power and now is socializing the individuals in the struggle of the society to exist and become stronger regardless whether the individuals exist or not. Ortega y Gasset, in his brilliant *Revolt of the Masses,* wrote: "Society, that it may live better, creates the State as an instrument. Then the State gets the upper hand and society has to begin to live for the State . . . the people are converted into fuel to feed the mere machine which is the State. The skeleton eats up the flesh around it. The scaffolding becomes the owner and tenant of the house." [18] Anthony Eden, during his visit to America in December, 1938, stated: "Man was not made for the State. The State was made for man. . . . We are living through an attempt to reverse his faith. After centuries of endeavor he is threatened by the State he himself created." [19] Albert Schweitzer wrote: "Society . . . it wants servants who will never oppose it. . . . Ethical progress consists in making up our minds to think pessimistically of the ethics of society." [20] When the Church and the Francos determined to destroy the Spanish Republic and restore the property and authority of the Catholic Church and pensions for a lot of military grandees, it didn't matter whether all the Spanish people were killed in the process so long as they died as good Catholics and not as Republicans—where, as Danielle Hunebelle, in her article *The Endless Crucifixion of Spain,* writes,—"Thinking is banished, reasoning forbidden and criticism is sacrilege." [21]

During the First World War, the German priest went into the trenches and cried, "Kill for God, for your Kaiser, and for the Fatherland!" Another priest was in the French trenches doing the same. Still another priest was in the Austrian trenches and encouraged the soldiers to kill the Italians, and the Italian priest was in the Italian trenches extolling and appealing to the soldiers to kill for God, for their King, and for their country. All Catholics killing each other for the sake of the same God. It is this same historical disease which has permeated and cancerated humanity by evolving societies which have become "living gods," personalized states of religious, economic and political conglomerates. It isn't enough that religion devaluated the time of a human being by throwing it into the hopes and fears of heavens and hells; the societies keep devaluating the human being's time on earth by making sure he doesn't find heaven but keep giving him plenty of hell. It is this fight for the precious value of Time that is the heritage of each individual and which, being sublimated, frustrated, twisted and withered by a thousand and one what-nots, is constantly being kept down from raising its head and breathing in the fresh air of joyful living that enlightenment and a proper rationalization of oneself and one's life might bring. The whole idea of Democratic principle is to accomplish this; if it has not been so successful as it should have been, it is just the truism that Democratic theory is just better than the best in human nature. Spinoza, writing on the purpose of the State, stated: "The last end of the State is not to dominate men nor to restrain them by force and fear; rather it is so to free men from fear that they may live and act freely with full security and without injury to themselves or their neighbors." Regarding the ideology of Democratic principles, John Dewey tells us: "Can we find any reason that does not ultimately come down to the belief that democratic social arrangements promote a better quality of human experience, one which is more widely accessible and enjoyed, than do non-democratic and anti-democratic forms of social life? Does not the principle of regard for individual freedom and for decency and kindliness of human relations come back in the end to the conviction that these things are tributary to a higher quality of experience on the part of a greater number than are methods of repression and coercion or force? Is it not the reason for our preference that we believe that mutual consultation and convictions reached through persuasion, make possible a better quality of experience than can otherwise be provided on any wide scale?" [22]

A social state that makes it possible for people to live peacefully, more secure, and in a more harmonious relationship, that makes possible the pursuit of happiness and the welfare of individuals on a practical and realizable basis, is not only understandable and acceptable, but absolutely essential and indispensable. Eric Larrabee writes that "a society that makes individualism possible is a new thing under the sun, and one that could endure would be our most admirable artifact." [23] But a social state that considers itself important to such an extent that man cannot seem to catch up with the pursuit of happiness and the people have little welfare but keep building up a higher, wider and more powerful state to perpetuate their struggles—then something is just "cockeyed" about such a partnership. I am afraid that by the time the State can evolve to the status of properly serving the individuals, there might not be any individuals left to serve. "In questions of social morality, more fundamental than any other particular principle held or decision reached is the attitude of willingness to reexamine and if necessary to revise current convictions, even if that course entails the effort to change by concerted effort existing institutions, and to direct existing tendencies to new ends." [24] The human being is gradually being taxed out of existence, and taxation has now become a science dedicated to the attempt to find out how people can possibly survive on what they have left. "Never a world so filled with the contagiousness of being alive, yet so difficult to live in!" [25] Alas, the world is really a beautiful, wonderful place (what else have we to compare it with?) *to live in, not to die for.* The world is but a whirlpool of whirling, conflicting societies arrayed against each other in a common, unceasing conflict in which the life of the individual is overlooked, subordinated and forgotten, though the individual is at the bottom of things and though the spark of life, the fire of all action, enterprise and progress, exists in the principle of individuality.

The evolution of instincts, the rise and fall of nations, the adaptations, variations and transformations from emotion to intellect and the compromises of both, intertwined, affirming and opposing, in short, the story of all mankind is also the story of any man's life. The impress of custom, of hypocrisy, deceptions, all sorts of subtleties and pretenses, innuendos and conspiracies, of hate and struggle, of stupidities clamoring for pride and prestige, private profit and public acclaim, medals for the dead and glory for the speech-makers, shelves and shelves of social strata, each shelf pressing upon lower ones, this is the story of civilization, the curricula of the socialization of the individual.

To a missionary, civilization means putting more clothes on primitives; to some women, the privilege of alimony; to a statesman, the power to rule according to rule; to a soldier, the right to kill according to ethics; and to the banker, the power to take your money according to contract. Alas, perhaps this would be the best possible world —if only we would let it alone! Or would we if we could? Let's see.

In the first place, before we begin evaluating, from the viewpoint of Time and the Individual in their relationships to Society, we should realize that it is elemental to understand that no society can completely fulfill all the hopes and affirmative "dreams" of all individuals alike, simply because all individuals may be similar but not exactly identical in all aspects and more often are radically different from each other in heredity, temperament, intelligence, experience, etc. Also, individuals within a group have different influences and reactions from that same group entity under the same circumstances in which they find themselves. Also, as a result of the 19th and 20th centuries' scientific and material advancement, and even though there are many peoples and tribal groups and even countries far or greatly removed from relationships with other groups or countries in general, yet communication, association and relationships between the peoples and the societies of the world have brought about a shrinkage of aloofness and afar-ness, so to speak, and people, as a result, know each other so much better and are to this extent less subject to propaganda and "patriotic" antipathies affecting these relationships and experiences. This entwining and mixing, small or great as the case may be, tends toward greater understanding and a minimizing of isolation, a gradual and slow decline of the fears, prejudices, barriers, and false concepts between individuals and groups, and with this a lessening of dogmatic or "frozen" societies plus an increase of fluidity and the flexible, regenerating sympathies and appreciations that only social intercourse and the intermingling of people could bring about.

They say that woman's greatest strength lies in man's imagination. In a way, a sovereignty's great strength, whether it be political, social, religious or economic, often lies, too, in the imagination of its people and in the people's ignorance of realities, so the people can be so much more easily worked up or duped into hating, even killing others considered by the particular sovereignty or power to be its enemies although they may not be the real enemies of the people. The new communication, travel and transportation, have greatly absorbed distances and time, and have made somewhat more difficult the process of propaganda to influence people because of this world-wide increase in social and educational intercourse between individuals of different countries.

I do not intend to convey any general idea that society and the individual are irreconcilable and unnatural companions. Man cannot and even does not desire to be alone, physically, biologically, and psychologically. Being free to be alone does not necessarily mean that one is free. *Man is a social animal,* to begin with, and society has evolved out of the natural leanings of the individual in which the desire to socialize, colonize, group or "get together" is a basic, unalterable trait of the biological make-up of *Homo sapiens.* "No man and no mind was ever emancipated merely by being alone." [26] The idea is to retrace and to recapitulate our social assets and liabilities, and to clarify two salient points that are self-evident: The individual's natural desire to socialize is nature's way of protecting the species through the traits of the individual, as is evident in other forms of animals and plants; and it was intended, I hope, through the natural mechanical processes of life and living continuums, that the process of socialization should enhance the life of the individual at least no less than it enhances the life of the group, instead of enhancing the group at the expense and expendability of the individual. The second point is that history does reveal an abortion of this process by the nature of events. Because of the very innate aspirations, weaknesses and intelligence of the human being, ontological as well as empirical, events brought about, unfortunately, the creation and furtherance of social and political entities that have become dogmatic, inflexible powers, asexual in nature, that exist and motivate an influence to submerge the individual in its many types of manifestations, into a devaluation of individual value and the life of the individual, bringing about a willing or unwilling sacrifice, because of these misconcepts, of the inalienable rights of the individual. This is especially evident in communistic countries and in those countries where there is the mere appearance of democracy but where a police state actually exists with a dictatorship thrown into the bargain. These are sacrifices only to increase the fastness of the social, economic and political Bastilles in which the individual so often finds himself imprisoned, yet built more often than not with his own hands. "The masses are always patriotic," wrote William Graham Sumner. "For them the old ethnocentric jealousy, vanity, truculency, and ambition are the strongest elements in patriotism. . . . The group force is also employed to enforce the obligations of devotion to group interests. It follows that judgments are precluded and criticism is silenced." [27]

Man, in his effort to find security, more freedom, peace, control over his environment and extend the longevity of his own life, built these Bastilles that have imprisoned him instead of securing him. Time is of concern here because the individual is a living thing, and Time is its "space" of life, while society is kept "alive" by the individuals that compose it. Where the society adds to the living "space" of time for the individual, then it is good. *Where the tendency is to lessen the value of life to the individual, to this extent the society is a negative and an evil factor which must be recognized by the individual if he desires, as he deserves, to live a fuller life.* Only the sacredness of life can stand as a principle for its own preservation. Each death is a clarion call for life, a reminder for those still living to live well, with honesty and courage to fulfill the fullest possible happiness for each precious hour. "It is inhuman to sacrifice one generation of men to the generation which follows, without having any feeling for the destiny of those who are sacrificed, without having any regard, not for their memory, not for their names, but for them themselves." [28]

To accomplish the possibility of a fuller and happier life the individual must *know* himself so much more and the social hive he lives in, so that he may be a free and unprejudiced thinker for his own good and so, in turn, for the good and peace of others. Good-will among mankind will come about, if it ever does, not from a dogmatic principle of form, belief, or self-denial, but from understanding and sympathy risen and cultivated out of the realities that face us. Man is healthy so long as he remains a social animal, dedicated to the normal and natural processes of enhancing his own life by coordination, not by compromising his freedom for the present and future and not by sacrificing his precious and irreplaceable years to a Molochian furnace of the past. Health depends upon body; body upon mind; mind upon everything else. The welfare of each one is held by a string around the equator. Man is a healthy animal so long as he remains a social animal and does not evolve into a *socialatom,* a mindless, blind, gullible, non-self-identifiable robot of a social Babylonian tower in which he has not only lost his tongue but the very essence of the free, normal, unfettered enjoyment of a happy and peaceful life.

Justice William O. Douglas, one of our great jurists and equally as great a philosopher, stresses this problem in his little book *America Challenged.* He writes: "The struggle of this century is between the two forces—for submission, subservience, and conformity on one hand, and for independence, individuality, and dissent on the other . . . Machines have to a degree moved men closer to the role of automatons. They move and act in unison; their tasks are more and more mechanical; the individual is transformed into a statistic." [29]

The difference between a social animal and a socialatom is that a social animal can still be a free individual with a free mind and body while a socialatom has lost, through a strange process of ossification, these precious rights of life by ignorant and blind conformance to rules. A man who lives by rules and makes no exceptions, becomes a rule to measure by, not to live with. A purist, to whom pure conformance for its own sake is desirable, is one who insists that the rest of the world should be as bored as he is. Not all rules, or accepted forms of a society are bad or good; the idea is that people, before accepting rules and deciding to live by them, should inquire into the nature of them, their origins and tendencies, their purposes, whether they are just, honest, good and wholesome, and whether they are rules and customs which are conducive to living in a naturally happy, unsublimated and peaceful community. Conformance, when it is supported by rational evidence, when the act of conformance gives proper meaning to our Time in this life and thus elevates and broadens our vision to a higher value of time for us relative to our perishability and transient nature, then it may be a good and wholesome conformance; such is a conformance based upon what a free-thinking individual has ascertained for himself, not something merely to follow because one's great grandfather believed in it. Actually when this occurs it really isn't conformance any more, in the strict sense, but doing things because we *know* what we are doing and like it. Blind conformance, if sincere, is like when goodness without reason gives money to a faker. On the other hand, reason without goodness is just thinking without a good reason. This brings us to the importance of a free mind. So refreshing and valuable is this sense of being free and I am so often imbued with the feeling of freedom, that I seem to be always racing with time to express myself.

A free mind recognizes that people are important firstly by what they think of themselves; secondly, by what others think of them; and of course, the wise man discounts them both as unimportant. What is important to the wise man is that the mind be used, not as a barometer of reputation or reflected egotism, but to live better and make the greatest possible use of time by the mind's function to translate, as best as it can, experience into meanings and values. Minds have their limitations in the same manner as other parts of bodies; one's potential is not unlimited or haloed with free will that "certainly knows" whereof it chooses. A mind, like the body itself and of which it is a part, is the result of heredity and experience, each the parent of each other. The firm believer in free will just didn't have to look for a job long enough. It was Dostoevsky who said that "there is a certain hell—comparable with none—in seeking work and finding none."

There are all kinds of minds, like everything else. I have seen foolish traits of my race, some better, some worse. I have met people who, for the peace of the rest, ought to remain ignorant. I have seen people who, for the protection of the rest, ought to be deprived of their atrocious, subtle intelligence. Then there is the mind of the neurotic who is so over-anxious that life should not escape him that he outruns it, with life behind vainly trying to catch up to him. And then there is the mind of the theologian who refuses to open his mind to the world of reality but tries to twist the world and experience to conform to the size of his brain. Then there is the mind of the dictator who comes to believe that every thought he has is a revelation of God, even though he would kill every living creature in the world to satisfy his sadistic paranoia (all dictators are sadists) and prove his twisted megalomania that he is really the "savior" of mankind and everybody's best friend by burying them under his feet. And then there is the mind of the scientist who becomes so purely scientific that his science has become so lofty and absorbed in objective and detailed technology regarding the parts of the human being as to lose sight of the principal value of the human being as a whole and as an individual. And there is the mind of the coward who has to become drunk out of frustration for not accomplishing something he hasn't got the courage to do when he is sober. And then there is the mind of the young, of the aged, of the savage and of the sage, the mind of the worker and of the captain of industry, of the chemist and of the cleric. All are different minds but the important thing is whether they are free and function properly, equitably, justly, wisely.

What is this *mind*? It is something that teaches you to learn from your struggles that it is your guide; from your shelter that it is your builder; from your happiness that it is your friend; from your pain that it is your sentinel; from your laugh that it is your jester; from your pleasure that it is your slave; from your remorse that it is your judge.

Self-criticism, self-guidance, self-judgment, are indications of the maturation of the individual and of a free mind. Criticism is naturally opposed to conformance, dogmas and absolutes, for the latter tend to an inertia of ideas and a stagnation of the intellect; while criticism is the field wherein lies the only possibility of error exposure, of ever finding out whether we are right or wrong as we go on our way in life. To many maturity never comes. To others it comes in a sudden moment of disillusion. With some it arrives with joy, to others with tears. Some receive it with despair, regret and retrospect. Others greet it with anticipation, wonder, and a ladder to the horizon. But when it comes, it stays, and the nature of its stay reveals the life one has lived. And it reveals how you have weighed the value of Time and whether you have used this incomparable preciousness to attain a satisfying and reasonably fulfilling life. Life is sweet and valuable to the one who can look above the turmoil and mixtures of the day, the intrigues of men and the whims of women, above the petty thrills of monetary gain and possession, above the maelstrom of human haste and strife, and calm in his own stellar sphere see the direction and the possible correctness of his own life.

This is the potential of the mind and when this mind is reasonably free and unadulterated with social and religious customs and rules to hamper its free expression, then the body of which it is a part can have a fair chance to be happy and follow a reasonably natural life, and what is considered by any mental therapist or psychiatrist as a *normal* life. This requires the mind to cultivate the intellect to function according to what has thus far been the natural, fortunate or unfortunate, acquisition of the mind within experience in general and its experience in particular.

What is *Intellectualism*? This word has almost become an ambiguity and of so varied an interpretation that it has almost lost even a general idea of what it actually is. It has turned into a subjective word, a perspective of different cultures, groups, clubs, societies, peoples, etc. I can only give you my own opinion of what I think it is or should be.

It is a pity that so often one meets with a person who appears or considers himself to be an intellectual, but who really confesses the lack of two things that, I think, should be evident of intellectualism. These are a self-introspective honesty and a simple open-mindedness to respect and regard the apparent limitations of the self and others in knowledge and judgment. People reveal their own self-exposed fear of inferiority, regardless of their knowledge, by either ridiculing others or in their polite arrogance and aloofness. Intellectualism should be neither cold nor negative. It is warm and positive. The intellectual is never a fanatic. He cannot be dogmatic. Inferiority and superiority are both negative and extraneous to the intellectual. Humility, sympathy, an open mind, the constant nurture and expression of the accepted principle of the fallibility of fact, value, and criticism, the transient perspective in all things due to the constant motivity of experience, and the judicial approach to people, things, and experience, all these are self-evident identities of the process of intellectualism. The intellect of a person is a process, not a prize, not a *fait accompli*; there is no medallion on the lapel of the intellectual. His mind and heart are open constantly and his mind and heart may bear a critical but never a negative attitude towards others. To the intellectual there is no such thing as the absolute ignorance of others because this is a relative variable concomitant with one's own limitations and probable misconcepts. We are all relatively ignorant to a degree. To admit the constant variable of error is also to admit the constant ertia of learning without forbearance and regret. The world moves on with people who are in intellectual motion, and this implies a tolerance, always vigilant, towards ideas and with surrounding peoples, things, and with life in general. The intellectual bows to the age of the world and the great stage of the future. Regardless how ingenious his oratory, the intellectual realizes that it is not possible for him to be or represent all the notes of the keyboard.

But of what purpose is all this mind and intellect? To gain a hold on meaning and value through a clearer discernment of the verifiability of certainty and the processes of certainty which is experience. Why meaning and value? To pursue and achieve, if possible, as great a degree of happiness and life satisfaction. The greater our value of Time the more we see how valuable and how meaningful it is to be happy. What is happiness? It is reasonable to try to analyze this most sought after and most valuable product of living.

Happiness is the measure of how much one desires to live. Happiness is a rational pleasure, and man to acquire it must be a rational being. Happiness is a thought arising out of some identity of physical experience. Born out of reality, yet its only reality is the idea of it. Outside of the mind, where conscious identity ceases, it is not real. Some are gay, some sad, some in despair, some serious, some humorous and some genuinely happy—all looking at the same thing. To an unhappy person even the finest palace becomes a dungeon while to a happy person even the simplest hut will always seem to be a palace. Thus the reality of beauty and possession depends upon who's looking at them.

The happy person digests Time so easily. His appetite is always good. He looks forward to the next hour like a gourmet anticipating another wondrous feast. I well remember a stanza from a poem written by a dear friend of mine, a hardly known but great mind, Robert F. Hester:

> "Ambition may aspire to fame—
> To luxury, or a lofty name,
> But life has never more supplied
> Than simply—to be satisfied!"

The happy person realizes, even subconsciously, that there is nothing so precious as Time, because in it he can keep refreshing his thoughts and actions. The happy man lives in peace because all values of things and experiences are scaled against the value of Time, and finding that this value of Time and with it the sense-value of his own life, is most important. All other things, profits and losses, as one sees them, become lesser relative values and factors. Anything that wastes time is a conscious loss. Pity it is that almost every one of us, including myself, wastes part of today because we simply take tomorrow for granted. As Llewelyn Powys puts it, "Every day, every hour that you breathe you experience a miracle. You are still free and alive. Brief as a rainbow your dream also will be. There is no clemency, no reprieve, no escape; no, not for the strongest heart deep mortiesed in life." [30] It is a pity that we take for granted our breath of the very next moment, millions of them hopefully expected, and yet we cannot buy even one of them with all the wealth of the world! Time is like a woman; it wants to be looked at and enjoyed. Even the hobo is a thoughtful philosopher who seriously considers Time to be more important and more valuable than money. It is silly to spend one's life philosophizing about things we could hardly ever dream of visualizing a true knowledge of and then realize, often too late, that while we were doing this we forgot to live. The key to wisdom is just to be happy; this is time best spent.

Whether it be with hoboes or aristocrats, young or old, men or women, a certain quanta of harmony is essential for the happiness of the individual and group. In the face of the realities of the world and the experiences of people, it is also essential that our knowledge of the social sciences be applied in every possible way to pave the way for this coordinating process of individual and social happiness. "As soon as we come to believe that the solution of the problems of human happiness will come not from religions nor from systems of metaphysical philosophy, but from exact sciences alone, the obstacles to progress will be removed." [31] Even applying our intelligence or reasoning power is not enough. "Not logic alone," wrote Benjamin N. Cardozo, "but logic supplemented by the social sciences becomes the instrument of advance." [32]

To the happy person Time is like good music, *hearing beauty,* which melts the world into a thought and a million years into a moment. The happy person doesn't bear hate. Hatred is the money of the weak. Oscar Wilde wrote, "When we are happy we are always good, but when we are good we are not always happy." [33] Or should we go by what George Jean Nathan felt about it: "When all is said and done, each and every man's philosophy of life, whatever it may be, is profoundly right so long as it makes him happy." [34]

The happy person "feels" the strength of his own thought and such thought is never one to be enslaved. Sympathy, the desire to appreciate and to understand, is born of happiness. Life to him is not a compromise of social negotiation or resignation to it, or as one who tries to make the most of a bad bargain, or a clandestine self-worshipping egotist who expresses his "smartness" by robbing his benefactors and friends. Or like the fellow who just "lets it go" for fear that a disruption of the *status quo* is just too much to be bothered with and it is so much easier to learn to be unhappy. He is not like the person who plans his life only to suit himself and at the same time expects other people's clothes to fit him. The happy man's life is more or less a balance of tolerances that remove conflict and bring serenity. This balance, of its own nature, essentially necessitates the dissolution or voidance of false concepts and false values, fictitious and superficial attitudes and mannerisms, the elimination of needless frustrations, repressions and inhibitions, the rise from the florid artificialities of conforming mannequins.

The word *Tolerance* is very often abused, misapplied, misunderstood, unjustly acclaimed. Even Jehovah, according to the Bible, is a very vengeful, jealous, punishing God, obviously fearful lest another god should grab a little love from some human yokel. Tolerance is something that is too often considered as a necessary subterfuge in our daily lives. The worker is tolerant of his boss although he may despise him for his wealth and authority. The suspicious wife is tolerant of her playboy husband because he gives her security, prestige and luxury. The beautiful young secretary is tolerant of her romantic old boss because of his gifts, niceties and an occasional raise in pay. The married young are tolerant of their parents so long as they do not forget they were only supposed to come for dinner. The salesman is tolerant of some power-drunk buyer only because of the order he might receive. The young Romeo is tolerant of his beautiful girl friend although she may be mean, stupid, dirty, and lazy. The priest is tolerant of the heretic only because the law now protects the freedom of thought and the priest got to him a couple of centuries too late. Let's not flatter ourselves for our kindness unless we are honest to confess that they are sincere, compassionate, wholesome, and mutual. Tolerance is born of understanding, a desire for peace, for mutual regard, of consideration, not of compromise, retribution or reward.

The very religious people, with all their supposed "brotherliness," are more intolerant towards others not of their own creed than ordinary good people. Intolerance is born of rigidity, of inflexibility, and there is nothing less progressive, less changeable and less tolerant than a religious fanatic.

To the extent that a person is happy, to that extent he is free in some way or other. To the extent that a person is unhappy, to that extent he is imprisoned in some way or other. To achieve a happy tolerance one must break down the bars, if necessary, that keep him imprisoned, cross the barrier of mere conformance and dogmatic subjections and go out into the fresh air of a free mind that can think in terms of how to live better and longer rather than how to suffer and die better. Living happily longer also requires a tolerance of the least number of regrets and remorses. This calls for an explanation of *conscience,* which is an ambiguous word and can only be evaluated or understood according to the content and potential of each person. "Conscience," states Dr. Friedrich Paulsen, in his *Introduction to Philosophy,* "is originally nothing but the knowledge of custom . . . [but] conscience assumes another form on a higher stage of development. Corresponding to the individualization of mental life, it here becomes an individual ideal of life, which even antagonizes custom." [35] God didn't hand out to each living thing the same quantum and quality of conscience. "Ideas, thoughts of ends, are not spontaneously generated. There is no immaculate conception of meanings and purposes. Reason pure of all influence from prior habit is a fiction." [36] Conscience is often a screen for evil as well as good, as a cover to beguile one's own guilt, or the composure that comes from being a town purist with greetings from people on the avenue because he is such a "fine" citizen. Even the road to hell, they say, is paved with good intentions.

The butcher does not necessarily feel conscience-stricken when he puts the knife to the steer's throat, but the steer, if he had a conscience, and perhaps he has, might have a few correctives to add if he had the time to say it. The soldier on the firing squad ordinarily doesn't feel his conscience smoldering because he is carrying out an order presumably created out of somebody's sense of justice. How about the good Bishop who blessed the English vessels before they left for Africa to fill their holds full of slaves? Did his conscience bother him? Hardly. The Turk who cut down with his sabre little helpless Armenian children and old people didn't feel his conscience hurt; on the contrary, he felt good because these deeds pleased his Allah. The Protestants who murdered the Catholic Macs in Scotland, I'm sure, felt good that they were serving their God; so did the Catholics who slaughtered the Huguenots in France; they, too, felt they were serving their God, and all their consciences were clean and very comfortable. Did the Indians feel conscience-stricken when they shot down the women and children of pioneers and then scalped them? And did the white gentlemen who sold them whiskey and rifles have a hurt conscience? Hardly. Nor is the conscience annoyed in any way that belongs to a thief who managed to pass through the customs with a bag full of hidden diamonds or another load of narcotics. Then there is the other kind of diamond trader who sends down thousands of Negroes, "freemen" slaves, into a large hole in the ground so that they can lay the precious stones at his feet and then lie down and die of silicosis. Does this trader's conscience bother him? Hardly.

Every conscience is *conditional* and *conditioned.* There is really no such thing as an inflexible, absolutely set conscience. The conscience of each individual not only varies in some degree from that of another individual, but every conscience is particularly conditioned and conditional. It is conditioned by the heredity, environment and its experiences. These three ingredients cooked in the pot of causes and effects called an event and a series of events bring out a certain type or status of conscience. It is constantly being conditioned according to what the individual needs more essentially or less essentially, what the person craves more or less, and how much tenacity he expresses on any particular principle of his conscience in ratio to the amount of desire exhibited on any particular thing he wants to have. Bank cashiers are usually selected for their jobs because of well-checked references regarding their reliability to be honest; the cashier may or may not be basically honest yet he knows that his records are periodically checked and that if he is caught he will go to jail and be publicly shamed and humiliated. A very fine genteel person will gladly share his water when there's enough of it to go around or when his life does not momentarily depend on it. But the same person, on a desert, in fear of dying of thirst, may strangle his partner so that he may get his hands on the meagre supply of water that just isn't enough for both of them. Or, for example, a person may not become dishonest if it involved a ten dollar bill but if it were a thousand dollars he may not be able to overcome the desire to get it. The conscience often works two ways: one way is not to look when its master steals, and the second way, to act as a phony justification to comfort the ego. The conscience, like the mind, of which it is only a manifestation, is like a man at the bar drinking; different people drink differently, more or less, and people vary in their capacity to absorb drink before getting intoxicated, but eventually enough drinking will get anyone intoxicated. In the Talmud, it says that there are three times when one can tell whether another is telling or expressing the truth: when he is drunk, when you come to him for a favor, and when he is angry. These words sound very wise and no doubt very often reflect a great deal of reality, but I have seen people feign anger only too well in order to bring across a lie, using the anger as a confusing element to make it appear the truth; I've watched people get drunk only so that they can lie really efficiently. And I've seen people granting favors as a way of disposing of a problem, not because they really want to help. Yes, the conscience is a very tricky thing, no less tricky than man himself and his strange ways of living. "We do not have to go to the interpretation of dreams to understand human behavior," writes Dr. Eichler. "The waking state hides very little." [37]

Many a tyrant started out as an idealist and a benefactor of the underdogs. Many a man lost his conscience whenever it pleased the lady, and a lady usually knows very well how to make a man lose everything, including his conscience. Many a well-heeled person can sit down in the parlor and chatter out long, windy expositions on the beauty and value of the conscience, but Jean Valjean, when he broke the window for a loaf of bread, just didn't have the time to think about it; he was starving and his life was at stake. We can continue endlessly with examples, but we can easily see that conscience is conditional to how much of anything a person can stand, require, or desire, in order to survive and without which he would suffer too much or perish. The extension of conscience is usually confined outside this circle of survival and want; inside the circle, it is usually every man for himself and let the devil take the hindmost!

How do you think the conscience of a child grows up when he was taught to ridicule and hate other children as "You dirty Jew!" or "You Irish harp!" or "You yellow chink!" or "You black nigga!" Do you think such a boy's conscience will grow up to be fair, intelligent, tolerant, equitable, just?

The conscience is merely the ability of a person, through his mental and emotional systems relative to the influence of external factors, to provide for himself a scale of conscience-value. This brings us now to an inquiry into the nature of "morals" and "morality," two words that have brought untold misery and unhappiness to people more than any other words I know of.

There have been and are many schools of thought, ancient, developmental and modern, regarding what Kant called his "categorical imperative," the mysterious "oughtness" in the "moral" nature of man which reveals to him what is right or wrong, and directs him to do right by some kind of divine compulsion and, if he were to do wrong, to feel that he has committed a violation of what he "ought" not to do. This would be just wonderful, if true, but it happens that it isn't true. In fact, it is so far from the truth that it remains purely an idea of *a priori* fantasy. We must try our best to face up to realities, things as they really are and not to beguile ourselves with futile dreams and desires. Albert Einstein wrote: "There is nothing divine about morality, it is a purely human affair." [38] Sir Leslie Stephen plainly stated: "The charge against the Agnostic, that he weakens his belief in morality because he brings it within the sphere of experience, is just as true as would be the same charge against the man of science, who appeals to facts instead of evolving the facts from the depths of his consciousness." [39] Morality should be based upon natural exposition, free intellectuality, and should be considered as all the other forms of social behavior.

What is particular to man does not prove in itself that it is universal; what is particular to a particular man does not prove or assure that the given situation of one man is indicative of a general situation, that is, of a group, nation, or of mankind. The reverse is also true, that what is indicative of a society is no indication of the same status of all the individuals, as individuals, within the society. This is not only simple logic but is plain ordinary common sense by any reasonable and diligent observation. A penitentiary may or may not be far from the Capitol, but whether the inmates of both institutions may or may not be similar or whether they are located in the proper places, cannot be taken for granted. By the same token, what a man may profess to feel or think does not or does necessarily coincide with what the society ordains or rules by. The same principle of variable and particularization goes on for groups in their relationships between themselves as individuals, and the same goes on for groups in their relationships between themselves, and the same goes on for nations, religions, social, political and economic orders. It seems, from observations of certainty within experience, that there is no universality or divine over-all principle governing the "moral" nature of man and his societies, regardless of the beauty of our oratory, the pleadings of our sermons, or our own clamoring prayers to the skies. Professor Joseph K. Hart gives us to understand that "Nature has no morality, nor moral leanings; it responds to 'bad' men as readily as to 'good' men—provided only that each has learned the technic of control. The rains of heaven fall with impartial favor on the just and on the unjust." [40]

Yet we know that the human being contains affection, love, friendship, principles of honesty, integrity and peacefulness, a sense of fairness and reciprocity, if not in most of us, at least with a number of us. The human being, while biologically a potentially cruel animal, is also a potentially friendly and lovable chap, as long as he is rubbed the right way. Cruelty doesn't necessarily mean criminal; what is meant here is that in the early stages of man's evolution his emotional system had to be attuned to his fight for survival, and he found hardly any love and affection facing him from the bleak, disease-infested, ever-encroaching jungle and the life it contained that constantly preyed and foraged upon him. The elements, the animals, the insects, the struggle for food and shelter, which readily included man eating man, and a thousand and one other natural enemies, were constantly pressing him, threatening him. Extended relaxation meant resignation to be swallowed up by the serpentine grip of the environment. Is it any wonder that man's basic emotional, instinctive, innate constitution is partly sadistic, partly masochistic, partly sane and insane, with a variable of a hundred possible complexes? The natural cruelty of the human being is the conditioned reflex evolved out of milleniums of fears and terrors. If any one is doubtful about this, let him see how a highly civilized city would behave if the police power, all laws and possible punishments, were removed. Sometime ago in the City of Churches, the headquarters of the Cabots and Lodges, a substantially Catholic city, the city of Boston, there was a police strike for a very short period of time. Thuggery, robbery, assault and hoodlumism reigned through the city and almost every store was looted and hardly an unbroken window remained. No wonder Walt Whitman, comparing the nobility of animals with the insincerities and cruelties of man, said: "I think I could turn and live with the animals."

The relentless and ruthless extermination of the passenger pigeon engraves the mark of Cain upon mankind—this alone establishes the cruelty of man far more cruel than all the gods this cruel creature has created to soothe himself with. If God chose man to be in his image, such a choice most assuredly originated in the self-absolving brain of man. No God could be so stupid. Civilization, the process of refining man's cruelty so one can be killed politely and by due process of power, has even increased the need for more cruelty. In Africa today the natives are killing zebras just to cut off their tails and they trap and kill lions only to cut off their paws and they kill elephants only to cut off their feet—these to become fly swatters and desk ornaments and waste baskets for the tourists—the tourists who were transported there by Civilization, and most assuredly tourists who believe themselves civilized and therefore "superior" and the natives mere savages. The pompous hunter with a gun, knowing very well that the animal he hunts hasn't got a gun, is far more cruel than the primitive who only kills for food and the animal's skin to keep him warm and with apologies. I remember a short time back when I was up in James Bay, Canada, where I fished the Nottaway and Broadback Rivers for trout; these trout became food for dinner. I remember, canoeing back to camp at dusk when one of the party suddenly saw a pair of peaceful little mallard ducks idling in the still water half hidden by the tall grass. He had his fill of shooting geese for the day but he felt suddenly triggered to shoot again. Here, indeed, was a great and powerful adversary! He asked the Indian to flush the birds, according to rule, and he shot down, within a short range, the male mallard with its beautiful plummage, bringing tragedy to a little family alone in the tundra of the North. This was needless killing. As we left the area and turned around the bend towards the camp, the sad scene of the small female duck wildly circling alone in the grassy water and the darkening sky is a picture I would like to forget, but I can't.

If anyone really believes that man is innately good instead of cruel, let him consider the fact that in spite of police power in every nook and corner, every village, town and state in our country, in spite of all the churches, temples, ethical societies, lodges, and millions of prayer books, the prisons are usually fully crowded and more jails have to be built to take up the increase of new criminals. Why should man be different than nature itself? "Dytiscus larvae devour animals of their own breed and size, even when other nourishment is at hand." [41] Man pretty well does the same. "Certainly when we know that men have not yet ceased from being murderers and killers for the past million years we have a better understanding," writes Dr. Raymond A. Dart, "not only of our human history but of our fellows' natures as well as of man's national and international madnesses." [42]

Moreover, the greatest amount of thievery is committed by employees against the people they work for.[43] Also we have to keep in mind that the greatest amount of cruelty and crime, legal as well as illegal, goes on without detection or arrest, and a still greater amount of cruelty goes on "within the law" in many forms and manners, from a petty confidence game or politician to working the stock market for suckers. If anyone really believes in the innate honesty of the human being, let him, on a crowded boulevard of a large city, drop a ten dollar bill and see how many people behind will pick it up and rush up to return it to him. There will be exceptions, of course, but the exceptions will be very few and difficult to find. The trait emanates from the deep-rooted self-interest of any animal to survive. In the human being it has evolved into a selfishness which tends toward cruelty, toward oneself and to others. To cite just an example or two, Dr. H. R. Hays relates that in "Australia, white frontiersmen were called bushrangers. They were the equivalent of the rangers of the old American West, and among them arose the inevitable outlaws and bad men. It is not surprising that such ragged colonial types, brutalized by legal savagery, were not inclined to be tolerant of a dark-skinned race of beings whom they soon branded as 'treacherous' and scarcely more human than dogs. In Tasmania, where thousands were hunted down like beasts and shot, by 1835 only 203 aboriginals were left, the pitiful remnant of thousands. The last pure-blooded Tasmanian died in 1861. On the continent they did not fare much better. Sheepherders, in order to clear the grazing grounds more rapidly, offered them, in apparent friendliness, cakes of flour dosed with arsenic, and thus poisoned off black humanity like ground squirrels. Other ingenious native-exterminators poisoned the waterholes." [44] What the animal cannot take or do with his prehensile limbs he now contrives to do with his brain. "Not a single moral precept—neither the golden rule of godmen nor the moral codes of good men—even when bolstered by the fear of hell fire or the promises of heavenly rewards, has ever brought about the slightest improvement in human conduct." [45] Felice Belloti, telling about the *Fabulous Congo,* says: "The reason for this comparative lack of crime among the savages, is not, however, to be sought in the natural honesty or good will of individuals but chiefly in the extreme harshness of the penalties and the impossibility in practice of the criminal's escaping punishment. . . . Actually among the negroes only one sentiment restrains them from crime—the fear of punishment." [46]

Some say that there are communities in the world where the people are amazingly and habitually friendly, even brotherly towards each other and any cruelty and crime, in any recognizable or more than a negligible percentage, is hardly known. In such communities, if investigated judiciously, one will usually find that the nature of the punishments or penalties, dating back over long periods of tribal history, including taboos and environmental limitations, were of so severe a nature that it created a fear-pattern which evolved and still persists, although the present descendants do not realize the real causes and origins of their culture-pattern. John Dewey cleared this point well when he wrote: "The idea that morality ought to be, even if it is not, the supreme regulator of social affairs is not so widely entertained as it once was, and there are circumstances which support the conclusion that when moral forces were as influential as they were supposed to be it was because morals were identical with customs which happened in fact to regulate the relations of human beings with one another." [47]

In such communities in which the fear-pattern was evolved, the penalty needn't be death; mere ostracism or exile from the tribe was tantamount to self-destruction, especially in those parts of the world where a member of one tribe was usually regarded as a natural enemy of another tribe, and subject to be killed on sight and often eaten. Also, if we find that certain communities or societies have cultivated or evolved tribal festivities, influenced by environment and religions, of a competitive or "prestige" nature, including offerings to the gods and temples, dances, group trances, group dependence upon the tribal witchdoctor or head priest, animism and ancestor-worship tied to cyclic resurrection or reincarnation, group dependence for food resulting from limited arable soil and collective tillage, then we will also find that in these communities the extent of communal friendship and mutual assistance are naturally furthered and patternized.

Even the belief in a god is the admission that man cannot cope with his own personal problems and most willingly grouped together with his fellow-men for common protection and fear-sharing. This belief in a god caused the reduction of the ego to a slave and a coward, two factors that can, by sublimation, bring about a certain cruelty and masochism in the person by his own fears, uncertainties, repressions, and dissatisfactions. Dr. LaForest Potter claims that "the human animal, when he runs amuck erotically, is the most lascivious beast on the face of the earth. Nothing that has ever been discovered or observed in connection with the conduct of animals—no matter how cruelly conscienceless and bloodthirsty these might be—can even remotely compare with the unthinkable horrors, the unmitigated degeneracy, the abysmal depravities of the sex-crazed human." [48] He informs us "by unimpeachable scientific authority that soldiers and monks, in particular, and others who have been entrusted with the duty of watching the dead, are frequently overcome with sexual excitement, during which they have gratified their horrible desires with female corpses." [49] Allen Edwardes, in his *The Jewel in the Lotus,* relates: "It is reported that by the regulations of caste in India, should a betrothed pubescent girl die before the ceremony, she had yet to be penetrated by her husband before disposal of her body. This was performed in the groom's house in the presence of the village priest just before cremation. Such a custom (intended to free evil spirits) may be traced back to ancient Egypt where virgin corpses were deflowered by priests or embalmers." [50] "The Magians (*Mujoosee*) were the first Parsic pederasts. They, fire-worshipping sorcerers, danced frenziedly in the nude round the flames at midnight, wailing eerie incantations and fornicating among themselves and with boys doomed to evil sacrifice." [51]

The fear of nature and the survival problem also were incentives to grouping and tribe-making. With some favorable constancy within the tribe achieving a reasonable security and peace, plus a maturing mutual-assistance program of a sort, it is easy to understand how people can learn, from the satisfaction of essential and basic wants, to like and be friendly towards each other. Albert Einstein comments that "A man's ethical behavior should be based effectually on sympathy, education, and social ties; no religious basis is necessary. Man would indeed be in a poor way if he had to be restrained by fear and punishment and hope of reward after death." [52] Let us say that man is cruel because nature forced him to be that way for sheer survival, and man could be good and the cruelty-weapon subordinated when he obtains the securities, satisfactions, freedoms, and peace of mind which are essential to a reasonably happy and healthy life. What comprises a reasonably happy life is for him a problem of analysis, and, in the more civilized races, the task of cultivating the mind to analyze experience with the view of reaching for higher values and a more sustained healthy normalcy and happiness.

Again we must remind ourselves to face up to realities. Civilization is an order of law *imposed* upon the majority by the few for the "good" of all according to the culture of any given civilization. Phillip C. Jessup emphasizes that "law is indeed a human necessity." [53] Cardozo, in his *Nature of the Judicial Process,* states: "Law and obedience to law are facts confirmed every day to us all in our experience of life. If the result of a definition is to make them seem to be illusions, so much the worse for the definition; we must enlarge it till it is broad enough to answer realities." [54] Lord Josiah Stamp stresses the point that "in society, the price of liberty is not merely eternal vigilance but also perpetual restraint." [55] Erwin Schrödinger hopefully goes further: "To establish the necessary order and lawfulness in the human community, with the least possible interference in the private affairs of the individual, seems to me to be the aim of a highly developing culture." [56]

Relax law for a moment and the majorities (whether they are right or wrong) will eliminate the minorities until no one would be hardly left. Assuredly, if there were no laws and governmental forces to prevent them, our Southern white nobility of "dear hearts and gentle people" would most probably grab every Negro and dump him into the ocean and thus solve, their way, the problem of integration. In business, the grape-vine philosophy is for one to try to do or outdo the other as you can expect others to do or outdo you. In politics it is the strategy before election of one party to outwit the public so much better than the other party. In religion it is the lamentations of the clergy to be louder than another set of clergy, and the prayer books of the religions are just loaded with exalted praises that take for granted that the Almighty is a blind, stupid old Egotist who just loves to be tickled by man's flattery and bribes of pretense and show. Even among sportsmen the sportsmanship doesn't usually extend to any overdose of altruism on the part of one to wish to see the other shoot a bigger bear or catch a heavier fish. And so on in all the little ways and byways of life. Because it is natural for people to think of themselves *first,* it never occurs to them that if only we could realize the good others could do for us, we would give them a fair and better break to begin with and enjoy a higher and finer degree of satisfaction with ourselves. And yet, paradoxically, because of this very fear for oneself, man has nurtured the psychological desire to mass together, create groups or tribes for increased strength in unity. He translated the greater strength and pressure of the group as his strength and his pressure; he grew in the emotional gathering of the idea that the group was inside him and made him bigger, more powerful. We often see the common example of how a man would act as part of a mob which he would rarely act likewise if he were alone.

We have seen in the chapter on the *Power* factor that there is an undercurrent, a fast and furious stream of seething pressures being pushed along by the generating processes of life and living which exhibits itself in the competition of man, men and his societies, to "push" up and extend, to *overcome* and to *overwhelm* something or other. In business it is called the spirit of competition, in politics the shrewdness or astuteness of the statesman, in religion the "sanctity" of the saint, and in society the incredible genius of the "leader."

Peace is something we always *seem* to want, yet when we get it we simply do not know what to do with it. Apparently, the complete or relative tranquility of peace becomes too monotonous, sooner or later, in its contrast to the activating desires of living. "So restless is man's intellect," writes Loren Eiseley, "that were he to penetrate to the secret of the universe tomorrow, the likelihood is that he would grow bored on the day following." [57] Unfortunately, most people would rather be aroused than tranquilized. As a result, no other creature has intended or tried to exterminate man except man himself. In my study of the history of man and his institutions I cannot remember one year of peace throughout the world; somebody or some country was at war, somehow, somewhere, just to keep the scorching flame of war from flickering out. Often I felt that war is so stupid a thing that the only plausible reason one can find for it is that the old are just too jealous of the young.

In man's primeval, slow, creeping climb up to the present

heights of civilization, he often forgets, in his transient escape from his insecurities, and rushes headlong into the grip of mob rule and violence, mass maelstrom, and social and political holy rollers. The grape has fallen into the barrel to find divine intoxication by dissolution. He expresses, almost in panic, his primitive fears and to overcome them expresses, in turn, his primeval cruelty whenever the conditions are conducive for its incidence. "The social instinct has been acquired by mankind too recently, and it is still too feeble to be a trustworthy guide in all conduct." [58] Many a reformer who had infinite faith in the natural goodness of people was killed by the very people he tried to free, befriend and even love. Jesus was crucified because his love for humanity was too great for men to bear or understand. Lincoln was ridiculed from pillar to post, from the pavements to the pulpits, and finally assassinated for his great humanitarianism. Thomas Paine, most probably the real father of our Revolution, was plagued to his dying day. Columbus was ignored by Spain for his explorations in the New World. Shelves of books can be written on the life and history of the truly great benefactors of mankind and the nature of the gratitude they received from their peoples. The pages would be too bloody and tragic. If God really made man in his image, all I can say is that it isn't very flattering either way and a sad commentary upon our origin. If it is more to the truth that man made God in man's image, then it is all very understandable why God is taking revenge upon us all these many thousands of years!

Yet, in spite of all this, we humans do have love and affection, friendship, sympathy, appreciation, compassion. How often I have enjoyed the warmth of a primitive's smile and the generous hospitality of his hut and shared his meagre food and drink. The handshake of a genuine friend, the extending kindness of the stranger on the way, the simple desires of people to please or to be nice, these experiences are legion. What wouldn't many people do for their horses, dogs or other pet animals? There are honest people in business who do consider very conscientiously the feelings and interests of the people they do business with and of the people who work for them. There is a softness, a desire for self-approval and external approval, a desire to "father" or "mother" other things besides ourselves. I do not mean, in this instance, natural father or mother love which is instinctively and biologically innate in people; I refer to the desire *to do good,* the enjoyment of an ethic, like enjoying good music or a fine painting, which aspires man to greatness of a degree and fulfillment within himself, the desire not to be alone, to give to and receive from others the warmth and affection that is so precious and which gives meaning and value to life. This is the *counter-current* which flows above the undercurrent of cruelty we have described previously. This counterbalance activity can be observed in and out of a hospital. In it many patients, utter strangers to each other but finding themselves incapacitated, helpless, suffering, feel sorry for each other, reaching out for each other's sympathy and giving sympathy as a form of strength to the giver, which is good and wholesome. Outside of the hospital, these same people, now healthy and strong, return again to the old way of dog-eat-dog and everybody is back in the arena of racing rodents again.

The *undercurrent* is cold, heavy, lightless and blind, powerful and relentless. Above it, the *counter-current* is green with the freshness of Spring, warm, throbbing with regeneration, much slower, more peaceful, yielding and throwing light, like a dance of love, on life's surfaces. It seems to me these are the two currents of a man's life. It all depends in which current one spends more or less time. It all depends on which current we find ourselves swimming with, very often much against our better judgment or helplessly forced to by pressures too strong for us to penetrate through or else fearful and helpless in ignorance of our own strength and potential. Fortunate is the one who finds himself sailing the top current where the stars, the sun and the moon can shine upon him and the horizon is there to wonder at.

As there appears to be an innate cruelty in man, so does it also appear that there is an innate potential of goodness in him. Vigilance and wisdom may find one or the other and often both in the same person. *Human nature is the one consociated paradox of compatible paradoxes.* Consistency and inconsistency are just parts of the perplexing, complex nature of man. This *counter-balance* of actions appears to be man's strange method of keeping interest alive and escaping from the growing monotony of anything; it is part of the restlessness of life and man's attempt to compromise even the dullest of moments with so much as a pilot's light to hold the flame of curiosity and adventure from going out completely. No reasonable person deals with perfections; only fools and thieves do. Life is made up of positives and negatives to complete a circuit. Each keeps the other alive and necessary. This counter-balancing expresses itself in social relationships. If you say "yes" all the time you will live to regret it. If you say "no" all the time you will probably die alone and lonely. Between these poles lie all the compromises in life. No man is so wise and lucky as to pick them all well. But we can try. In trying, judgment should rule, but never resignation or submission. It's even very normal, and wonderful, too, to let loose once in a while, so long as no one is hurt, no heart broken, no life destroyed, no health impaired, no one cheated or deceived. We know we live once, die once. We are entitled to a few "irregularities" just to prove that a straight line is of no real measure or relative importance until it goes off at an angle, even if for a short distance. "Living is a devious affair; nothing seems to develop in straight lines." [59] If life has to be a continuity of errors, let them be happy and peaceful ones.

Then, strangely, we find people who are good one moment and surprisingly cruel the next. There are people who wouldn't twitch an eyelid to buy a ten thousand dollar painting or sculpture in their sincere appreciation of the fine arts, and yet their employees can lie begging at their feet for a trifling raise to eke out a survival. There are people who profess religiosity, are sweet of nature, seemingly noble of mind, peaceful, of an understanding, patriarchal temperament, in the public or social eye just adorable and wonderful, but yet who, shrewd and cautious, connive and steal from their own trusting friends and from those who are their own close associates and benefactors. It reminds me so much of a little anecdote from the Talmud (Midrash):

> "When Rabbi Johanan ben Zakkai fell ill, his disciples went to his house to visit him. They spoke to him, saying: 'Master, give us your blessings.' He said to them: 'May it be His will that the fear of heaven be upon you in the same measure as the fear of those who are of flesh

> and blood.' They asked: 'Only so much?' He answered them: 'Would that it were so much! For you must know that when a man intends to commit a transgression, he says, 'If only no man sees me!' "

The vigilance and wisdom we have stressed before are shared in the effort of man to use his mind to investigate, inquire into these natures so that he can possibly filter off the bad and retain the genuinely good. But if the mind really wants to do this it must, in the first premise, be honest with itself by lighting the way with analysis and reason. One must have the courage to face up to realities, and when he really does this he will be surprised to find out that it isn't really bad to look at them. The more he will look at them honestly and fearlessly, the more he will learn to like to look at them and will not compromise for anything less than that. He may even find the road ahead to be brighter and happier. The good in man is like bits of gold sprinkled lightly over the vast sand dunes of human existence. Truly, the goodness of the mind and not all the material goods in the world, can filter the mountains of sand and uncover the precious chain of value that might bring peace to the people of the earth.

Hence, the meaning of morals goes deeper and wider than the mere skin-deep veneer that society is so often painted with. In the natural morals of life there is no desire for tragedy or dying. A moral that necessitates the subjugation of a person from normal desires of love and affection and the natural, biological requisites of life, is not only an abortion and a miscarriage upon both the individual and the society, but it is plainly a criminal assault upon life and the value of Time within that life. If God created life, as they say, surely He must have created life for the purpose of living, not for a constant dying before death.

It is strange that a substantial part of the European or Western idea of morality consists in anything that, because of some abnegation and self-denial of some kind or degree, must envisage a certain quantum of frustration or "holding back," abstinence, sublimation or asceticism, which should result in a "respectable" discipline resulting, in turn, in some degree of suffering and loss of freedom. Why morality must be identified with unnatural, unnecessary and unrealistic self-denial and suffering must be only known to the saints and prophets of two thousand years of European history and misery who, if they lived today, would be most probably confined to the nearest insane asylum or to some school to eliminate or soften, if possible, their hate of life, of humanity, their illiteracy and self-cruelty. The greatest travesty upon the minds and bodies of men and women is the "dark ages" idea that people *have to sacrifice* something or must suppress or deny themselves, in order to "purify" their minds and bodies, to "cleanse" their souls of the "devilishness" of the normal and natural requirements which make up a reasonably happy and healthful life. This puritanical, traditionalist trait of Western culture, in particular, has not only vulgarized the finest concepts of human feeling and expression but it has wrought untold misery upon millions, if not billions, of people over the centuries and if this is the best that the immaculated gods can do, I can do without them very nicely. The monk on a bed of straw who hides from a woman, in fear of pollution, or the prude who can't stand people kissing and necking each other, are both simple idiots who are really sick people and for whom we can only feel sorry. The affixing of the idea of sex life as something sinful, impure, etc., to the religious theme of Christian morality, makes impossible any maturation of its followers. They say we are all "creatures of habit." Most probably they are right, but when a habit becomes more important than happiness, than life itself, then it is high time to dump the habit, and if we can't, then we should see a psychiatrist or go back to elementary school.

Morality should not mean that the sex impulse is a bad impulse. On the contrary, it means that the sex impulse is something that is fundamental, that it is indicative of good health, normal wants, that it is basic in life and to the perpetuation of life, that it is clean, precious, sensitive, and that it should not be disrespected, vulgarized or lowered by the abuse of its sacrament. The purpose of the sex impulse is to maintain life; the purpose of love is to make that life a beautiful consciousness, an enjoyment, an appreciation and a scale of value within it. They are distinct qualities and entities that oppose each other, and it is because of this very opposition that they are fused and attracted, as the ends of a magnet, into one unity and purpose. The world exists by opposition, by attraction and fusion, and it is these very things that make the world go on, that make man try to keep perfecting himself into a greater, better, finer, and happier life.

Passion may seethe as the torrid sun, yet the sun has its specific performance. Were it not for the sun, the earth being in fact a dead and cold planet, all life would immediately come to an end. The sun maintains life; so does passion. Love soothes as the moon, and the moon, too, has its performance. Passion possesses; love cherishes. Passion craves; love calls. Passion, within itself, is the attraction of opposites; love, within itself, is the tie of equals. Passion is the curricula of nature; love is the understanding and appreciation of nature. Passion is the preservation of a kind; love is the intelligent ascendancy of that kind. Passion is the course of natural preservation; love is the meaning and value of this preservation. Passion denotes an identification of nature as a self-preserving system of opposing forms; love denotes an identification of nature as a self-progressing system of uniting forces. Passion is an automatic process of natural energy; love is a conscious process of reflection. Passion is objective life; love is esthetic life. Passion is the expression of weakness for the absorption of strength; love is the expression of strength for the absorption of weakness. Passion dies by cause of its own volition; love increases by cause of its very expression. Passion is the matrimony of the body; love is the matrimony of the life within the body. Passion is the worship of beauty; love is the worship of the ideal by which the beautiful should express itself. Passion seeks body equilibrium; love seeks life perfection. Passion is essential to mating as a particular; love is essential to life as a whole. Passion seeks the body in things; love seeks the personality in things. Passion is diencephalic; love is cortical. Passion is the means of life; love is life itself. Passion builds the house of life; love lives in it. Passion is the criterion of life experience; love is the criterion of life happiness. Passion is the instrument of music; love is the music we hear. Man is born with a heart, not with an incantation.

What is this *love?* It is stronger than any code of belief or law. Belief makes him a hypocrite. Law makes him very often a slave, as there are good and bad laws. Unlike the windows of the temple of creed, the windows in the temple of the heart are crystal clear and pane-less. Creed looks forward to death. Love looks forward to life. In the Temple of Love no one gives the bride away; there is unity, not an offering. How to define love? Love is inexplainable as life is inexplainable, and yet we live it and know that it is in us, that it is ourselves, and that without it we are nothing.

The years are like the leaves of a tree. Sooner or later all are gone. The leaf, in itself, expresses the story of existence. It comes up through the roots that live hidden in the soil, and it goes back to the soil that holds the roots. The coming and the going are the same. Take love away and life, too, turns into such existence, a cosmic drift, just coming and going little leaves, senseless, blind little leaves that come and go as a matter of event.

Life is always in the ultimate; the next moment is already the present. Where there is love, life is the moment. Where there is no love, it is just a knownless drift, coming and going little leaves, *sans* the beautiful, *sans* the real, *sans* the worthwhile, endless little leaves that come and go within the soil.

Mankind's history casts a tragic shadow over the word *love.* No word has been so misunderstood, confused, abused. The mirror of the past reflects an awful picture of what society has done in the name of love. Hundreds of thousands of Egyptians toiled their lives away to erect a tombstone, in the name of love. Carthaginians gave their children to be burned in an idol furnace, in the name of love. The early Christians became a lion's repast, in the name of love. Monks went into the desert to bury themselves in solitude and filth, in the name of love. Wars, crusades, massacres, bloody towers, guillotines, burning "witches" and swinging gibbets, dying columns of men, heart-devouring Toltecs, machine-guns, world conflagration and broken hearts, all in the name of love.

Once an individual *rises* to that status where he realizes the finer, truer and better meaning and value of life, of the good in unity, of the value of his time, he will not need laws, beliefs, prohibitions, to force any goodness. He will become good of himself that rewards itself and which is the foundation of proper and wholesome human conduct and morality. The history of the heart is the story of splinters gathering themselves slowly together; it is like the hunt of Isis for the segments of her love. The fulfillment is far from the end. It seems the last man must sit upon his brethren, and, coming out of his trance, realize his mistakes in a solitude that only complete extinction can release.

I hope some day history may portray a different picture than it has of the past. Men have worshipped and "loved" everything under the sun and have rarely understood anything. But Time waits for no man or his prayers. He must be up and doing, seeking the meaning and value of things and out of *knowledge* may come, if it ever should come, some lasting happiness. "A new Renaissance must come," pleads Albert Schweitzer, "and a much greater one than that in which we stepped out of the Middle Ages; a great Renaissance in which mankind discovers that the ethical is the highest truth and the highest practicality." [60]

Love says: *Know thyself*! Love, to be love, is based on a knowledge of things, based on the realities and certainties of things, and this knowledge comes as a result of using our minds, not our fears. Love is built upon *Simplicity,* for this leads to candidness and understanding, not a catechism of complex rules only fit for slaves or a habit of double-talk that makes it almost impossible for us to see ourselves and others as we really are. It is built of *Sincerity,* for this brings truth, confidence and integrity. It is built of *Courage,* for this assures the realization and continuance of the ideal. It is built of *Patience,* for this leads to a reasonable, just and well-poised attitude in all things of love. It is built of *Giving,* for love is happiness and without happiness all else is worth nothing.

Form is usually misleading and fictitious. Appearances approach but alone they cannot stay. Those who do not possess the personality and vitality of love are but empty shadows, waiting for the twilight of life to take them into it. Falseness, intrigue, deception, diplomacy—these things are complex and complicating. Love has no secrets. Love is an open book. It has nothing to hide. The great trouble of many people is that they are usually trying to be what they are not, instead of not trying to be what they are. Pretense has many suits to wear, changing its dress, form or manner to match each occasion, from tears rolling down profusely in agony to laughter enough to burst a steel-banded barrel.

Common sense is the art of thinking out what to do wisely instead of first doing something to regret later. One may be wise enough to understand life; it is another thing to live it wisely. In life there is only one obligation we cannot change and must meet, the contract to die sometime. All the other clauses in this contract allow us to make changes until then. That is why love is the deepest of convictions as well as the most beautiful emotion that moves the possessor along the hard and cobble-filled highways of life. Thus, too, love has the highest of courage and endurance. When poverty comes in by the front door, love stays. It does not fly out of the window. Thus, too, in the relationships between men and women, a man must think as a man, a woman as a woman; when one thinks like the other, someone is either going to be disappointed or get hurt. Like Kahlil Gibran said, eat but not from the same plate, drink but not from the same cup. This is what I mean when I say that people must face the realities of life, both fortunate and unfortunate, and turn them into fortitude and goodness. Love is built of patience, a waiting enjoyment that we hope never ends. I think the most ardent and patient fishermen in the world are the Frenchmen who fish the Seine. I have watched them for days, in the early evening and in the early morning, in the rain and in the fog. I have never seen any of them ever catch a fish. This is what I mean by a "waiting enjoyment."

Love is never an obligation, a contract to perform, a social convention, a signature in a lawyer's office, a property settlement, the luxury of opulence and flatulent security, a sense of weakness for one to serve and the domination of the other to be served. No one should, better still, never should even try, to dominate or possess another person. *Love must breathe freely* to keep its bloom and express itself of its own volition to keep its fragrance. A man is never a failure so long as he possesses the principle of love, expresses friendship to have friends, nor is he poor as long as he values the day he lives, nor is he sad as long as he has love in his heart, regardless of anything, anywhere, anytime. Love brings wisdom by actually fulfilling the value of Time. He who remembers yesterday and realizes tomorrow will become another yesterday, usually tries to know how to live wisely today.

Were you ever at the ocean shore, and, watching the sun turn the tumbling waves into violet foam, fall asleep listening to the ceaseless swishing of the waters? Then you will know what I mean when I say that love possesses a ceaseless bond. Did you ever, stretched out on a carpet of grass and gazing up at the clear sky, wonder at its inexplainable simplicity? Then you will know what I mean when I say that love is inexplainably simple. Did you ever study a little child at play? Then you will understand what I mean when I say that love possesses an unconscious sincerity.

Love is the bloom of wisdom. It is the titan of the worlds, for its strength can mold a man into a god and a woman into a goddess. It is the one miracle that nature has been kind to give us.

Life is a brief song. There is a rhyme in that song, that song composed of volcanic fury and the cool lagoon, the towering redwood and the blade of grass, the soft waving field of wheat and the tangle of the jungle. There is a rhyme in that song composed of the suffering and the pleasure of the flesh, of the mind, with the *allegro* of the wind and the *penseroso* of the calm, of the frozen ether and hot strata in the core of the earth. *It is the undercurrent and the countercurrent,* the ever-balancing counter-balance. There is a song composed of the bubble of the baby's cheek and the pain of the gray and the dying, composed of the beautiful and ugly, the sad and the glad. That song sings in the sway of the branch, in the sweep of the plain, in the echo of the forest and in the downpour of the rain. There is a song composed of the sparkle of the brook's ripple and the sparkle of the star, composed of the work of the day and the rest of the night, the sweat of the sun and the romance of the moon. The flower in its beauty gives its fragrance to the world; the wise man in his wisdom gives his heart to the world and finds peace within himself. Both are rooted in the same principle. A little seed, a little earth, a little water, a little warmth, and a flower is born. A little thought, a little expression, a little hope, a little understanding, and a lover is born.

The wise man need only study a flower and in its nature learn the beginning, the way, and the end of his own life. He learns that, as the flower, he should open his heart to the free air, send its fragrance to others and bring fruition upon himself, so that other hearts may be born and nourished. He learns that in closing his petals upon himself he will only dim and smother to decay, like a gnarled bud that cannot open and yellows to die, even though the stem be strong as iron and set in the richest soil in the world.

The eyes of lovers are full of the colors of life. Beauty goes in parade before them from the first warm, vermilion streak of dawn to the cool moonlit glimmerings of the evening. To them each thing is a thrilling spectrum through which they see that it is good and great to live as they do, for the heart is the lens that fuses the shades of the day and the night into a singular aurora of stereoscopic space into which they can enter, a field of enthusing and inspiring colors that makes it worthwhile for them to live and want to live another day. True, indeed, he is the most friendly who can listen to a bird all day long. He is the most peaceful who can find repose in a little song. He is the happiest who can find wisdom in a flower, and he is the most contented who can bathe his mind in a rainbow shower. Love is the way.

Like a beautiful song softly being born of bow and string, stirring across our hearts to its just awakening and expression, so does love stir our hearts to the real meaning of what life really is. In the evening, at rest, with our eyes upon the cool darkness, we reflect upon the heat and activity of the day. So, too, in the evening of life we reflect upon the hurry and waste of youth and pine that we should not have traveled so fast. Youth is something we do not usually value when it is with us, and cannot buy when it has left us. The beauty of October is the mellowness and rich colorings of the fallen leaves—something the leaves have little reason to appreciate. Old age is something we all strive to reach—slowly. The stay is but a moment, even though one can live a million years and it is love that transforms this moment into a meaningful, realistic joy. It, like Persephone, returns with the Spring and the moisture of love can be seen on the petals of every flower in the morning. To realize that we are always just one day old—for the present is all we really have, the past being gone forever and the future an expectation and a hope—this is to put upon life its highest value and greatest practicality. Time, the genie in your Alladin's lamp of life, will come to you if you call him. We are born anew each morning and yesterday's life has died with the night.

The night weeps for the day, the mother for the child, the old man for the running boy, and all the world is moist in the morning. It is the continual dawn of new reflections and courses. A new evening comes, another's eyes gleam in the joy of her heart, another child runs, another old man muses, new lovers clasp their lips beneath the bowers in the sweet quietness of the evening, while the moon, keeper of secrets, smiles down with love streaming through every crevice and across every path, well knowing that at the end of the coming day she will return again with her everlastingly changing shadows, the shadows of ever-ebbing, ever-recurring life that have so much to tell.

Foolish people take tomorrow for granted. Tomorrow is never here. We always live in the present. The next moment is again the present. It may find us gone.

Love takes life at its face value. Love does not ask you where you were before you were born or where you intend to go after you die. It says, "I'm here and I stay. You only come and go." It does not say, "You will live forever." Such is the blind wish of misconcepts. Love says, "You will never live at all unless I am with you!" The heart is the watch of life. Love says: "I know what you are, that I am your measure of happiness, and because I know that some day you will have to stop, I will stay behind to say that once you kept time."

Time is a vast affirmation built of negation. Love is the affirmation. Within Time exists light, yet it does not see. Within it exists sound, yet it does not hear. Within it exists matter, yet it is not material. Within it exists conflict, yet it is totally passive. Within it exists peace, yet it is forever restless. Within it fly the furies of the universe, yet it is deathly silent. Of itself it is measureless, out of nothing and into nothing; it is truly the eternal nothingness. Therefore, do not array the number of calendar years you have existed; count the moments you have loved. Without it there is no happiness, because love, as a principle of living, is the guide to the meaning and the value of the self, of life, and of Time. "We know that it has made us, and what we are without it remains meaningless." [61]

What is so brief, so much in haste, forever in quicker pace, as this quibbling, feverish round of man? His moments are like the rush of the wave, to disappear in the foam between the rock, and return to the calm. Out at sea the wave does not see the rock; it laughs as it rises out of the distance; it rolls along indifferently and carefree as it descends again in the distance; it dances with the silver sparkles from the moon, swooping up and down, rolling on, inevitably to the awaiting rock—a new wave is born at sea. Like the horizon, when we come to it we have just begun. We end where it begins. It begins once more; some one else goes out to meet it. What does it mean? It means that life is limited to the individual. *Therefore it has greater value.* Albert Schweitzer reveals his ethic: "My life carries its own meaning in itself. . . . The idea of reverence for life . . . and I produce values." [62]

The value of life is not like the value of a commodity. It is like the value of a song. It can only be enjoyed while we hear it. Commodities can be exchanged, made over, bought and sold. Not so with life. What is done *is* done. Only the new can be different, the next moment can be lived differently. Though a man own all the wealth in the world, he does not possess any more than his own life. Though a man pray to a dozen gods, he does not possess any more goodness and any more godliness than he contains within himself. Though a man hope to live a thousand years, it will never hold back death for a single moment. When one is old, the sun seems so cold even on a summer day and the night seems to come too soon, and when it comes it seems to last forever.

A wise man said, "Life at the longest is not so very long." [63] Emerson said that "the years teach much which the days never know." [64] Lin Yutang philosophizes, "The goal of living is not some metaphysical entity—but just living itself." [65] Epictetus, that wise old Greek, once said, "In life you ought to behave as at a banquet." [66]

People can never measure the value of life until they know its definiteness, its composition, its certainties. Most of them, unfortunately, never realize its length until they are in their death pangs. That they are born and die—this is definite. Here is a specific length of time. It is limited; the end is certain. The fact that they do not know when that end comes makes the length ever more valuable. Why waste this precious moment dreaming for eternities, or building towers of power to entomb oneself, or covering oneself with artificial concepts of tradition and convention solely because one thinks he "must" follow them regardless whether they are good or bad? Angels never made aeroplanes.

A bundle of nerves, a bundle of emotions, and a sensory junction where experience lists the reflexes for future courses, this is man. Love seems to be the happiest culmination of these reflexes. Senses lead to experience, experience to reflection, reflection to reason, reason to morality, morality to love. This is the ladder of life and every human either climbs up or stays down.

Love is the activating appreciation of the knowledge and acknowledgement that life is specific. What will it help a man if he wears a jeweled symbol and showers praises on his god, so long as he realizes little of the value of life but merely possesses the jeweled symbol and mumbles prayers, dreaming the illusion that these will make him live forever when he does not even attempt to live the moment he really possesses? Can man make things out of the ground, whisper words that grow and die with time and say that these things will bring him eternal life?

This moment is the present. It is our material, the plaster for our model of life. We are the sculptors. The plaster untouched today dries tomorrow. Let us mold it today. Our knowledge is our skill, our hearts the inspiration. Let us build our model to the concept of happiness resting upon the firm pedestal of equity and goodness to and for ourselves in accord with the experience of our being, of the world, and of our scale of value in the evolution of all life. May it be a symbol of love that can justly say, "I have thought; I have lived; I have loved; I am satisfied." There is no pay envelope waiting at the end of life. Life must be enjoyed. Its enjoyment is the only reward.

Let us love and be happy today, conscious of the errors of yesterday, conscious of the opportunities of tomorrow, conscious of the truth of the dawn and of the twilight. Love today, conscious of your place in the ever-passing audience before the drama of life, to which we never return in giving our place to others. With such consciousness, each day will be our only day, our greatest day, a day in our song of life, an added chord in the symphony of human happiness. Thus, let us love and live today, for tomorrow we may die. Ah, to get up healthfully early in the morning, breathe in strongly the fresh air and say, "This day is all mine!"—this is the greatest wealth in the world.

To summarize regarding the self, the society, and Time, we have seen that the importance and life-value of the individual should not be subordinated to the absolute will of the society; that each individual has the right to accept from the social order what is really good for him and to reject what is bad for him; that the social order should never be so powerful or influential that it becomes a dogmatic status freezing the potentials of change, although actually nothing is static completely, and cultures and societies change like all other things, subject to necessity and adaptabilities; that man, in trying to achieve his finest and highest form of happiness and expression finds it in the esthetic principle of love as a way of life. What is this principle, how is it detected or recognized?

When you feel good in doing something for a friend, for a stranger, for one who is stricken or helpless, for the unfortunate and the forsaken; when you can pat a dog or enjoy a cat nursing its young; when you admire a flower on the wayside and stop to think about it; when the wind fills you with pure freshness and you are grateful, and no cold is cold enough to remove the warmth of your heart and no wealth rich enough to remove the honesty of your mind; when your canary greets you each morning with a kiss, and the moon reminds you how lucky you are to enjoy good health; when you can admire the beauty of youth and romance and listen quietly to the wisdom of the aged; when you feel exalted to see the birds build their nests in the Spring; when you can worship among trees and consecrate yourself humbly in their shade; when the ceaseless murmur of the galloping falls and the twisting, rushing, swaying stream thrill you with their movement and symphony; when you can face everyone and say, "I have tried to harm no one, and I won't let anyone harm me, for the world is beautiful and life is too precious, and there is enough happiness and love in the world for everyone"—then be grateful and joyful and know that life has been truly kind to you, that you are walking in the light and have found the true meaning of peace, you have used correctly your little measure of Time, and you have found and live the principle of love.

In order that the principle of love should not be overcome or broken down by misjudgments and misconcepts, it is important that it be protected by reasonable vigilance. It is only logical to surmise, where anyone or anything does not tend to lessen our ignorance about the *realness* of a particular thing or event but instead, on the contrary, leads or tends to increase an *unreal* idea or judgment, that we are liable to be victimized by the acceptance of a misconcept regarding it. What is a misconcept? It is the acceptance, knowingly or unknowingly, of a non-existent status of content or relationship concerning a something which may appear as real or unreal to us or which we accept as real even though we may know it is in *actuality* unreal. As the space of anyone's life is Time and if the fullest possible use of this Time necessitates a premise based upon verifiable and meaningful values of any subject or object of inquiry, it is reasonable to assume that it is far better to try to detect, wherever possible, various misconcepts that have grown up in our society, and when discovered and exposed, to replace our misconcepts with concepts based on what they really are, and not what they have seemed to us or what we would like to make them appear to us notwithstanding. How do we know that it has become a misconcept instead of a concept, or a concept instead of a misconcept? By investigation and analysis, by applying whatever intelligence and experience we have, hoping that both may be sufficient and competent to accomplish the desired ends. Concepts and misconcepts are legion in every society and to try to list or enumerate them would take volumes, but let's peek exploratively into one or two.

Although far removed from esthetics or the principle of love, let's take, for example, the field of advertising and public relations in our own country. While, in so great a specialization there are no doubt honest people, as we find in almost every other field as well and whose only intention is to bring a legitimate product or service to the attention of the consumer, yet we must admit that the greatest proportion, especially the unscrupulous and greedy ones, moves on the premise that the average intelligence of the average adult American is either that of a jackass or that of an eight-year-old child. While any reasonable person can easily understand that presentation of a product or service and some education about it is important and necessary, and highly desirable, the advertising field has become a fetish and the advertising specialists are becoming a priesthood dedicated to the principle that if you lie long enough and loud enough about anything the people will eventually believe you, and you may even get to believe it yourself. In this way it is a partial take-off from traditional religions. With color, splash and a million variations of show, innuendo and vagary, the hordes of society are led into the modern new Temple of Display and No-Money-Down! A sponsor, even though his product may be worthless, misleading and deceptive, as long as he has a buck for advertising, becomes a trustee in this new temple. It also becomes a substitution for the social vacuum created by the gradual dying out of colorful religious and social festivities which made life in the past less boring for the primitives. Instead of public relations being a channel to serve and enlighten the public, it has become a way of getting the public to serve the sponsor. The "temple notices" are the billboards that have cluttered up our beautiful country and hide the natural art from every road and highway. No doubt the advertising specialists are already planning to discolor the moon with signs, and I fear the day when a billboard will separate us from the sun. Vance Packard tells us that "a still more ingenious entrepreneur has begun offering admen the chance to plant their messages against clouds and mountaintops." [67]

The general premises of advertising and public relations are gradually creating a "display philosophy" and a "sales pressure culture" in the business, social, political and even the religious life of the country, and which is superficial, unilateral and indifferent to the real good and interests of the people. While medicine and drugs have performed miracles in saving lives, yet we are most probably the most drugged people in the world; many of these drugs are such that our bodies can't fight them back and so we need additional drugs to fight the poisons of the first batch. As I stated, public relations is becoming part of the modern replacement of old traditions; both are based on "take what you can get," "never give a sucker a break," and "dog-eat-dog" approach. Besides, it is unnecessarily wasteful. Advertising does not add a single atom of value to a product but plenty of atoms to its cost. Vance Packard again states that "money spent for advertising in the United States has risen considerably faster than total sales have risen." [68] If the costs of advertising were narrowed down to legitimate needs of the public and of business, the cost of distribution of the general economy would be much lower to a great extent and thus create a greater possibility for the people to consume more of what they produce. "Marketers in general have been subjecting the consumer to a barrage of selling strategies that has rarely heretofore been matched in variety, intensity, or ingenuity. Millions of consumers are manipulated, razzle-dazzled, indoctrinated, mood-conditioned, and flim-flammed." [69]

American business seems to be more concerned with what they can sell rather than with what the people should buy. Anything that misleads or takes advantage of a human being detracts and subtracts so much value from the Time essence of that human being. To produce for the sake of production, to sell for the sake of sales, to impose purchases beyond the point of necessity or absorption, to create material wealth for the sake of surpluses and economic problems, to keep making luxurious non-essentials for more profit when too many Americans haven't seen the first bathtub, this is to put upon labor and management the mark of idiocy. We will yet be lost in the machines we invented to serve us. Acceleration and accumulation can reach a point where they will only be markers for our wasted lives.

Take another example of a misconcept—the caste systems. I do not refer to those in India, Bali, Japan and the other countries. Goodness knows they're just loaded to the gills with all sorts of castes. The caste system must have originated out of the idea of some primitive (most probably a shaman, medicine-man or witchdoctor) that other people are stupid enough to be persuaded to serve him, and could be actually conditioned to believe that not only are they lower in animal grade and quality but that they were born with the destined role to be his slaves or followers. If we go back to primitive times we find the original social organisms or community "executives" were the priests or witchdoctors, who also started the first aristocracies together with the later kings and military leaders or chief warriors. These "leaders" not only began the caste systems of the world, from Bali to Brooklyn, but also started the early drives for individual and group power and aggrandizement which impelled the early conquests and contributed to the origins of warfare.

However, I am particularly more interested at the moment in the caste systems in our own country. The castes in other parts of the world are mostly predicated on religious, military or cultural traditions, a scale of deification, a reincarnation status from Brahman to Pariah. In our own country the castes are not predicated on religion and superstition. Our castes are based on wealth and position. The more money a person has, so much higher is he in the general esteem; the less money a person has, it seems, so much less respect is tendered him. One of the sad and false misconcepts in the United States and other highly Westernized countries, is the very unfortunate situation that the governments and economies, social and "prestige" values, the religious and the successful revival meetings, have made the dollar the barometer of a person's importance in his scaled position in the particular field. The worship of the dollar means that a person becomes valuable not so much in what he can do for others, but in what others can take from him. The increased tendency to create standards by which people are evaluated in terms of the dollar instead of the dollar being evaluated in terms of the people, has caused throughout the world a rising rebellion on the part of most of the people who haven't got any money.

Democratic principles disapprove this, but democratic practice reveals it. If one hasn't got any money at all, he is liable to be locked up as a vagrant, the American pariah. If one has millions, his or her picture gets into the newspapers whenever they take an aspirin or go to Paris. This is, it seems to me, a grotesque misconcept of the true value of a person and an insult to the real dignity of the human being. In terms of the value of life to a person, in terms of the irreplaceable value of Time in the life of that person, all the gold in the world cannot possibly match it in value and meaning. Money cannot breathe, think, laugh or cry, but people do.

If we put our searching lens a little closer, we can easily

verify this caste system based on wealth as really the same basic root which created caste systems all over the world, regardless of the visual veneer—the basic root of *power*. Money is power in our culture. The Brahman represents a power in India because he has more power than the common people in his position as intercessor between the peasant and his god. The Raja in Bali has power because of his higher standing in the divinity-value before the gods and also because of military strength, signified by his kris. Fears produce castes; the fear of the gods and the fears of insecurity. The witchdoctor is a higher caste because he has the power of magic which the ordinary savage doesn't possess; the savage fears the sorcerer and this fear is the source of the other's power. When people rebel, it merely signifies that conditions have become so intolerable that their fears are temporarily and compulsively thrust aside in the struggle to live. If we reduce all the complicated mixtures and strata of human cultures all over the world, we will most probably see that it boils down to greater or lesser powers, and these variations of powers bring about, by conditioning, the caste systems of the human race.

So, when you peruse the pages of the Social Register, hold in mind that these listings came about not because they have more or less brains than you or I; or that they are morally better or worse than you or I; or that they take more or less baths than you or I; or that they are wiser or more ignorant than you or I; or that they are more honest or less crooked than you or I; but because they have more money than you or I.

"There is a tragic alliance," writes Albert Schweitzer, "between society as a whole and its economic conditions. With a grim relentlessness those conditions tend to bring up the man of today as a being without freedom, without a self-collectedness, without independence, in short as a human being so full of deficiencies that he lacks the qualities of humanity." [70]

Thus, by recognizing misconcepts and identifying them, we are able to verify the difference between superficial factors and human value factors. True progress, and with it the peace of the world, is based on the realization and meaning of a proper appreciation of human values.

Pick my left pocket of its silver dime,
But spare the right,—
It holds my golden time!

OLIVER WENDELL HOLMES
A Rhymed Lesson

FIVE

TIME AND THE *EGO FACTOR*

The *Ego* of man is his peculiar *inner* twin brother, the little fellow with a thousand faces and names. One thing it always wants to do,—and *want* is the unconscious expression of itself—it wants to *be* and stay *being,* no matter what. The "no matter what" is the history of mankind.

Many thousands of years ago man saw himself "perpetually" surrounded by almost everything that could destroy him at any moment. He saw others, like himself, destroyed by the elements, by the ever-constricting coils of the living, moving-in jungle. Never could he find reasonable peace of mind, feel really secure for the night or the coming day. His was the constant, continuing resistance to death, the ever-persisting struggle to live. *Homo sapiens,* our species, is probably about a million years old. How far back are buried the origins of our own species, of the pre-tool periods of the predecessors of mankind, we do not know. Even today there are many tribes and communities living under the same or similar conditions as people lived ten and perhaps twenty thousand years ago. The Ego was as persistent then as it is now. The force within him that *could not* yield was the Ego. So long as he had an ounce of strength, so long as he could bear the deepest and sharpest pain, so long as he could hold out against hunger, thirst and exposure, he could not yield. That "little man" within him carried on, relentlessly. That's how he survived. The *Ego cannot* yield; it knows only one thing—*it wants to be.*

The paleolithic savage who tore away a bit of the jungle with his bare hands and teeth to make a little clearing for himself, and the well-tailored banker in Wall Street today eagerly passing the ticker tape through his anticipating fingers to make a fortune, are different expressions of the same drive.

This is not only the nucleus of man's determination to live; it is the same nucleus of all forms of life, animal and vegetable and combinations of both. In the more basic and elemental way, in the structure of the atom, the Ego may well be compared with the nucleus or *nuclear center,* the Proton, the positive charge of life with its dancing electrons compounding and moleculing its structure in the same way as experience exhibits itself in the varying, relating, mixing continuums of environmental pressures. It may be that the first Ego was the Proton, attracting to itself, for stability, growth and extension, the *negative* or pressure electrons that originally polarized the first spark of the living cell. "Habit is interpenetration of organism and environment: habit is equilibrium between the living animal and its world." [1] In the moleculing processes of cellular life may yet be found the elemental ertia that eventually led to the evolution of the social instinct—the helpless, unconscious *banding together* of cells out of the sheer nature of submissions and resistances, to evolve forms or chemical activities for protection, growth, stability and potentials of variation for survival. "When two atoms are held together by sharing electrons, the force that joins them is called a *covalent bond.*" [2]

To survive, the Ego or nucleus submitted, resisted, or formed to variations, to *adapt itself* to new and *modifying* conditions or situations. Resistance or submission, or combinations of both, brought gradual or sudden changes. Development, although imperceptibly slow, was constantly conditioned by the external relationships continually *emerging* and pressing the internal organism within whose center sat enthroned the "little man" *determined to be, no matter what.*

The Ego seems to *sense* only the present; even the fears of tomorrow or the regrets of yesterday are syphoned off into the present so far as the Ego is concerned. It doesn't know or function any other way—because *life is the present, neither the past nor the future.* The mind may become either the master or the slave of the "little man." In both cases conflict is generated, conditioned, patternized, and habituated in mild or acute neuroses. The compromises of living and of life make up the nature and extent of these conflicts. The almost impossible trick is to try constantly, vigilantly, by all the means at our disposal, *to know ourselves* as realistically and as honestly as possible, either by introspection or self-contemplative analysis, on one hand and within, and by constructive criticism, guidance and analysis from and through others, from without. The more dogmatic we make ourselves feel, the less are we able to flex or move to invite and encourage self-enlightenment and self-improvement; the more rational we are the more we are able to be receptive to knowledge and the therapeutic, constructive criticism from the outside world. Rationalism is the neural means of the Ego for natural selection and by impounding it through dogmatic attitudes man merely creates a superficial and blind barrier against himself and against a free-flowing state of *naturalism,* which is the "protoplasmic pond" in which his Ego or nucleus "floats" in living experience.

The wider the division of conflict between Mind and Ego the more are we unhappy, unstable, mentally and physically sick, and in turn, cause misery to others. The narrower the conflict between Mind and Ego, to this extent we are happier, mentally and physically more healthy, and we are desirous of spreading happiness to others. The former tends to conflict and tragedy; the latter brings us closer to a self-gratifying, livable peace of mind.

When the mind follows a line of thinking that goes along with the elemental and basic behavior and expression of the Ego—the determination to serve its own preservation and the continuity of being—then there is a more harmonious, conscious or subconscious, alignment of self-satisfaction. The mind evolved as the *outer protector* or *neural shield* for the Ego, the "little man" encased in a body which it has developed and which it automatically wants to preserve because *It* exists only so long as the body it lives in, exists. When the mind thinks in terms which appear or are "sensed" by the Ego as against its basic nature and interest, an *anxiety* is born and *worry* is the conscious sense of this internal disagreement between the master "little man" and his frontman, the mind. When this disagreement continues for long periods of time in confusion, that is, when the Ego and the mind are no longer sure of themselves and of each other, and move into a state of incompetency and lack of courage to make decisions or changes and to activate these decisions—*Stress* is born. Continued stress gradually affects the entire organism and psychosomatic ailments set in, manifesting themselves in many ways, even to the point of bringing death to the entire organism.

With this in mind, let's go back again, tens of thousands of years ago, to primordial periods of human history. When the Ego, through its mind, and the mind through its senses, felt the overpowering forces of nature pressing it as a constant, continuing play day in and day out, the mind—the "eyes" of the Ego—in its efforts to counterbalance these "super-nature" forces which controlled existence and which were beyond the control of the Ego and of which it was itself just one manifestation, evolved the social instinct, out of sheer necessity and survival pressure, to group with other Egos for psychological and physical strength to stand up against the forbidding walls of external forces. Originally, the primary cause of the origin of the social instinct in the human animal, as well as in other forms of life, may have arisen from the *survival* pressure or the mindless, naked process of the Ego, and the social instinct in *Homo sapiens* may have developed as the mind developed. In the other animals, like the insecta, the social instinct may have developed by the transfer of the individual-ego to a group or colony-ego and each individual organism became a helpless but detached part of a group-body and this group-ego expressed itself, by the instinctive processes, through the group-pattern and behavior.

But let's get back to *Homo sapiens*. The development of group or tribal life must have given each human a greater degree of security, which was primary, and much more was obviously accomplished in common protection, food-getting and family building. Kinships, clanships, matrilineal and patrilineal, semi-yielding expressions of the Ego toward grouping, gradually arose and "cultured" themselves into the tribal patterns. Totems, taboos, and tribal protectives developed out of millenia of patternization and group living.

However, in spite of all this and with the development of the mind, something seemed to have happened somewhere along the primeval way that shaped the destiny of the human race up to modern times. As the tribal mutual-assistance pacts continued to stabilize more or less into a reasonably secure and fairly serene existence, considering the habituated or cultured resistances to opposing obstacles and favoring benevolent factors, the savage mind, through its dreams and fantasies, began what it considered as a "realistic" explanation of death, the one irresistible, eventual occurrence it dreaded, could not overcome or break through, even with grouping. Here is where the fiasco might have happened. The Ego—the "little man" who wants to *be*—was very receptive to any idea the mind would contrive or discover that would give it an "out" on the greatest fear of all—extinction through death. The mind, seeing in dreams, visions, drunken stupors, comas and general mentalizing fantasies of departed and still living people, animals and experiences, obviously established these apparitions as certainties of the present. Spirits and souls were born. Fantasy became, to the primitive, "reality" and what he "photographed" in his imagination became "objective" experience. Professor Oliver L. Reiser wrote: "Consciousness depends upon cerebral oxidation . . . conscious perception resembles neural photography." [3] This neural imprinting process upon the hominoid mind could only come about as a result of his *focalizing* his environment, his group, the world about him and his observations and experiences within it. Wen Kwei Liao wisely states: "The individual is essentially a product of the community, and yet may by chance become a guide of it . . . no matter how much freedom the individual might express, in the process of expressing it he is determined by his community. Because his knowledge is simply a gift of his age and he thinks in the light of what he knows." [4]

Ancestor-worship, or the religion of the *manes,* was evolved because the Ego, wanting to go on living, past the door of death, evolved in turn the fantasy, which appeared real and logical to it, that the soul is transferred from one ancestor to another, or in some other form—the development of the idea of reincarnation, *Karma.* The *breath* became the symbolic manifestation and identification of the spirit, soul, etc., as live people breathe and the dead do not. Here's where the unfortunate Ego could have been innocently and logically misled by the mind, later defrauded by the establishment of priesthoods who found it very profitable to trade in spirits, which solved their own economic problems, and who specialized in the magical powers that preserved one's soul and represented its interests, as intercessors, in the new world of unseen, spiritual forces. The Ego found itself a new desire, because it now turned away from its activity of wanting to live *now* to the new and greater mirage of wanting to live *forever!* Thus the Age of Magic was born and its later offspring, Religion. The objective present was *etherealized* for the phantasmagorial vision of eternity. The Ego, thinking through its mind that it has finally overpowered the most powerful force in nature, let go of the bone of life in its mouth, to drop it in the watery vision of its own misconception. Thus the Ego, the self-preserving nucleus of human life, forfeited its coveted and strongly guarded value of Time in the present for a timelessness which cannot possibly exist, not even in the fullest perception-potential of the human mind. "With the progress of civilization the emphasis of religion shifts more and more to the hereafter, because death remains the one dread against which man has no hope of permanently protecting himself." [4a]

What followed then? It is obvious to see that with the centralization of power in the hands of the few—wizard, sorcerer, witch-doctor, priest, king, leader, chief, etc., the primitive became a religious, social, economic, military, political slave. Complicated and ever-modifying and varying patternization brought about structures from which there was hardly any escape. Not all was dark and gloomy. The people loved it. Social life, castes, kingships and kinships, filled the people with enjoyable and colorful activities and emotional stabilizers which otherwise would have left a boring vacuum of isolation and misery, general dissatisfaction and cause for rebellion. Religious ritual, sadisms, ceremonies, festivities and reverential dramas and spectacles, brought deep comfort, order and peace, to an otherwise chaotic, confused, "exiled" existence. Economically, through submission and resignation, it supplied channels and new fields for labor and usefulness, created empires, temples, cities, and built the wheels by which the various races moved on progressively to modern times. Politically, it cultivated nationalisms and patriotisms, self-protective defensive countries and offensive, warring communities, whichever fitted its needs, and brought about the various divisions, peoples, and national entities of the world. All in all, the cultures of the world, from the Aztec to the Zulu, from the Cro-Magnon to the gentlemen of Downing Street, rose from the general development.

But the price of human life was heavy, awesome, staggering.

The patriotic Jap boy who, hearing the magic cry of *Banzai!* actually craved to die as a hero, the Hindu ascetic who religiously sits on spikes all his life to "purify" his soul, the society matron on Park Avenue who flourishes in a false philosophy of appearances and insincerities, and the unfortunate man or woman who frustrates his or her life through prudery and accepted superficial, groundless, misery-making conventions, "morals," and social mores, all have been victimized by the development of misconcepts of the mind that to "live forever" or to "live properly" is better than to live at all. To these people Time has been substantially lost. Fortunate is the one who finds his way back to it and regains a fresh hold and a realistic value on life. Very often the happiness one longs for is the road just beyond the puddle in front of him. Fortunate is the one who has the courage to jump it and carry on.

The scientists tell us that this world is probably between five and seven billions of years old, and it is also possible, they say, that the world may continue for another or even longer period. "The development of life," writes Irving Adler, "appears as something that just *happened,* without any design or purpose. It started from the accidental mixing and combining of chemicals in the primitive sea. But the direction of development it took was not all accidental. It was influenced by the natural preferences the chemical elements have for each other. It was built on the basis of carbon's ability to form long-chain compounds. In the later stages of chemical evolution it was also directed by the effects of natural selection. The rule of survival of the fittest guided evolution toward the development of more complicated and more efficient organisms, and finally towards the emergence of intelligent beings." [5] . . . *"Life developed out of the chemical mixtures that were in the sea over two billion years ago."* [6] "This evidence which we carry in our bodies," writes Richard Carrington, "of our debt to our universal mother, the sea, may also help us understand that the brotherhood of all living things is not just a sentimental conceit but one of the most indisputable facts of science. . . . If it were not for the watery covering of the earth, and the complicated processes that have taken place in it during more than 1,000 million years of evolution, neither you or I nor any other living thing would ever have existed." [6a]

Dr. Earl A. Evans, Jr., sums it up for us thusly: "The study of fossils shows the presence of primitive life some 1,000,000,000 years ago. The earth itself was then five times as old. . . . All the evidence indicates that life originated in the absence of oxygen and that the addition of oxygen to our atmosphere was brought about by photosynthesizing organisms. . . . The green chlorophyll of the photo-synthesizing plant and the red hemoglobin of our blood are chemical cousins; the malaria parasite uses sugar for energy in the same fashion as mammalian muscles. This similiarity indicates the common origin of all living organisms and provides valuable evidence—especially when we study variations—of the evolutionary process. . . . The evolutionary sequence has the beauty of logic. It can be traced through succeeding eras of enormous duration: the oxygen-free, hydrogenous atmosphere of the primitive earth permitting the synthesis and accumulation of organic substances as the precursors of living organisms; the interaction of these to form self-duplicating systems leading to the depletion of the organic molecules in the ocean and to the appearance of photosynthesizing organisms; finally, the emergence of oxygen-requiring organisms—each step irreversible and essential for evolution. . . . We feel we are more than science can analyze. Yet every attempt to separate the psyche from the physical body has failed. If we modify or destroy our molecular fabric, we modify or destroy its psychic properties. The scientist concludes that they are but different aspects of the whole human." [7]

When we think of this for a moment and also look at the night, the starlit sky and gaze upon our moon and silently wonder at the immensities of space with its billions of stars, suns, moons and planets in our very own galaxy, just one galaxy among millions, the experience of living becomes, in itself, a wondrous thing. But there *is* more, the miracle of sight, to see these things, and to bring the pages of daily and nightly pageantries to our minds, to fathom our thoughts deep into this well of the mystery of the universe and out of it to build an altar of contemplation, appreciation, and the compassion to understand ourselves and others who stand before the stage of the universe, fortunate to breathe, to see, and to think how wonderful, just wonderful, it is that we can experience these events in time.

Not that we are an insignificant, passing little life that comes and goes within a brief few years; this we can observe only too well. But of *ourselves,* while we live, there is no such observation except of others whom we see as beginning and of others whom we see as ending. But *while we live we are not insignificant!* We need not bend to the earth and cower, as trembling slaves, before this cosmic spectacle. *We are all parts of it,* like anything else, which make up its totality, its grandeur, its mystery, its continuity. By measuring ourselves in finity we have reevaluated the universe and compressed all of its potential and its beauty within *our* time, within *our* hour, within the moment we *now* breathe. We sense this revitalizing force of appreciation as we begin to realize more and more that we are not alone, *never can we be alone,* that we stand not outside the universe looking in but we are inside looking at the reflection of our own nature. This is the miracle of experiencing the essence, the flavor, the aroma of living that a true value of Time can give us. This is the miracle that a sound body, mind, and a healthy Ego, by intelligence, courage, love and coordinative unity, can produce. Through an appreciation of life in general we can come to appreciate our very own; through an appreciation of our own lives we should learn to appreciate the lives of others. Through our hearts we can reach out with our hands and gently touch the stars in the heavens as we gently touch a flower on the wayside. "Nature herself is one vast miracle transcending the reality of night and nothingness. We forget that each one of us in his personal life repeats this miracle." [8]

It is in the nature of things to make mistakes. No human being can be so perfect as to give himself and others a positive assurance of an errorless life. Santayana said that "every phase of the ideal world emanates from the natural" and is therefore a concept of error or certainty based upon and within experience. A growing intelligence, in itself, denotes a correcting and maturing attitude in things. This attitude indicates a courage to look at realities more clearly and which brings around a finer, more sustained maturation of an individual. This is what A. Powell Davies called "the full and free acceptance of reality." [9]

Integrity towards oneself and to others become fastened to one's character. Maturation also does not see error as something to create prejudices about or to heap abuse upon it, or to provide oneself with a scapegoat, or to sit and cry about yesterday's failings or to boil up the pressures of hate and animosity. Instead, he sees these things as natural consequences of living and realizes that only by applying our minds more correctly and cultivating characteristics of compassion and understanding, can we gradually and possibly diminish these negative things that cause misery and unhappiness. Understanding is one of the necessary pillars of wisdom, which, in turn, acts as a basis for selectivity in human and worldly relationships. As stated, by a rising and realistic appreciation of our own lives we can first begin to appreciate, concomitantly and spontaneously, the lives of others, and this is conversely true. Understanding leads to appreciation. A proper barometer of values, based on the realization of the incalculable value of Time is the first step in appreciating life; this is our measure of man and of men, and all the things that men have created and may yet create.

The history of mankind can be compared, in a way, to the life of an individual. The helpless child sees in his father and mother two divinities who "must" look after him and provide him with the security, affection and devotion instinctively necessary for him to survive. In a similar manner the primitive, the child of mankind's life, took to the idea of gods and goddesses for the same reason. The whole phase of supernaturalism is but a passing phase in the history of the human race, and part of the development of the rationalism of the human being. Maturing rationalism and its offspring, the sciences, is the gradual growth to mankind's manhood. All forms of reality, even including the fantasies and dreams of people, are understandable and are things of analysis and observation. But if we are to see at all we see with open eyes, not with closed ones, depending upon our imagination alone to mislead our "little man" with unreal though savory pneumas of make-believe; these belong to our childhood. It is tragic that so many thousands of years and billions of lives have been spent *running away from life and anticipating death* to carry them on the wings of fairy-tales to heavenly bliss. It may be thousands of years more before mankind may leave the Age of Magic and Religion behind, but gradually it may possibly come about. Slowly the "little man" with a better and wiser mind, may see that there is greater glory in living happily the all-too-brief years of his existence than not to live at all. Such a transformation may slowly bring about, in turn, a new philosophical approach and analysis of ethics and conduct, and a new synthesis of the content of life and its appreciation. It will no longer be based upon what a man "ought" to do but what a man *can* do to achieve happiness, peace, and longevity. To begin with, let's realize the most realistic observation of all:

Tomorrow morning tens of thousands of people all over the world will not be breathing the air they are breathing now. They will be dead. Be understanding and grateful, then, if you are reasonably healthy and comfortable. Appreciate the reality that you are *still* living. When you arise at dawn, open your window, close your eyes, breathe in the fine, fresh, living air, and say, "It's wonderful, just wonderful, to live *this* day!"

"The basic principle of ethics, that principle which is a necessity of thought, which has a definite content, which is engaged in constant, living, and practical dispute with reality, is: Devotion to life resulting from reverence for life." [10]

Let this be your "prayer" of mortality.

This occasion can only delight the Ego. He deserves it. He needs it. Let's continue with our quest regarding our "little man."

In savage and primitive times a human being found out that that in order for him to achieve a reasonable quantum of peace, satisfaction, security and happiness, he was "persuaded" whether he was willing or not, to realize that he "must" do certain things in order to find and accomplish these benefits. With the rise of barbarism and the early civilizations and with it parallel growths and intricacies of magic, religion, and their offspring, the social and feudal customs oozing out of this religious maze, there arose the "dark ages" ethics that were predicated upon the fantastic idea of a universal "moral imperative" that a man "ought" to do this or that in order to remain a "moral" person and without which he could never achieve peace of mind, happiness and general life satisfaction. Besides, he would be dumped into a pit of eternal fire for a *forever* roast!

Today we have found that it isn't what a man "must" or "ought" to do, but what he *really can* do to find the new and earthly content of what morality and ethics *actually* are, and by ascertaining and verifying their natures and contents, to try to see *possible, reasonable, justifiable, gratifying* forms or methods of conduct to learn to live by so that the security, happiness and peace of mind we are seeking, can be achieved in a possible, reasonable, justifiable and gratifying way.

The *possible way* is the limitation of probabilities—the essence of inquiry and investigation.

The *reasonable way* is the government of mind potential and its subsidiary arteries of the sciences, logic, and the unknown quantum.

The *justifiable way* is the cultivation of the highest possible values and meanings of things of experience and of the relationships within it, of which we are a part—philosophy, law, societal direction and culture.

The *gratifying way* is the art of appreciation, the new esthetics, the nature and the adoration of the beautiful, love—love, the *finitesimal continuum.*

The possible way may be found by the reasonable way, which may find for us the greatest justification in love.

Each of these should, like Cinderellas, fit the slipper of Time. Value is the fit and the slipper, the *space* or existence of Time itself. First, let us ascertain the nature and the direction of a *possible way.*

The nature of possibility depends, of course, upon what we are seeking. It may be possible for one to use his imagination to any degree regarding anything and everything his mind is capable of imagining. In this instance, his imagination seems to be of two parts: The reality of the imagining itself, and the probability or improbability of finding its *real* verifiability within certainty and experience. It doesn't follow, however, that what may appear to be a most fantastic or "impossible" dream today is "absolutely" beyond the possibilities of present or future experiences. Very often the dreams of yesterday laid the groundwork for the engineer of today; yesterday's poet often inspired the scientist of the present; and the painter of legends and symbols paved the way for many an architect of today's structures. The fire of inspiration and the emotionally expressed hopes of the ancients unquestionably unrolled the carpets of ideas upon which the progress of mankind walked firmly and securely as it advanced to us from the path of the centuries. The problem is to try to understand the limitation of probabilities within the human mind that can leave us to express the fullest potential of imagination, idea and ideology, and still be able to retain its probable realization, no matter when it may be realizable at any given time of present or future, and which may be fairly and honestly considered as a possible part of, and related in essence to natural phenomena and thus possibly realizable within the experience of what we call existence or what Lucretius called *in the nature of things*.

Here the theologian might enter and say that the idea of a god or the idea of a universal moral imperative, or the idea of a supernatural, mystical, intelligent, purposeful, omniscient force, presently empirically not verifiable, lies within the outer circle of possibility. Descartes's assertion, he can further claim, of "*Cogito ergo sum*" also implies the possible existence of a god by the very existence of the idea. In the first proposition he is stretching or *diluting* the idea *beyond* the outer circle of probability, outside natural phenomena, and at the same time refusing to admit the evidence of experience that proves it to be fallacious, meaningless, and unverifiable. In the second proposition, Descartes's idea of "I think therefore I am" proves, in a reasonable way, the existence of the one who is thinking it but it doesn't prove anything else simply because of the idea of it. I *can* think I am right now in Tahiti but the only transportation to Tahiti I can have here is the dreaming, thinking, or imagining, which, while it may be true or untrue in terms of reality, remains unverifiable and uncertain so far as *establishing* it outside the mind and *realizing* it as an experience of internal factors between the mind (or the person possessing it) and the external factors relating to each other, in the course of living experience.

Even the most astute and conscientious scholar who may be studying the history, culture and living conditions of a certain country or people, could not have the fuller perception and realizable knowledge as when this very same person combined his study of this country or people with the actuality of seeing, experiencing or "touching" this country or people with his sense as a coordinative, collating experience.

Even the most systematic processes of extrapolation regarding any subject or object could not, in the reasonable and more verifiable sense, be soundly and logically carried on unless the object or subject has become, in an experienced way, a related external factor to the senses or internal factors of the observer. This is so because life is a series of related continuums which comprise experience. The essence of existence is relationship and it is in this very factor of relationship that one relative can possibly verify another. There isn't one particle in the universe which is not related to everything else in the universe. The closer the relationship the greater the potential of perception; the greater the perception-potential the more *possible* the realizability of true or verifiable concepts.

This does not imply, in the least, that a person hasn't the right, or a good reason, to study anything or anybody without actually experiencing a relationship with it. It does imply that in such an instance the knowledge or ideas thus acquired become the related factor and the perception-potential is thus necessarily and obviously limited to this relationship.

So when the metaphysician, theologian or Utopianist, deduces concepts of ideas, by fantasy or inference, which cannot be reasonably related to existence or within the sphere of natural phenomena (there really isn't any other phenomena, anyway), he is eliminating *ipso facto* the possibility of a realizable idea, and is following a line of attempting to touch the horizon line as a directional end, behind which he drags his life, intellectually befogged in vagary and self-delusion, when all the time he could have developed the *reality* of the horizon as a *means* which is everywhere and to which he can reasonably and empirically apply his poetry, his art, his hopes and dreams, and even his fantasies, in a *naturalistic, possibly realizable way.*

This is what is meant when we say that a sound possible way extrapolates and implies, within the nature of experience, a limitation of probabilities. But this limitation extends *to all there is,* whether we have realized it or not; it includes all the "mysteries" that science presently acknowledges as the presently unknowable but still a *possible knowable and a realizable possibility.*

Therefore, in any attempt to weigh our ideas as a possible

way so that our experiences and the fullest and highest possible value and use of Time might be truly fused, it is necessary for us to understand that both experience and Time, which identifies it, exist only within existence, and *Existence is our limitation.*

The limitation of possibility and probability is, therefore, a range of projection and knowledge-potential that may be limited to man by the nature of the universe but becomes unlimited within it by the nature of man.

The commonly accepted Oriental philosophy of attaining unity, or that the purification of soul or idea becomes dissolved, so to speak, with the "Great Universal One" is an attempt to solve the problem of survival by "joining" with the "endlessness" of the "absolute wholeness" of the universe. This process of purification, discipline, ritual, absolution, contemplation, etc., by which the Ego of man hopes to attain this final "bigness" or unity with the Universal Whole is, it seems to me, a processed dilution or wasting away of the very life he is trying to expand, thinking it will be as large, as strong, and as enduring as the universe itself. What *actually* happens, I think, is that man's idea that "if you can't fight them, join them" is not the road of greater strength but one of weakness, submission, slavery, and absorption. By such trying, no matter by which method or by what fancy name we call it, we attempt to buy eternity with the money of hours we have in our pocket. If my life could be metaphorically considered as a drop of water I would rather try to understand that while I am just *one* drop, the great ocean of existence is made up of similar, many, *many* drops like myself, like a long chain of existence-drops which make possible what *appears* to be a phantom of wholeness and which implies by its very hypothetical idea to be motionless, directionless and unrelated rather than what in actuality it really is—a constant, restless, manifesting vortex of creating, integrating and disintegrating relationships, a constant diffusion of its parts by the nature of its content and by the nature of event and experience.

Thus, by refusing to dilute our few hours of existence into an eternal nothingness, and by the realization that our lives are, in essence and in actuality, no less and no more valuable than the rest of existential life, we attain a status of appreciated value rather than a stasis of depreciated value. By raising the value of ourselves in relation to our own selves we are raising the value of the universe to ourselves in relation to our own general scale of value. The greater the value of ourselves to ourselves by ourselves, the greater the value of the hour increases. All the value of the cosmic process is drawn from all points outside of ourselves *to* us, to dissolve itself in *our* essence of Time, in the very moment we live now. Instead of striving to join the cosmos to become *it,* the cosmic "whole" joins us, which it does anyway by the nature of relationship, so that *whatever we are can be it.* To know that, whatever we are, the cosmic "wholeness" can't leave us in its inexorable chain of diffusing relationships, the full possible value of its possible "wholeness" becomes no more or less valuable than the value we place upon ourselves.

What does all this mean in simple words? What does all of it mean to us regarding a proper directional course, the *possible* way, in order not only to find a basis of ascertaining reality but also of finding a basis upon which this reality can be translated and transcended to us into meanings and values?

It means that it is not possible for us to accomplish this if we stretch our hopes and dreams beyond the sphere of possible existence or beyond the cosmic limitation. *What is measureless cannot be measured.* Where there is no possibility of measure there cannot be the possibility of Time and where there is no Time there is no possibility of comprehensibility (the identity of an existence) and of value (the appreciation and meaningfulness of this existence). By understanding the limitation of probabilities within natural phenomena, as even the *concept* of supernaturalism is part of natural phenomena, we give ourselves (our lives and intelligences of limited content and extension) "unlimited" possibilities within the entire scope of what we term the universe. By "unlimited" we do not mean the possibility of the human being to acquire total knowledge and total experience, which is impossible within our very own limited means of acquisition and the reality of our brief personal existence, but it does mean that this total knowledge and this total experience are still within the potential of our domain, and it is up to us to *absorb* as much as we can and desire to, to make our lives for the better or for the worse according to our knowledge, our patterns, our behaviors, our hopes, our capabilities and our good fortune in the constant arena of experience to which we are tied and related to and to its natural ways of accident and the mechanics of living and existence.

No polity, no culture, no religion, no philosophy, is a "must" in the life of any individual. Each person has the right of expression that can change the cultures of the world. Each person has the right of idea that can change the opinion of the world. But with this inalienable right is also the concomitant appreciation that all others have the same right; the common protection is the protection of each. Within the nature and the extent of this appreciation lies the possibility of peace, the happiness and the security of the human race. Whether the people *actually* desire peace, happiness, etc., is a question for the psychiatrist and the sociologist, and is another matter that requires special, separate consideration and analysis regarding the geno-psychic tendencies underlying the mass-mind subconsciousness of peoples and races, and perhaps, of life in general. Man is egocentric and geocentric in nature and what twists and turns these two forces take, often very paradoxical and inconsistent, are inviting mysteries for the psychologists of the present and future to fathom.

However, let it suffice for us here to realize that when the Ego let go a long time ago of his mortality to achieve "immortality" he left, to the extent of his fantasies, the domain of verifiable reality and if we are to be able to re-evaluate Time and our lives within it we must return to the *real* world and restore to the Ego its natural right of life preservation, life enjoyment and life value. The possible way, then, is not through fantasy beyond the limits of probability but *within its limits,* where exist measure and the possibility of value and appreciation. The important thing is that we recognize that Time does not exist of itself but solely as a relative factor and related to our own lives, its length, width, and depth. It is a constant variable because our lives are constant variables. If Time is to be maintained in our measurement of ourselves, as a *real* and *understandable* factor, then we are compelled whether we like it or not, whether it complies with our cultures and religions or not, to accept it as our identity of natural phenomena, of which we are a part, and not as a "fantasy measure" of an endlessness which the human mind is incapable of perceiving. So long as Time can be applied, as a directional guide, by its acceptance as a *determinable* factor instead of an indeterminable hypothesis, then it creates itself as a possible way and becomes a self-identifiable premise and an essential attribute in our analysis and re-evaluation of Time.

While we are on the subject of possibilities and probabilities, many people are misled into thinking that the philosophy of *materialism* concerns only that which is more or less "objective," which we can see, touch, weigh, of substance like earth, a chair, a cell, water, our bodies, etc. Materialism concerns itself with natural phenomena, regardless of their state, form, or knowability. It does not recognize any premise that *all* there is is more or less divided into two principal parts, the physical world and spiritual world, the natural and the supernatural. It concerns itself only with the knowable and the probable rising out of the knowable. Man's inquiring mind and nature have created the *sciences,* which have become instruments by which man has thus far been able to "perceive" beyond his own physical limitations of perception and, as a result, has been able to establish and identify forms of reality which man cannot see or touch and has been able to make rational concepts in his analysis of various transient "states" of "moving reality" which is the constant ertia of all existence. Whether it is the microscope or the quantum theory of relativity, man has and will continue to pursue and attempt to understand *objectively* what has been for thousands of years considered to be subjective and the "dualistic" counterpart beyond his touch or reach. If man's intelligence can advance, so can his sciences which, after all, is the natural, knowable child of a natural, knowable parent. Materialism, therefore, is an attempt to dissolve out of existence the dualistic concept of centuries of metaphysical speculation that matter and spirit are apart from each other, as the body and soul, the tree and the tree-spirit, the natural and the supernatural. Many able scholars and scientists, feeling the full impact of the *new physics* of Einstein, Planck, Bohr, Schrödinger, Kramer, Slater, and others, men of science who have aided substantially in heralding the *atomic and nuclear age,* feel that the concept of materialism should be modified or rectified in the same manner as the classic physics should be modified or even discarded. Professor Werner Heisenberg, who was awarded the Nobel Prize for Physics in 1932 for his brilliant work on the quantum theory, contends that "we cannot escape the conclusions that our earlier notions of reality are no longer applicable in the field of the atom and that we are dealing with weighty abstractions when we set out to define the atom as what is truly real. Modern physics, in the final analysis, has already discredited the concept of the truly real, so that it is at the very starting point that the materialistic philosophy must be modified." [11]

Reality is not an object of identity alone, but a process of identity, and the process is natural phenomena; the *realness* of this process is not any less real because of man's innate limitation of possibility, and though the process self-identifies itself as a limitation of *knowable probability.* Long ago materialism accepted the premise that energy and matter are forms or states of each other. Materialism is merely a generalization indicating that the realm of all probabilities lies within the field of natural phenomena, regardless whether any of these probabilities can be recognized or established by visual or observable analysis or whether they can be statistically, mathematically or theoretically identified by the space-time quantum theory of relativity—the base premise of the monistic concept of the universe. Man, in his helpless humility and in his knowledge of his own identity as a specific limitation, may come more and more to understand that while the identity and perceptional verification of his own concepts is an indeterminable limitation of possibility unto himself as a man, the statistical, theoretical, conceptual verification of the *science of nature* becomes a determinable factor of limited probability. Science is the "seeing-eye dog" which makes it possible for the limited vision of man to "see" beyond his own natural limitations and to be able to possibly pierce the barriers that may unfathom the riddles of existence and explain the totality of experience, and to open wider the range of probabilities ordinarily "impossible" for men to reach or encompass while he remains and will always remain a small segment of this totality of nature.

It is not that the philosophy of materialism is to be modified but rather our concepts of its realities that are to be modified so that the observational or visual and the "basic" or analytical realities as elemental particles or atoms, can both be synchronized, from different lenses, into a fundamental, verifiable, focalized interpretation of reality and the nature of the process by which this reality reveals itself to us. There is no need of the general philosophy of materialism to modify itself every time science moves ahead and extends the frontiers of knowledge; it is sufficient that the philosophy of materialism recognizes that the sciences are concerned and could only be concerned with reality, whatever the forms revealed to us, as attributable to manifestations of the knowable, probable sphere of experience of nature and existence. Any probability within the limitations of experience is by the nature of itself a possible objectivity; science is the means by which this possible probability can possibly make itself known to us. Science is man's specialized media of observation, experiment and analysis of natural phenomena and regardless what roads we take to identify it the totality of the *potentia* of experience can only fall within the arena of natural phenomena, which is, regardless how you call it and regardless of any and all forms it takes, materialistic. There is existence or there is no existence; so long as there is any identity of existence, no matter by what process this identity takes place, it is in the nature of itself materialistic. We cannot etherealize experience and the nature of its processes, but our sciences, by their analyses of these processes, may come to verify and evaluate the transient knowables that are all part of the general process by which experience and existence identify and express each other into an identity of one process and which is the monistic basis of all reality, and this reality, by the nature of itself, is one of *limited probability.*

This leads us to the second part of our quest—the *Reasonable Way*. The word obviously means a way of using our reason or reasoning in order to establish realities, that is, verifiable, determinable materia, and by applying this acquired knowledge and experience to the processes and faculties to reason, to attempt to establish, in turn, a series of identities and the identities of the relationships involved, that could possibly further, by these experiences, additional knowledge of what is true or untrue, proven or unproven, real or unreal, so that these states or events of established realities can, by arrangement of our reason and reasoning faculties, bring about what we know by the common word of *education.*

Of course, education doesn't necessarily mean formal teaching or learning in schools or by private teachers, or reading a book in the library corner. As long as people live and think, the process of education, no matter how meagre or highly specialized, goes on relentlessly. As long as we can live we cannot escape from experience. As long as we can think, this experience is translated and retained through education of a lesser or a greater degree. So, education is not a goal, a failure or a success, a degree or a destination, but purely a *process* of cumulative experience. *Rationalization* is the evidence that this process is actively and expressively operating. As long as we rationalize, whether wrongly or rightly, we are recognizing the continuing process of education as a *parallel nature of living,* not apart from it but *within* the monistic framework of our lives.

Thus, any judgments, ideas, observations, patterns, forms we create are created *a posteriori,* that is, as a result of experience and not *a priori,* that is, a judgment or idea created antecedent to any experience on which it is hypothetically based. Rationalization cannot concern itself with "spheres" outside natural phenomena because, to the best of our knowledge based on experience, experience doesn't seem to exist outside of it. Theoretics of the sciences concern themselves with probabilities or inferences based on previous experience or acquired knowledge and are methods of deduction or induction rising out of indicative processes of reasoning. Contemplation, meditation and introspection, are reasoning processes rising out of the desire for self-knowledge through the general knowledge of the relationships and interrelationships between the nature and experience of ourselves and the nature and experience of things outside of ourselves. All these are empirical forms based on experience. Even anyone's concepts of heaven and hell, of angels and saints and demons, are still formed and processed by the limitations of experience. Thus, all expressions of the arts, whether realism, expressionism, impressionism, surrealism, abstract, imaginative or photographic, are still expressions born out of the experience of the nature of people, things, and the play of relationships and events.

Logic is mathematical, provable reasoning, but in the broader sense, many things may appear logical but not reasonable. Very often logic can become the quicksands of reason and many things brought down by logical thinking can bring about a *reductio ad absurdum* which can nullify the very purpose of reasoning itself. But we are not concerned here with the intricate and almost fathomless depths of logical formulae but primarily with the more or less reasoning capabilities and attitudes of the more or less rank and file average human being.

As stated previously, as long as people live they are experiencing, and this experience is a form of education, whether it comes from working in the sewer or in a bank, whether in the field or in the classroom. These variations and degrees of experience are a matter of relatives, but in all these manifestations of experience, so long as a person is alive and conscious, his concurrent quantum of focus and the nature of the experience will be able to eventuate in some resultant inference, no matter whether this inference is actually identified or not. Whether it be fantasy or knowledge, mere superstition or pure science, series of inferences are evident. Whether any inference is true or not, that is, whether it can be verified as knowledge, is quite another story, in itself, and again requires delving into the intricacies of epistemology, and this is not our purpose here. Suffice it for us to realize that the powers of reason—plus quanta of heredity, environment, experience—can free a person or enslave him, make him a pauper or a prince, a demon or a friend, a fool or a sage, a hating bigot or a kind-hearted atheist, a greedy miser or a soft sentimental humanitarian. It all depends on the ingredients and the cooking to *determine* the dish that can be served. As we all know, one can have what may be fairly considered as "rich" ingredients but have unwise cooking and result in an awful meal or we can have "poor" ingredients and wise cooking and result in a fine, happy banquet. Good ingredients and wise cooking are ideal but the types and experiences in variation are legion. In China, they say, if a person in a previous life deserved a good merit, then he is born "lucky" in this life. As we have no way of proving a pre-life merit system, we just have to be satisfied that a person, when born, is "lucky" if he contains good ingredients. The same goes for his environment and his experiences. *In the rational and scientific sense we can understand and appreciate the circumferential limitations of each human being and the catalystic potential between material and process.* As a result, it is more, in the true sense, what life makes of a person rather than what the person makes of it. To say that we are the masters of our destinies and the captains of our souls is certainly a very poetical and nebulous attempt to ascribe to a human being the omnipotence of a god, and even if a god existed he still would be limited to his own nature, and his process of experience would be limited by the containment of himself. That's why man created male gods and female goddesses, cat-gods, snake-gods, wind-gods, and fire-gods, and a million other gods!

What does all this mean to us? It means that by the realization that all human beings are not born equal and that all experiences are not the same, we are able to create for ourselves a *tolerance* of judgment and from it an appreciation of values *reasoned* from the reality of things. Man was not always the same. "Our minds may stagger at the idea of the hundreds of thousands of years that separate living men from their Australopithecine ancestors; but this steadily unfolding story shows that the origin not merely of Tierra del Fuegans and Bushmen but of all living men goes back through these brutish, blundering phases of prehistoric man to that ancestral stage when human brains were no bigger than those of living microcephalic human idiots or of living apes." [12] And as of today the variations continue in degree. "If nature cannot reproduce the same simple pattern in any two fingers, how much more impossible is it for her to reproduce the same pattern in any two brains, the organization of which is so inconceivably complex." [13] Even "instincts are by no means fixed, in no way inflexibly responsive to certain outer stimuli, but very elastic and adaptive." [14] "Heredity is never absolute; resemblances are never quite perfect . . . heredity itself changes . . . variation is but another aspect of heredity . . . variation is a primary condition of progress." [15] "It has become increasingly obvious that an understanding of the evidence concerning heredity is fundamental to any philosophy of life." [16] "No two humans have ever set out with identical equipment in mind and body. Every face is stamped with individuality when it issues from the womb. Every baby brings into the world a pattern on its finger tips never seen before." [17] "The very protozoa, the one-celled animals, possess individuality. Beneath the microscope it is seen that no two of these miserable minute jelly blobs always behave the same under identical conditions." [18] "Every human life is a unique adventure." [19]

When I was in Macao, a small Portugese settlement on the China coast, containing mostly Chinese people, I saw little Chinese children between four and eight years old, squatting in small groups outside their little shack homes, making fire-crackers. They usually work about eight hours each day, and, according to my Portugese guide, they can earn an average of seven cents each per day (in U.S. money value). They live in slummy huts not far from the border of Red China. In my own life I was forced to start working and earning my keep when I was about five years old. I know what it is. But as I looked down at them and my heart wept for these tiny human beings, I also seemed to say to myself: "There but for the grace of luck I could be squatting over dirt, making fire-crackers for seven cents a day!"

That's why in most of the world, where humans have to struggle for survival each day the peoples believe in "luck" and hope that the gods will make them "lucky." As a matter of fact, if a Japanese or a Chinese should unfortunately pick an unlucky prayer slip at the temple he ties it to a tree and gently gives it right back to his god, with no thanks.

On the other hand, while it is verifiable that all things are not born equal, nor can possibly experience all the same events with the same reactions, it is also verifiable that many humans, as a matter of fact the great majority of them, are denied, for one reason or another, through circumstances or limitation, either by political, cultural, religious or sociological impress of one kind or another, an equitable and reasonable right or opportunity to change or better themselves. Here is where the reasoning power of the human being can be intellectually and scientifically nurtured and developed to produce a just and a more rational approach to a consideration of these factors.

Another fundamental concept in the use of the *reasonable way* is the realization of the difference between believing without reason, and observing and verifying things through reason and objective experience. In primitive times the savage, even in an instinctual manner, used his intelligence through his senses to survive and to develop and extend his inventive potential by using his reasoning faculties to overcome as much as possible his own physical and environmental limitations and barriers. He also used these very same means to develop his family, group, and tribal life by the mutual-assistance of numbers to gain individual ends for survival. If this primeval rational process continued unbroken this would be a happier and a more peaceful world today. Perhaps. But what happened? As we have seen in preceding pages, he *tripped* on the idea of souls and spirits. We can rationally understand how he came to fall into this trap but that doesn't make the trap rational or the mistake less tragic. The long darkening era of *belief before reason* began and it is still going on. This gradually patternized into a *credo quia absurdum est*—I believe it because it is so unlikely!

As we have seen, by entering into this new phase of the prehistoric beginnings of the Period of Magic and Religion, man became a vulnerable slave to the witchdoctor, later the priest, king, emperor, etc. Originally the necessity of sacrifice, austerity and forbearancc, necessitated in turn, a submission to the priestcraft or kingship within the tribal group and area. The actual result of the earliest taboos and magic gave birth to the earliest development of powercraft, power-politics, eminent domain, transition from private to public property. While it is purely conjectural, it may be possible that the earliest human sacrifices came about as the king's and priest's attempt to depopulate the growing and expanding group to cover economic disaster or land deficiencies and limitations. It is also very probable that when human sacrifices increased and the supply of victims lessened and dwindled, to avoid tribal genocide the idea arose in the minds of the leaders to capture victims from neighboring tribes, and this brought reprisals, one of the principal origins of *war*. The chief priest or king thus discovered a vehicle for increased power and property, besides a solution for satisfying the gods—not only one of the origins of war, inasmuch as the primitive nature of man does not indicate, from the evidence, an instinctual desire for destruction or for property, but also one of the origins of power politics and aggrandizement against the rights of individuals and groups. Besides, primitive man was too busy struggling to survive, and to satisfy his immediate and always-pressing needs took up so much time that he simply couldn't think of power or property aggrandizement. This human trait even expresses itself in people today who are too busy with their own lives, occupations, and problems, to see through or protest vigorously when their rights are naively invaded. Thus, strangely as it may appear to many people today, the origins of religious belief are also the origins of war, twin developments of the same ghost—the trap of fear into which the prehistoric human Ego fell.

What does this mean? Simply this. The submission and submergence of the Ego to dogma instead of maintaining a free developing mind to express itself, unencumbered and without needless fear, caused the greatest and most tragic period in human history—the enslavement of the mind and the devaluation of life, the fear and hope for a future life, the exploitation of his labor for the benefit of the ruling few, and the destruction of his life by sacrifice or war by his own leaders or by the enemies that were generated in the human arena of relationships. The enslavement of the mind was the worst of all because without an unfettered and free-flowing rational faculty, the human Ego was helplessly lost, without direction, measure or value of life. Without a *reasonable* attitude toward life, it was impossible to have a sound philosophy and value of Time. It's our mind that tells us constantly that we are alive and to the extent we suppress our minds in lieu of blind belief, to this extent our minds cease to exist and to this extent we are devaluating the value of Time, because only through our minds do we conceive a scale of value. For this reason the philosophy of an open mind, always willing and capable to change its convictions by the acquisition of new verifiable evidence and the steadfast adherence to a rational attitude, can never injure us; it can only help us to build fortitude, respect for ourselves as human beings and not exist as the mindless caterpillars. "Freedom of inquiry, toleration of diverse views, freedom of communication, the distribution of what is found out to every individual as the ultimate intellectual consumer, are involved in the democratic as in the scientific method." [20]

When we begin to use our reasoning faculties we begin to

see the world, the universe, as the "property" of each and all individuals, that freedom is a *natural* right and the only way to keep it is by preserving the same right for all others. By reason we begin to see the light that no man, king, priest or government, has the right to impose his or its will upon the people, or defraud them, or to exploit them or allow others to exploit them, or use them to kill each other. Thomas Jefferson so wisely stated: "American independence . . . may it be to the world what I believe it will be (to some parts sooner, to others later, but finally to all), the signal of arousing men to burst the chains under which monkish ignorance and superstition had persuaded them to bind themselves, and to assume the blessings and security of self-government. That form which we have substituted restores the free right to the unbounded exercise of reason and the freedom of opinion." [21]

The sanctity of life implies the natural right to the freedom of expression, the natural right to the benefits of one's labors, the natural right to be left alone in peace and to enjoy life without impressment or regimentation. Governments and dictators took over from the kingships and they have no rights except to *serve* the people who constituted them. Thomas Paine, one of the great liberals of all time, stated: "Government ought to be as much open to improvement as anything which appertains to man, instead of which it has been monopolized from age to age by the most ignorant and vicious of the human race." Dr. Gordon W. Allport, President of the American Psychological Association, wrote: "The individual is more than a citizen of the state, and more than a mere incident in gigantic movements. He transcends them all. Striving ever for his own integrity, he has existed under many forms of social life—forms as varied as nomadic, feudal, and capitalistic. He struggles on even under oppression, always hoping and planning for a more perfect democracy where the dignity and growth of each personality will be prized above all else." [22]

By reason and the reasonable way our book of knowledge is never done, our door to new ideas never closed, and the impact upon our senses that so long as life exists for us the *unknown quantum* always lies ahead, like the horizon—there for us to reach to, but *living* to get there. By reason and the reasonable way we come to learn, through a better and truer evaluation of life, a true and greater measure and value of Time.

Now we come to our third quest—the search for a *justifiable way.*

Thinking and thinking justly are two things very frequently confused. The act of *thinking* is a natural process of the human being and one doesn't ever have to read a book or go to school or have any set of ethical, religious, social or political opinions in order to think. *Thinking justly* is quite another horse of another color, cultured and bred in a special manner in order to bring about certain directional attitudes, an analytical method for self and the outside world and of the relationships between them, and the selectivity of a constantly reflecting scale of values rationally processed to bring about not only an equitable way of looking at things and experiences, and which is a parallel fit into the meanings brought about by the science of verification, but also and primarily a *method* of thinking that attempts to build, out of these observations, a guide for happiness, honesty, and peace of mind. Bear in mind that it is a *method* and a *process,* not a set of standards which vary, change and are generated throughout the various peoples and cultures of the world.

Of course, Gray in his *Elegy,* said, "Where ignorance is bliss, 'tis folly to be wise." This takes in the implication of an already established—*ipso facto*—condition of ignorance, which if established as certain, can never be either a state of bliss or folly about the particular premise involved. As stated before, *pure and total ignorance in a conscious state does not exist.* Ignorance and wisdom are constantly moving, increasing or diminishing factors operating from and within experience. A person totally ignorant at any given moment is either soundly asleep, unconscious or dead. However, it is more reasonable to assume that Gray intended to convey the meaning that a *relative* amount of ignorance *conditioning* a state of bliss concerning a something is better from the value scale of actual living, than a relative amount of wisdom which, though more realistic, brings about a state of folly or unhappiness.

The main thing we should hold in mind is that while thinking is an automatic, natural process that comes to all persons, more or less, thinking in a certain methodic or logical pattern is an acquired process of learning which, when habituated and cultivated as a way of life, affords some protection of a rational attitude in one's thinking about oneself, of others, and of things in general. Yet all of this, even if it were the most perfectly achieved process of thinking, would mean nothing unless it had a certain scale of value and meaning. What good are all the great scientific achievements in the world today unless they *mean to and actually do* further the peace, health, and happiness of the people? And of what good is it if one scientist struggles to bring about a longer span of life for people while another scientist is working hard and furiously on a new vehicle of mass destruction? Very often intelligence itself has the unfortunate tendency of defeating its own purposes. The human world seems to operate on a subconscious but obvious activity of self-betrayal and wanton disregard of those things for which intelligence has originally shown ways of preserving and bettering. Fire and water, in their various forms, have been most attractive to us even though both are antithetic to each other. And yet isn't it strange that we, too, all too often keep making fires with our left hand and keep putting them out with our right? Pure science can become so lofty and absorbed in objective and detailed technology regarding the parts of the human being, the outside world and all sorts of ingenious devices, as to lose sight of the principal value of the human being as a whole animal and as an individual.

Before the era of magic and religion the primitive thinking was basically very sound, elemental as it was, no doubt, because its principal purpose was the pursuit of security, comfort, and peaceful general satisfaction. There were enough fears and dangers constantly lurking about man without creating additional ones. Apparently, from the objective evidence thus far revealed, there seems to have been no warlike spirit or tendency or the thought of destroying other human beings. Even during the early periods of rudimentary magic and group shamans, whether it was in the Upper Paleolithic, Mesolithic or Neolithic periods and before the establishment of organized religions in any way similar to what we have today, there seems to be hardly any evidence of internecine warfare as came millenia later. The necessity of war, murder, and even mass genocide, came about as a result of the emergence of the era of more specialized priestcrafts and nationalized religions. The need for sacrifices, the establishment of the priestcrafts and later kingships, the power of eminent domain for the priest, pope or king and the subsequent containment or walling in of villages, towns, cities, empires, caused the evolution of widespread and more specialized forms of warfare and destruction. We have previously stressed this factor but it is worth repeating for emphasis. The various nations in the world today have taken over from their predecessors of power and sacrifice, the idea of glorification and social, religious and economic needs, emotional and objective, of war and all its sisters and brothers of destruction. Whether mankind can undo the thousands of years of slavery, religious morass and psychotic power, is problematical. So long as it continues its present path of submergence to these terrible ghouls, all the brilliant ingenuities and achievements of science and scholarship may be sacrificed at the altars of Mars and mankind may be the continual sacrifice. However, what may happen to mankind in general in the possible centuries ahead is a matter of speculation, hope and conjecture. The ways of man are unpredictable and masses of people have often been trigger-happy and eager to throw themselves into a general melee of chaotic, purposeless destruction, geno-sadism and the weird satisfaction of suffering and death. Here there is the clear example of the destruction of the Time-Value by genocidal hysteria and social, religious, political group paranoia. H. R. Hays comments: "As anthropologists grew more sophisticated they discovered that the primitive's mind was not so different from that of his civilized brethren as it had at first seemed. When the same scientists turned their attention to contemporary civilized society they discovered that modern man was far more primitive than he cared to admit." [23] Charles Duff contends perhaps bitterly that "a crowd is not only a gathering of people but a *state of mind* representing a mental unity. In the history of human culture it is a cruel fact that *the quality of this state of mind has always gravitated toward the level of the lowest intelligence and the lowest morality of the individual unit.*" [24]

It is, therefore, left for each individual to understand that if the proper decorum of value is to be restored, it must primarily emerge from individuals who can free themselves from the group-chain gang and begin rebuilding for themselves a free expression of their thinking faculties. "One basic feature," states Dr. Joseph Ratner, "initially characterized all modern liberation movements: they advanced against the social and intellectual tyranny of Church and State, their oppressive authoritarianism, by appealing to the superior integrity, nature, authority and power of the individual." [25] Bertrand Russell comments: "Collective wisdom, alas, is no adequate substitute for the intelligence of individuals. Individuals who opposed received opinions have been the source of all progress, both moral and intellectual." [26] Dr. Henry C. Link states that "the individual is the only foundation on which any social order may safely build." [27] Every great deed, invention, discovery, all the wondrous works of brawn and brain, art composition, all the amazing things of which the human being is proud and which differentiates him from all other animals, all these great achievements of thousands of years of struggle and persistence, were the results of the intellectual freedom of the individual and not of the suppression of the individual. Whenever he was suppressed, progress was halted, art blinded, literature falsified, science blocked, and he retreated or stood still. Every song that has been written, every picture that has been painted, every figure that has been carved, every great creation, every poem, all these were attempts of some individuality. John Dewey puts it plainly: "Impulse when it asserts itself deliberately against an existing custom is the beginning of individuality in mind." [28] Clyde Kluckhorn adds: "A stable world order that takes account of new, wider, and more complex relationships can be founded only upon individual personalities that are emotionally free and mature . . . Demagogues and dictators flourish where personal insecurity is at a maximum." [29] John Dewey states again: "When social life is stable, when custom rules, the problems of morals have to do with the adjustments which individuals make to the institutions in which they live, rather than with the moral quality of the institutions themselves." [30] . . . "There can be no conflict between *the* individual and *the* social. For both of these terms refer to pure abstractions. What do exist are conflicts between *some* individuals and *some* arrangements in social life; between groups and classes of individuals; between nations and races; between old traditions imbedded in institutions and new ways of thinking and acting which spring from those few individuals who depart from and who attack what is socially accepted." [31] J. H. Denison clarifies this point well: "The great problem in social life is the question as to how far it is necessary and right to sacrifice the individual to the group, or how much of his own liberty each man must surrender for the sake of unity. When the individual is entirely free, there is anarchy. When he is absolutely controlled, there is tyranny. There are two types of tyranny: that exercised by one man, or autocracy; and that controlled by the group, or communism. Both aim at complete suppression of the individual." [32]

We have read about the Reformation and about the Renaissance. This new phase of the recapture of a free, rational and intellectual process dedicated to the principles of peace, security, happiness and the enthronement of the Time-Value on an empirical basis to accomplish these principles, could verily be called the *Restoration*. With practically and probably more than ninety percent of the world's population as conformists or steeped incurably in cultures and religions which have almost obliterated individuality and individualism, most of the rest of the ten percent who may claim to be free intellectuals but unfortunately are probably neurotics and inflated egos of a hundred different kinds, the small balance remaining who are really free individuals, in the reasonable sense of the word, can only view the possibility of this great Restoration as a dim, dubious, remote dream of the future. I am afraid that the human race might probably destroy itself long before any such thing can even be a faint realizable hope of a few rational minorities. The possibility of this wholesale genocide of mankind is increased as the industrialization and "Westernization" of Eurasia, Asia, the Near-East, Southeast Asia and Africa, continue to move rapidly onward with the eager aid of the profit-seeking Western countries who eventually may prove to be the first victims of the early phases of this worldwide cataclysm of destruction. These Asiatic races, just recently emerging from the past fastness of depressed orders and from the mud-baked villages, might turn out to be the trigger-happy ones with new destructive playthings like atomic bombs. Besides, their value of individual and quantitative life is not so concerningly worrisome or important to them. To them the elimination of life means more living room and less competition, and more power for the survivors or victors.

Personally I feel that, even so early as today, Russia has more to fear from the expanding power and demands of Communist China than her cold or warm war with the West. Russia and the United States can go on for years with a colder or warmer war of face-saving, public relations conferences and diatribes at each other, well secure that neither wants war, so long as each country's home economy is prosperous and producing, even though they operate on an unsound economic premise. Russia may be happy that the United States has the hydrogen bomb, not China. So long as the *status quo* continues Russia may be expected, short of a revolution within Russia, to remain the titular head and main office of the world Communist movement. Russia may be vigilant and suspicious of the United States but, I think, she does not fear us. I think she fears China and will use every diplomatic artistry to keep China as a bosom friend but looking up to Russia as the acknowledged Parent-Mother and Guardian, from whom China can reasonably, with face, take orders in actuality if not in form. Ralph Linton in his *Tree of Culture,* makes this keen and sharp observation: "If history repeats itself, China should be able to dispose of the Russians and become a world power in its own right within another couple of hundred years. The idea that the Chinese are simple, friendly, non-warlike people is far from the truth. China has been a world power during several periods of her history and has spread her conquests to an amazing distance." [33] It may come sooner, one way or the other.

However, before we wander too far from our premise and lose ourselves in the labyrinths of contemporary politics and world problems, it may not be amiss to analyze the meaning of progress as part of an attempt to find a justifiable way as a higher and better value for individual and group. We should try a rational brake to restrain the use of rationalism itself as an excuse and defense for our own transgressions and guilts. In our own Bill of Rights the meaning of liberty or freedom is the power not to injure other people in our own pursuit of happiness and security. If we can only apply this single premise to an observation and verification of any content of progress, we will have gone a long way in the reevaluation of experience, individual and societal direction and behavior, that can be called within reason a justifiable basis for thought and action. If the nations and peoples of the world would take this attitude seriously in their meanings of liberty and freedom, and the right of sovereignty over themselves, the peace of the world and the general safety of the human race might have a fair chance of being secured. What we are seeking here is not a "moral" basis which is dependent on the subjective perspectives of any person or group but a practical, more natural and living method of action that can parallel the realities of history and the possibilities of human restraint based on the premise *that which is justifiably good for the individual should be justifiably good for the group,* if a self-sustaining and sound basis for social action is to be achieved.

The term "civilization" itself is often a misleading and vague denominator from the standpoint of a rational justifiability. It all depends on the actual results, not on the veneer of appearances and surfaces. If one should dress up a donkey in a tuxedo and high hat, that, in itself, is not going to change the pattern of the donkey's behavior other than to make him unwieldy and uncomfortable. On the same premise, if an unlearned man would put on an academic gown and spectacles, this, in itself, is not going to make him an intellectual giant. Nor can paint make a woman beautiful; neither can perfume, by itself, make a person clean. Nor can a man in a beautiful, expensive office with a tremendous desk and a gorgeous secretary mean, in itself, that he is a great man of industry and accomplishment. Nor does it make, by itself, a man in a uniform beaming with five hundred multi-colored medallions into a military genius. These are outward, observable tokens of status without verification and without value analysis.

So it is with the idea of *Civilization.* To me Civilization is more present in a happy, peaceful, non-aggressive, healthy society, even if primitive, than in a highly complicated, mechanized, ultra-gadgetted super-duper modern society in which the people are generally miserably "free" slaves of sedentary labor, rich or poor neurotics, full of pretensions and problems, and where people and pretensions are kidding each other in a constant stream of superficialities, percentages, graphs, and all sorts of false values and selfish intentions. It's like wearing a beautiful diamond ring and being delighted by its dazzle and exuberance, on one hand, and worrying constantly how it's going to be paid for, on the other; this leads to the creation of a slick-happy neurotic with one half of his face smiling in admiration of his ring, and the other half a sorrowful, fearful face of doubt and anxiety. "The essential nature of civilization," writes Albert Schweitzer, "does not lie in its material achievements, but in the fact that individuals keep in mind the ideas of the perfecting of man, and the improvement of the social and political conditions of peoples, and of mankind as a whole, and that their habit of thought is determined in living and constant fashion by such ideals." [34]

Of course, I like many gadgets myself and I enjoy them to work with, play with, and they make my chores easier and more satisfying. If reasonably applied and not over-extended, gadgets can make life a heap more comfortable, interesting and enjoyable. But alone they will not give happiness; they can only help us, even exhilarate us, but the basic happiness and life satisfaction that are fundamental cannot be obtained or attained alone from outward sources but must be synchronized and polarized by a harmony of the individual and the world around him to fulfill his innate and basic emotional needs. "A society overendowed with gadgets cannot long escape the knowledge that gadgets are not enough." [35] Gadgets are all right so long as people do not became gadgets themselves. This harmony conducive to happiness and peace is the true aorta through which truly civilizing influences can flow. "This much I have learned," writes Harry Golden, recalling his youth and home life on the lower East Side of New York, "We were happier when mother emptied the drip pan under the icebox." [36]

If a civilization is superficially rational by its outward appearances while, in fact, it is shallow and irrational in its subsurface content, then such civilization is a down-grading from primitive cultures, not an up-grading. I think I would be better off as a happy Balinese who dances under the stars by his banyan tree, with peace of mind, and no anxiety, warbling his chuck-chuck in the monkey dance, actually accepting the good and bad fortunes of living with grace and fortitude, even accepting the terrible demons and blissful tales of Ramayana as an excuse to dance, sing, laugh and be merry to avoid a boring existence of rice consumption, than to be a world-wise smart aleck on 42nd Street who knows Wall Street upside-down, has ten millions in the bank and internally hates himself for living and who hasn't got peace of mind because he got most of his money by conniving against others and values everything by its outward veneer of service and shine rather than by its inward qualities of verifiable and enjoyable sincerity, honesty, and integrity. People may be what they also appear to be, but *appearance alone is not a conclusive premise for verifiable value and meaning.* That is, if we are really seeking a justifiable way for our judgments and directions. So if we are to observe and evaluate correctly whether a certain status is civilized or not, we must take into account the content or interior structure of it, the body under the skin, the mind under the face, as well as other factors such as the outward and exterior framework of a person. To judge alone from outward appearances, impressions and veneer, either of a town in Illinois or of a village below Agung Genung, would be a mistake, in my mind, and would not justly lead to a meaningful analysis of the places involved. What I state here is simply not going to change the world or its people, but at least I can try to see what's really going on with my eyes wide open. Let's try to analyze this problem of appearances and impressions a little further.

There are many people who go through life on the very obviously false philosophy that they can attain happiness, find friendship, and even cultivate love on the basis of impressions and appearances. Still others go even further; they believe they can quickly or gradually get people to love or want them, think more of them, approve them, and even to expect them to worship them because of an irrational, over-emphasized generosity towards these people. If a person genuinely hopes to find the finer and truer values in life that can possibly make for less regrets, less disillusions and disappointments, that could possibly cultivate a more sympathetic, deeper, appreciative and an honestly compassionate attitude towards oneself and to others, then the philosophy of impressions and appearances is not the road to take. By trying to fool others, even subconsciously, one can only fool himself.

A person may say that even if one "buys" love, devotion,

physical or esthetic pleasures, even "knows" that his "friends" are false and just make-believe but still enjoys their company, nevertheless, so long as he obtains, in his way, a sense of satisfaction, relief, or happiness, then one cannot deny, in fairness to the individual, the right to such experience and satisfaction. After all, he can say, who is the one who can irrevocably set up standards of approach, of values and meanings, which, after all, are practically based on perspectives and opinions?

The answer to this is that it is a prerequisite of liberty that a person has the right, or should have, of doing what he pleases so long as no one else is prejudiced or injured thereby. He can throw himself off a cliff provided his body does not injure the cherry tree below or he can eat popcorn night and day so long as he doesn't call the pharmacy in the dead of the night for a bottle of castor oil. Or he can drink himself to death in a nightclub or he can make love to a prostitute whose trade it is to sleep with her customers, and pretend to himself, sitting on his own throne of conceit, that she is emotionally disturbed because of her surrender to his captivating charms. Yes, people can do what pleases them, even if they recognize their acts as superficial but satisfying. If a person is happy being an idiot, then idiocy is justified as *his* road to happiness. However, we are here concerned with an *attempt to think* so we can possibly reach a sense of evaluation and as a result of which to possibly attain a more realistic, a more meaningful sense of value that can possibly lead to a happiness or satisfaction that has not got imbedded within itself the seeds of disillusion and regret. It is not a question of setting standards. Life is too fluid and flexible and changeable for all that. The idea is to provide oneself with a *way of thinking* in terms of *better, more sustaining* and *less regretful* experiences. Less regret within a certain quantum of experience enhances the value of meaningfulness of the Time essence within it and gives birth to a peace of mind and body that is fundamental to a grounded satisfaction in living.

There are certain things one just cannot buy. Not even with all the wealth of the world. And these things will be found to be finer because they rise, not out of gratuities or favors, but out of the hearts and minds of people, out of understanding and the weighing of genuine values reduced from the relationships between people. There is a big difference between leading the Ego through its mind to the transparent, calmer waters of life experience so that a genuinely sincere and relaxed attitude of appreciation, affection and sympathy, can be obtained, on one hand, and the muddied waters of misdirected egotism that can only fulfill the stupid desire of self-inflation which cannot transcend into other minds to draw the respect and admiration of intelligent and sincere people.

Very often these impressions and appearances take form or are exuded because of the auto-intoxication of power, authority, position, opulence, social or business caste systems calculated on wealth, religion, race, orders, clubs, etc. No doubt many a man automatically feels bigger because of a button in his lapel or because he lives in a bigger, more expensive house or is able to buy the more costly things. People who are so fortunate as to have money have the right to, and should enjoy their wealth and buy expensive things if they desire to, and they have the right and should have better homes and have as many comforts and luxuries as they desire to. There is nothing "immoral" about wealth or poverty. And there is nothing I can find wrong in joining a club or a society where the principal incentive is to bring people together, to find friendships, to better themselves for social enjoyment or education, or both; these things make for a healthier, more normal and congenial relationship. But there is a difference between enjoying one's wealth, and making friends by using wealth as a blind or bait to impress people or making friends by judging the *outward* coat rather than the inward lining. If a person has only his power, position or opulence and uses these things to gain the support or submission of people, he will never really get it. That's for sure. He may get them to "yes" him all he wants, to serve him, to prostitute themselves for him, to beg of him and to yield to his whims, to silence their normal free expressions, but he will never get them to sincerely love him or regard him.

Money buys services, not people. Gold leaf may gild the lily but all it accomplishes is that the flower ceases to breathe and an artificial creation is manufactured by a false and deceptive value. I feel sorry for the people who are compelled, by economic or other necessities, to serve such a person, but I pity this person more, because his servants may still be free in their hearts and minds, but he has become a slave unto himself and he has imprisoned himself in his own cell of false and deceptive intoxicants. Inflation does not increase value; it merely increases size, a weaker and thinner size that ultimately bursts, like a bubble, with the first puff of the fresh air of intelligence and rational self-analysis. People who want to face realities and to reach a state of peace within themselves and with others, are not those who beguile themselves and others with deceptive and false values of impressions and appearances. These false values are transitional and temporary; they are founded on the premises of what we possess, show, exhibit, fears of power and enticements of favor; founded upon what we have, not what we are. Until a person learns to give of himself, he cannot fairly expect people to give of themselves to him and the values he seeks but cannot buy. What he can only buy is the reflection he seeks, of a bigger self that is projected against the screen of his own self-deception and he will find his refusal to face reality and true values will keep blocking him, wherever he goes, from the hearts and minds of people.

The great stumbling block, I think, to the regeneration of a better and more honest concept is the unyielding self-intoxicating over-confidence that he is really "smart" and those who are willing to be accepted and judged for what they sincerely are, are called "peasants." This "smart" person wastes away his life in two principal illusions: the illusion he creates about himself and because of this, creates illusions about others. This is the snake-pit for the one who surrounds himself with the fog of self-contortion and deception, of false dignity and fictitious pride and affectation. He never enjoys the fresh air of self-honesty or inhales the true fragrance of love. Love is impossible to him. He never knows the enjoyment, the essence, or the discovery of true peace and life satisfaction. He never knows who his friends are because he has not really learned to be a true friend unto himself.

Another identity of this person is that he is usually or primarily concerned with pleasing himself and never hardly thinks of pleasing others. He wants others, of course, to do his bidding and to please him and he thinks this is a natural and proper way for people to react towards him, and which he honestly thinks is fair and justifiable, just coming to him. He never thinks or *feels* to do things which would not only please others but which would be considered good sportsmanship. If he does go out of his way for someone, the act is usually combined with some purpose of pleasing himself. He is rarely reciprocative, and when he is reciprocative, it isn't a wholesome spirit of mutuality but one that is pressing him with guilt and the urgent desire for relief.

This person belongs to a confused, restless and bitter class. They are the poorly rich or the sickly powerful, the ignorant poor or the ugly one with peacock feathers stuck in its backside, or the fearful cleric bathing in haloed phrases that he hopes will bring in more money, the little "big" business executives, arrogant and domineering because people grovel at their feet catching the dollar bills being blown out of their pockets, or the sartorial perfectionist who cannot be seen unless his tailor and beautician have stamped their daily approval, the smart gentlemen who "kid" everyone and think their manners, slick style and a few fancy words will conceal them; or the lazy poor student with an encyclopedia under his arm when he takes a walk, or the dirty wife who powders her nose but forgets to wash her face, thinking that the fragrance of one will "outsmell" the other. These people walk the plank to fall into the sea of false concepts and twisted roads that never can lead to a happy life. Happiness is something one cannot find only with himself; it is a stocksharing profit plan that originated hundreds of thousands of years ago in the satisfaction and comfort the social instinct has provided the human being with. While this social instinct is deeply rooted in the very core of the human constitution, it remains a flexible, transcendental *movement* so that it can breathe and grow; an egotist is one who cannot find it possible to transcend his own opinion or see things in another light. The introvert is one who cannot understand what the other people are doing on this planet, while the extravert is one who feels that the only reason the other people are really here is to see how he runs it. If such a person ever goes for advice, the answers are already stuffed in his ears. These people are usually very snappy to others, dictatorial, and anxiously desirous of telling all that can be wrong with you but all this is merely that no one should take a good look inside of them. Beneath it all lies the bared guilt of inferiority, fears of ogres that really do not exist, fears of others because they fear themselves, the insecurities of their own fantasies and illusions, and the constant ordeal of making appearances and impressions. This is the art of never being able to be truly relaxed. Time to them is a whirlpool of restless moments; they fear the next hour and find no companionship of a peaceful mind and body. The clock to them is a constant challenge and threat and their plan is not to enjoy it but to overcome it, which of course they never do. They walk backwards, following their own shadows, slaves of their own illusions.

What does Time mean to these people? Nothing but a problem, a wall to climb or another sleeping pill to do away with it. They waste life by wasting Time. If they realized the importance, the value—*the sacredness of the Time essence*—the realization of its synonymity with life and living, the brevity of its uncertain measure and eventual end, they would realize the foolishness of feeling the way they do and they would create natural fortitude, strength, honesty, and the courage of character and conviction. Thus they can find peace within themselves and the possible happiness they should seek and enjoy.

Therefore, the justifiable way necessitates, due to the nature of human nature and human experience, a process of self-discipline and decorum without becoming dogmatic or rigid, without lessening the visibility towards realities, and the flexibility to change, learn, and change again if need be and whenever necessary, to achieve the possibility of greater and finer satisfactions in life.

In the broader and social way this discipline and decorum necessitates the *law* and its enforcement agencies. Contrary to the blissful phrases of the optimists, it appears that dishonesty, kleptomania, deception, masochistic and sadistic tendencies, are usual factors in man's nature, that is, it is reasonable to say that under normal, natural conditions, a human being could be easily dishonest or sadistic if the opportunity was there and the fear of being punished or ashamed nonexistent. Professor James Harvey Robinson states that "not only is the whole human race derived originally from wild animals, but each boy and girl enters the world as a wild animal." [37] Robert Briffault comments, also, that "the human world has risen out of savagery and animality, that its dawn-light shines on no heroic or golden ages, but on nightmares to make us scream in our sleep." [38] Dr. Raymond A. Dart, the great scientist-adventurer of the missing link, reviewed this point well when he wrote: "Perhaps it is because man has grown so self-centered and clever and progressed so rapidly during this century that it is mentally difficult, if not revolting, for civilized people to look back upon their primate past. . . . Their lowly origin offends their newfangled ideas of what is fitting to their present estate. Yet if some footling human beings had not been courageous enough to forget their social status and study in deep humility the lowliest living creatures as well as the primates (and also their dead bodies and internal parasites, as well as those that crawl over them and the foulest diseases that destroy them) civilized beings would certainly not be able to live as they do today. The study of the repulsive and the filthy is usually more vital to our welfare than that of the attractive and the clean if humanity is to obtain the understanding needed for survival." [39]

People will most assuredly rush to watch a prospective suicide about to jump twenty stories to his death and enjoy watching the event much more than attending a wedding or a concert. Recently a plane rising from Idlewild Airport in New York lost some of its landing gear and circled the field for hours. Probably fifty thousand people came from every direction, clogged the roads and pressed forward in mobs over the field area to watch the possible crash of this plane containing over a hundred people. That the plane came down safely and without incident must have been a terrible disappointment to these mobs that expected a great explosion and the incineration of its occupants.

I realize that many will feel insulted, irritated, and even feel injured at this remark, but the public and private records prove it again and again very convincingly. On the other hand, this should be no exposure of those who are genuinely and naturally honest, good, and kind. On the contrary, it may be a word for those people to be vigilant, alert, and realistic in their approach and appraisal of the relationships between them and the world. Shakespeare wrote that "a fool and his money are soon parted." If the world were basically honest the fool and his money would be fairly safe and together. It is, truly, a sad and pessimistic commentary on the human race but we must face realities. Take away the law, the fear of punishment, penalty and public exposure, take away these for a single week and the world will lay in worse ruins than Carthage!

Whether law is just or not is a problem of another order. Before we can rationally say that a person has "violated" a law we must establish the justification of the law in the social order involved. If the law is not just, then the problem of whether this is a reasonable status of violation can be questioned. Who has the right to judge the law? Every individual. Whether he is right or wrong is for the individuals or jury to decide by their individual, collective, or group expressed actions, selectivity and judgment, and even here we come up against the fallibility of human judgment. Whether they are right or wrong is also a question of opinion. Therefore *law* is not a sacred end unto itself but should be considered as a pragmatic process or vehicle to be modified, replaced to suit the best and basic needs of security, justice, peace and the general welfare of the individuals and society. Life, constantly in flux and subject to change, being in itself a manifestation of flowing and forming forces, cannot be justly subjected to a concept of unchangeable dogma or status.

The enjoyment of freedom also implies the obligation not to injure others or to deprive others of freedom, as we have noted in our own Bill of Rights. What has all this to do with our philosophy of Time? It means that the best and highest possible use of Time necessitates the reasonable assurance of a minimum regret and remorse, and, in order to further this end, we should face and judge realities of human nature so that we can avoid getting hurt, or prevent by reasonable vigilance, without lessening our own desire for compassion, friendship and association, the pitfalls and the disillusionments that await those who make the mistake of creating the concept for their own unwary pleasure that others think and act always as they themselves do, without first making some rational attempt to verify their concepts from experience. Also we must face up to and we must be fair to emphasize the equally realistic observation that there are many people who, because of their upbringing or sense of fairness and intelligence, are basically honest and trustworthy under most circumstances. It is, therefore, reasonable to understand that equitable law, to assure individual and group political, social and economic decorum, the general welfare and the pursuit and sustenance of happiness, is an essential and necessary arrangement in human order and relationships.

The greatest test of wisdom is to uphold *Equity* regardless of the established or accepted rule. Here is the courage and the persistence to keep salvaging and maintaining the defenses of the minorities and the protection of the majorities from the minorities, even to a single individual, the rights and judgments that are considered, in the light of simple justice, to be fair, tolerant, ethical and honest. It was upon the basic principle of Equity that the main pillars of political and social democracy were firmly implanted. Laws may be just or unjust, obsolete or premature, over-penalizing or too moderate, preventative, harassing, or ineffective, but Equity remains the same quantum because it is not *the* judgment or rule, but a *way of judgment* based on the essence of impartiality and fairness, without preference or prejudice. Equity implies the principle that justice and honesty must go hand in hand, and that where the law cannot fulfill this intention then it is still in the process of Equity to fulfill it. Once justice and honesty are parted, both are lost. Equity takes into account the fallibility of law, as very often the law can become a chain that enchains itself, unfortunately, and its very links often become loopholes. Not so with Equity, which is primarily concerned, not to maintain regulation, a rule, an ordinance or a fine, but with maintaining a principle to which anyone can appeal to so that *justice,* fairness and honesty, can be upheld as a living, basic source and process of democratic action. This marks the difference between a true democratic state which calls itself a republic and a police state which also calls itself a republic.

As in the principle of Equity, so in many other things it is important to understand and evaluate people and experiences in terms of the highest possible ethic that can aid in eventuating the greatest potential of peace and security, instead of lesser and more questionable principles that may sow dragon's teeth for the future. One of the sad and false concepts in the United States and in other highly Westernized countries, and I have even found this very widespread in the Orient and other non-Westernized areas, is the very unfortunate situation where the governments, economies, the social and prestige values, even the religions and revival meetings, have the dollar as the barometer of a person's importance. The worship of the dollar means that a person becomes valuable not so much in what he can do for others but more so in what others can possibly take from him. Money is meaningless unless it is attached to something more valuable. What represents value is the test of a man's philosophy. The worship of money is so evident from the general principles and practices of too many in the political, professional and business fields in almost every nook and corner of our country, especially so in a concentrated form in the newer states where, in order to do business or buy property or get a tooth extracted you simply need a battery of lawyers to protect you from these marauders on the purse strings of every new comer to their dens. Of course, there are honest and fine people with integrity and good intentions almost everywhere, but these are the exceptions and one has to have an extra fine comb, the FBI files and the Congressional Library, to filter them out. The increased tendency to create standards by which people are considered in terms of the dollar instead of being considered in terms of the people, has caused throughout the world a rising rebellion on the part of most of the people who haven't got any money.

Such a situation brings about a false value and a twisted, unethical tendency in the affairs of people which is not conducive to peace and happiness. All the intelligence, experience, material progress and all the great advances in all the sciences amount to little if they do not bring a basic, genuine *gratification* to the peoples which can possibly make for real peace and the lessening of conflict and hate. What good is civilization if most of the people of the world are steeped in poverty, in want, and without any reasonable security for the simple essential needs for even elemental living? Any progress, therefore, must be judged, not only in its objective usefulness and potential in the relationships between it and the people but, of more importance, it must be judged by its potential to inspire and achieve a gratification of the desire to produce good, to cooperate and to appreciate, to cultivate our vision and our venture into esthetic values, meanings and expressions, that can help fulfill the inner and emotional needs of people, their basic securities and wants, and to achieve a reasonable safe tolerance of compassion which may possibly by actions and accomplishments ensure peace of mind and justice for as many of them as possible.

Silence is golden only when it "speaks." So it is with Time, which moves silently away from us when we least regard it. It comes to us, in its fullest and finest way when we use it to express and translate our experiences into beautiful and worthwhile things and thoughts, and to cultivate our sense of appreciation and understanding which gives constant moving color and balance to the moments we live.

We have noted, in the chapter regarding the Social Factor, how the social entity, like the religious or cultural entity, gradually submerged the individual who became in turn subordinated to the "welfare" of the social and political state and that now and then when conditions became intolerable or inoperable the individuals rebelled and formed other types of social states more to the contemporary satisfaction of the people. In making this repetitive observation of social history we must take into account the parallel observation of many situations where individuals, benefiting from the freedoms and wider flexibilities of economic, religious, social and political mobilities, have used these media not for the benefit of other individuals or for the welfare of the state but purely, and often very cruelly and sinisterly, for their own personal satisfactions, power psychoses and self-aggrandizements. The dictators, tyrants and despots, are only the top, more easily observed veneer of this class of people. Below the "top bananas" of these cruel and vicious psychopaths are legions of lesser ones infiltrating the substrata and tributaries of human life and experience. It is for the good of humanity that very few of these paranoics reach the pinnacles of power but when any of them do they only spread tragedy over the lands and lay waste human life and property. It is also for the good of humanity that most of them are limited, in their eager subconscious drives to overcome their own guilts or inferiorities, but one can easily detect them when they express "authority" at the wheel of their cars, getting drunk and "telling everybody off," carrying on petty tyranny in business, in their homes, in their professions, fulfilling the desire of sadism by torturing people in a million little ways, torturing and even maiming animals, and destroying things of beauty. The list is almost endless, and terrifying. No animal on earth, whether it walks, creeps, flies or swims, is capable of the bestiality of man. No animal on earth has as much intelligence as the human being for the specialization of destruction, waste, floundering, and hate. Dr. Herbert S. Dickey, in his famous *My Jungle Book* relates about the treatment of Indians by whites: "Kindliness rewarded by meanness, hospitality repaid in treachery, decent treatment reciprocated by murder. It is almost always so when Indians and whites meet in the deep fastnesses of the jungle land." [40] Cruelty often exhibits itself in the attempt of one race to be master or exploiter of another race. If the Indians were the "civilized" people in power and the whites were the "savages" no doubt the Indians would not be any more or less cruel.

It should be noted, also, that it is naturally instinctive for an individual or group to group with others because in this way a person's Ego or sense of self becomes bigger, stronger, "wiser" by *accretion*. Man, knowing that alone he stands insignificant and powerless before the all-overwhelming giant of Nature, now grows defiant and confident, not only to overcome its present reality of life and death but willingly joins a mass psychosis to *pierce through* it and beyond into the ego-accepting idea of immortality. Of course, what actually happens is the very reverse. Unless he is cognizant of sound and logical values and meanings, which implies a logical and equitable interpretation of the value of Time, he is lost in the emotional quicksands of social mores and regimented oblivion. The oblivion which was descended upon people by religious organizations throughout history is now being taken over by Communism. Both rule by the sacrifice of life for the "future" and accept a value of insignificance for life on earth except in so far as it operates for the power and growth of the "State."

The West is changing the face of the world with gadgets, telephones, typewriters, motion pictures, strawhats, autocars, appliances, Fifth Avenue clothes and umbrellas. The Communists are not changing the world; they're just taking it over, lock, stock and barrel. The Christian says, "Let us teach God to your children and the Church will have them for the rest of their lives." The Communist says, "Let us gather all the souls in the world and we will own the world!" Alas, a world of slaves! Both the Churchmen and the Communists are dogmatic, self-imposing absolutes that require blind or imposed faith or adherence. The West, depending upon reason and freedom, will eventually lose the freedom to use reason. The West is *modern,* does not take into account the evolution of religion and how it came to overtake the misdirected Ego of man.

We were born free; we should live as freely. When we do die let us not die as slaves to anything except to the freedom which we were wise and fortunate to enjoy. The ideals which induced Karl Marx to strike out for the exploited masses are no longer the ideology of modern political Communism. If Karl Marx lived today and saw Communism at work in Russia and China, he would become a conservative Republican and filibuster every liberal Democrat until he dropped dead from exhaustion. Bertrand Russell remarks that "no political theory is adequate unless it is applicable to children as well as to men and women." [41] No political or social system can more or less permanently endure wherein the free expression of the individual is curtailed or suppressed. Such a system is a police or military state, holding in power those who impose their will upon the people by force or fear, or both. Such a state can only be in power so long as their power-hold continues and will eventually crumble from within. In the meantime generations of humanity and millions of lives are enslaved, exploited, and misled.

On the other hand, in our own country we have the regular conservatives who may be still fighting wars against the Indians; conservative Southerners who insist that the Civil War is not over and that the Union Army will yet be beaten back into Pennsylvania; conservatives of a hundred kinds who mentally live centuries ago, with minds so frozen and rigid that they resist every new day with emotional force. They forget that the conservatives draw salaries upon what the liberals took chances. They forget that most of the world is poor, living little better than the animals in the woods, that as the world population explodes more and more and as a result increases the number of the poor, that unless something is done to help them to help themselves, the fighters for democracy and freedom will not be strong enough to hold back this colossal tide of human want and frustration. The conservatives just wouldn't be around.

Of course, this is no affirmation that all the liberals are the wisest, that they are not subject to error and wrong. This is no affirmation that everything new is good and acceptable and every old thing is bad and obsolete. This is something that entails inquiry into any special or particular thing, to be analyzed and judged. There are many things of old which may be good and will always be good and there are many new things that are probably unjust or unsound. Life moves constantly, and the intellectual and political life of a nation cannot stand still. It also moves forward or becomes stale and needless. Albert Balz, in his brilliant book *The Basis of Social Theory,* comments: "The cultivation of mind implies the cultivation of individuality. If this cultivation brings in its train unsettlement and upheaval, then one can believe only this, that further growth of mind can provide the compensation. The continuous play of interaction among variant individuals, between these and the bulk of society, the free play of mind upon the cultural tradition, is a precondition of growth." [42] The importance of this growing mind in the world today and the actions which this mind impells might still save the world from Communism and free the peoples from poverty without losing their freedom. It is upon this road that Democracy must travel and travel fast if it doesn't wish to become just a temporary free breath in the history of the long road of mankind's tragedies.

The governments of the world are often in the hands of third-rate hangovers from the regurgitating wells of power politics, well-heeled dictators who keep filling foreign banks and vaults with their loot, or oratorical floor-walkers who know too well the science of practical politics and how to get elected for another term. Of course, here and there a good man appears but as soon as he is found out to be an honorable man, he usually goes the way of Caesar. If a politician can't play ball, the team gets rid of him. Very few fine and capable people enter the field of politics because they don't wish to wallow in it, get burned for their good intentions, or be "framed" for the good of the party. One reason for the woes of mankind is that those who possess fine talent with intent of honest analysis, projection and judgment, write excellent articles for special magazines which very few people read while the hams and the robbers rule the world.

What does all this mean to us in our quest to understand the Ego, the individual, and the relative factor of Time? It means that if the leaders of the world today were wise and honest servants of the peoples, they would get together for a general review and inquiry into the nature and purpose of government and society and thus attempt to reclaim the rights of individuals and reconstitute the duties of government towards the individual. Governmental organization, public regulation and private suppression, are substantially working towards creating robots and slaves out of human beings. Rebels become outcasts or neurotics. Centuries ago conformance was insurance against being stoned or burned at the stake. Today conformance is insurance against losing one's job or being disliked in the neighborhood, economic insecurity and social ostracism. Perhaps to a lesser extent, the same disease of obscurantism still prevails to a certain degree and this gives birth to many kinds of creatures in our social organism: yes-guys, liars, deceivers, behind-your-back expert sneaks, paraders, mannequins, applauders, and an awful lot of just saddened people who have simply been much too intimidated to express their true feelings. In each case, the individual has lost sight of the full impact of the value of Time within his life.

Very often courage comes with old age or a doctor's revelation of a dreaded disease, but then it is too late to recoup that which cannot be replaced or brought back. It is this threat against the individual spirit that made Thoreau go to his little pond and prefer the birds and the trees rather than the superficialities and falsehoods of people. Fortunate is he who knows the difference between enjoyable solitude and parrot-society. It is these things that inspire artists, philosophers and doctors, and many other kinds of people, to go into the jungle land, preferring to live among primitive people who are not sufficiently intelligent to be insincere, rather than to suffer within a concrete world with stone-hearted, self-mauling technologists wallowing in confusion and in a mechanical vortex of automatic controls but with little control or purpose in their lives. There must be some way, I hope, to realign the material progress of the world with less loss to the values and meanings of life itself. Unless this can be reasonably accomplished then mankind will be swallowed up in the vortiginous mess created by its own hands.

Let us build our hopes that this may not occur, that man may alert himself to the *presence* of his own life, its brevity and uncertainty, endowed as such by nature, its irreplaceability, its wonderfulness and the real parade of miracles that surround him each day and night. Let us hope that there will come men of good heart and wise minds who will show the people the path of peace and the end of conflict. Let us hope that there will continue to come other men and women of honest hearts and free minds to keep freeing the peoples from the self-enjoined chains of superstition and social obscurantism so that people can enjoy the day and the coming day and fill each precious hour with happy living. Let us hope that teachers with knowledge and courage will keep coming to pave the road for the young to walk upon so that they can see what is real, not in the ever-distant future but all around them as they go through life. The clock is our daily witness to our deeds and our hopes. The greatest hope is that we can begin expressing our hopes *now* and living it within the *present* hour.

Out of the earth to rest or range
Perpetual in perpetual change,
The unknown passing through
The strange.

JOHN MASEFIELD

SIX

TIME AND THE *CULTURE FACTOR*: PART I

Now we come to a phase in our quest for a philosophy of Time by attempting to examine and review the nature of human cultures, some historical background of the acculteration of the races and religions of the world and the processes observed in the making, and finally the value of these relationships between the individuals and the cultures, and what they can possibly mean to our analysis and value of Time in the life of the individual and the societies in which he finds himself a part of, willing or not willing but nevertheless a part.

What is *Culture*? There are here, as in so many other things, various interpretations and differences of definition and meaning. It isn't really important in order to understand its observable form, organs, functions and behaviors, to enter into a long ethnographical or anthropological treatment of the subject. For our purposes, holding in mind our *alma mater* of Time, it may be fitting to outline, very briefly, the present evidence available to us by the investigational and analytical sciences concerning the human race. As a guide and determinant of the evolution and mobilities of human cultures, it may be wise, perhaps, to keep in mind that there appear to be two prime reasons or principal pressures that kept up and which are still operating: First is the always immediate need of water. "The cradle of all living things is water. Sired by light, life was conceived in water. And in its amniotic fluids most of life is still carried." [1] Next is food, and then the other securities to overcome or ease the struggle to exist; and secondly, the constant attempt to satisfy, consciously or unconsciously, the basic emotional needs and drives of the human animal. Herbivorous animals will be found usually where there is the possibility of grazing, water, etc.

Predatory animals are found where the prey is usually available. Trees and grass grow where there is sufficient rainfall, moisture, or irrigation. Large fish follow the smaller fish and the smaller fish follow the still smaller ones. Man, too, had to follow this course in order to survive, although today a man can live in a desert and the mailman, plane, train and truck, can bring him all the necessities and comforts of life so long as he has the money to pay for it. But throughout history man didn't have all these services at his beck and call, and it's only a short way back when money or coins were invented and began to be used. The further we go back we find that humans stayed in one place if the environment supplied them with these two principal needs, more or less. If it did not or if they were in short supply, or if the population grew to numbers far in excess of being able to satisfy itself, they either moved or were pushed out, or they pushed others out. This *pushing* is the history of the emergence, growth, evolution and movement of human cultures.

In order to fulfill these needs, not only were people pushing other people around but they pushed the forests back, leveled mountains, changed the courses of rivers, created lakes or dried them up. They kept pushing in all directions that yielded to their adaptabilities. Where a group or people stabilized themselves into a favorable *modus vivendi,* they created confederacies, empires, kingdoms, religions, feudalisms, castes, etc. Then the pushing expressed itself *within* the organizations themselves as the drive for power and property expressed itself by individuals, groups, religions, political or societal entities or philosophies. However, before we should make any attempt to analyze the value of Time and the history of human cultures, let's try to make a rapid and brief survey of the emergence and nature of these cultures from prehistoric times to the present.

To begin with, it is accepted by all the sciences that man emerged or evolved from the same stream of animal life and by the same principles as other forms of animal life. The human animal is an animal with clothes on. No one really and seriously believes in the fairy-tale of Adam and Eve, the Serpent and the Apple, unless he is a fanatical believer in the religious stories that make this genesis part of their program. Such a person would most probably still believe that the world is flat and that the angels hang out the stars for us when the sun goes back into the barn at sunset. There are literally thousands of fairy-tales explaining the origin of life and the world. Some are beautiful and good, others ugly and cruel, depending on the people and the culture that created them. The Judean-Christian story is not only ugly but also cruel and nonsensical. The Chinese will tell you that their deities created the world with China as the pivotal center of the universe. The Greeks would have told you the same. To the Balinese, Bali *is* the world. "Every great people of antiquity, as a rule, regarded its own central city or most holy place as necessarily the centre of the earth . . . St. Jerome, the greatest authority of the early Church upon the Bible, declared that Jerusalem could be nowhere but at the earth's centre." [2] The Jewish people picked their spot, inherited from the Babylonians. The Hindus selected a little peninsula off India. The Eskimos "know" that the world was created by their Sedna, a goddess who lives beneath the seas and who controls their supply of fish and seals. Almost every American Indian thought and believed that the world was created with their chief's wigwam as the exact center of the cosmos. And, like the brilliant Greek playwright thought, if a donkey were able to talk he would tell you that the center of the universe is exactly where his tail hangs from. Man is not only egocentric and geocentric, but also cosmocentric. Professor James Harvey Robinson comments: "Few of us take the pains to study the origin of our cherished convictions; indeed, we have a natural repugnance to so doing. We like to continue to believe what we have been accustomed to accept as true, and the resentment aroused when doubt is cast upon any of our assumptions leads us to seek every manner of excuse for clinging to them." [3]

Man evolved and was affected by the same things as other animal and plant life, and continues in the same experiences today. Sir James Jeans states that "we know that man is an absolutely new arrival on earth; he has possessed and governed it for less than a thousandth part of its existence." [4] Because the nature of life and existence consists of non-static potentials and materials, so also does the nature of man from his origin until today, reveal a story of constant mobility, slow or rapid, adaptational or mutational, gradual or accidental, but regardless of the nature of the change, change it was and still is. All the worshipped gods and goddesses in human history cannot change this, but the gods and goddesses themselves have changed like everything else. It is a continual narrative of absorbing or being absorbed, of infiltration, modification, or resistance. It is utterly impossible to stop this fluidity and ertia of change. Movement and motion, accretion and dissolution, or both, go on constantly no matter in what degree or where. There is no such thing as a static or completely stationary culture, society, or individual. Whether parts of this process, depending upon circumstances and events, were kinetic, catalytic, or whether they were due to adventitious incursions or endogamous inbreeding, whether any occurrences or cyles were deleterious or beneficial, didn't and doesn't matter to existence itself. The littoral people, the desert nomads, the circumpolar vagrants, the tropical jungle cannibals, the invading Aryan hordes from the Caucasian steppes, all these and a thousand more, spread and moved over the earth subject to the problems of food and water supply, shelter, security and gratification, like other living things and subject to the climatic problems and limitations like all other forms of life.

As stated previously, scientists have thus far revealed that pre-Sapian remains indicate clearly the existence of human ancestors and human cultures going back six or seven hundred thousand years, and most probably much earlier. Remains of *Homo sapiens* prove that our present species may probably be a million years old or more. New investigations may yet prove the age of the hominoids much earlier and older. The prehuman animal learned how to make fire, crude tools, and spoke a language to communicate with each other. "In discovering language, man discovered his own mind." [5] "Language grew up out of earlier and simpler kinds of speech." [6] With the gradual emergence of *Homo sapiens* the art of tool-making increased and spread, and languages grew bigger and better. With the art of language came the discovery of the mind and each evolved and furthered the other. An animal has to eat and drink water; therefore the prehistoric humans stayed or roamed wherever they found these. No one thought of raising food, that is, planting or farming, or to raise cattle. It took countless millenia for this idea to sprout in the minds of our prehistoric ancestors. Most of these revolutionary ideas came about by accident, many came about by the sheer problem of urgency, like facing starvation and extinction, and many came about as the result of chains of extrapolated series of ideas and events in the daily and accumulated experience of these people. Only about six or seven thousand years ago we find the so far earliest evidences of planting to grow food, according to present knowledge. Man, equipped with a hypertrophied brain box supplied him in the Pleistocene period, didn't become an inventive genius intelligent enough to make the crudest tool until long periods of thousands and thousands of years later! The body of man did not become human because of the brain alone; the brain of man experienced and grew because of the various physical and neural impressions that evolved because of the usages of man's body in its long trek and constant struggle to exist in changing climates under many changing and environmental conditions. "Thinking cannot itself escape the influence of habit, any more than anything else human." [7] Morris R. Cohen clarifies further: "It is only in the interaction of man and his environment that the basic elements of history can be found. . . . The essential fact is that the environment of every human being and the context of every human act contains human and non-human elements inextricably intertwined." [8] Of course, the actual historical period of man reveals the gradual rise in his potential for inventive ideas, diversified tool-making, and the emergence of the scientific method in man's rational attempt to better himself and his fellow-creatures or to overpower and destroy other fellow-creatures who stood in his way. The basic and elemental tools in use today were not unlike and very similar to the tools used by prehumans before the dawn of history.

As it has been pointed out in a previous chapter, there is an innate cruelty in man that dates back to periods when he walked more on four legs than on two. The prehistoric human animal never knew the meaning of or the need for pity or mercy. There was nothing humane about him. "No animal," points out Dr. Potter, "not even the most bloodthirsty carnivora, could ever find it possible to commit the atrocities that students of psychopathy know are only too common among humans." [9] Unfortunately, in animal life we find too often that the good and affectionate and peaceful animal perishes and the cruel and stronger survive.[10] Humans in the jungle still roast monkeys alive by holding them by the tail over the fire and in the "great" cities men go about calculating and plotting to rob the poorest and the ignorant, or to cheat their neighbors and friends who innocently trust them. Eat or die was and still is the basic preamble of life. The eater didn't give a hoot how the eaten died. Even today humane societies have to fight bitterly against theologians and matadors in order to get laws enacted that would at least provide a less painful death for the poor animals who give us their lives so that they can land on our plates with smothered onions and mushroom gravy. "Monkeys are not as mean as men," writes Dr. Herbert S. Dickey, "nor are they so vicious. I wonder if they are less intelligent as a rule. Certainly they possess generally much more intelligence than many men whom I could name." [11]

To gratify himself man is innately and fundamentally indifferent to the woes and sufferings of others. Let's face it: we can only agree with Clyde Kluckhorn that "man is a domesticated animal," [12] and view the reality, as Fred Hoyle does, that "the human being is an animal and that at root he lives like an animal, controlled by exactly the same natural processes as other animals." [13] The great changes in esthetic, humane, ethical, equitable, and compassionate treatment of man and other living things came about by the rebellion, persistence and brilliant battles of the few against the many, not the many against the few. Slavery has existed as far back as history goes, and still persists in direct and indirect forms. Only a mere hundred years ago our Southern gentlemen of laced cuffs and cordial manners couldn't just understand why it was wrong to "buy" a human being at an auction sale and thought that Lincoln was just a crazed crack-pot because he couldn't see the "justice" of their "paternal" culture and order. Alas, man has changed little in thousands of years, but change goes on relentlessly, even if it means merely the changing of the scenes, the dress, language, the arts, and the veneers of the progressive, mechanistic and material innovations and luxuries of modern living. Today, if a cultural inventory could be fairly taken of our world, we would find that *today* there are places and people who think, believe, live and die, like almost every phase of human existence and development since the beginnings of humanity, and in some isolated places live like the prehumans of the Eurasian ranges.

The historical period of the human race goes back a mere twelve to fifteen thousand years ago according to the scientists, a trifling fraction in the evolution of man. In China, where the earliest historical records are revealed, the country had already achieved inventions, comparable to modern ideas, about nine thousand years ago. The Mesopotamian region, parts of Africa and the Southeastern Asiatic perimeter accomplished most of the principal inventions and tooling between five to six thousand years ago. Steel was probably invented in Southern India. Coal was used in China thousands of years before European gentry realized its use. The earliest wheels came from Sumeria in Asia Minor. The plow probably came from India, perhaps from the Southeastern section. The loom, they say, came from somewhere on the Asiatic continent, making substantial progress in Turkestan and the Near East. Different parts of humanity moved slowly depending on migrating invasions of new continents and the resurgence of old cultures, and the emergence of new cultures like those in Japan, North and South America.

Before we trace the "pockets" and "streams" of human movement, let us keep in mind what the essence or genus of culture really is, as a directional guide to understand these emergences, changes, mobilities of peoples and races.

The question again faces us: what is *Culture*? Every anthropologist and ethnologist, archaeologist, philosopher, sociologist, may have his particular way of interpreting his view and analysis of it. To me it seems that Culture denotes the acquired, accumulated and habituated sum total of experience which people have become accustomed to live within, for, by, and about. While this is obviously a very broad and encompassing view, the basic causal and effectual dynamics of this process are the prime needs to subsist and secondly, the need for power and gratification (whether it be money, property, prestige, wives, or art collections); thirdly, the need for "endless" or "continual" life after death; and fourthly, the need to satisfy the insatiable hunger to overcome peace and stability by the restlessness and strange psychoses of the human emotional system with its oversized brain, too good for its own good. All of these factors take defensive and/or offensive patterns, forms or actions, depending upon the particularized experiences of any people within its own patternizing behaviors and the environmental limitations and possibilities of its almost curvilinear tendencies of settlement, diffusion, and expansion.

According to the scientists, the nucleus of humanity is a historical probability located somewhere in Mongolic Eurasia, although there are scholars and archaeologists who contend that Africa may have been the birthplace of the human race; Basil Davidson states that "tools found in Uganda are the oldest tools ever found anywhere." [14] Very recently in Tanganyika, Africa, at a spot called Olduvai, Dr. L. S. B. Leakey found the skull remains of a pre-human male which may be the earliest evidence yet found, and supports the growing thesis that the earliest habitat of hominoid types of animals was in Africa rather than in Asia. "I call him *Zinjanthropus,* or East Africa Man," writes Dr. Leakey. "He lived more than 600,000 years ago. . . . The possibilities are vast, and Olduavi is just beginning to reveal its secrets and unfold a history sealed for 600,000 years." [15] However, so far as our subject here is concerned, it doesn't matter just where the first hominoids originated from so long as we understand that man-type creatures roamed the world for hundreds of thousands of years and not just since 4004 B.C. as calculated by Bishop Ussher. On the evidence that great hordes of protohominoids and *Homo sapiens* congregated in the Asiatic mainland, one arm of this prehistoric mass of humanity may have spread out to the south and to the southeast. Another might have swept into the vast plains, ranges and valleys of China; a small finger made its way from the north into Japan and the northeastern regions of Asia. Still another could have fingered itself down to the southwest, Asia Minor and Africa. Another long and widening branch swept over the Caucasian steppes to the Balkans and west into the continent of Europe—*the drang nach Westen*—from Mongolia to the Thames. From Mongolia over and across the Bering Straits, then probably land and ice-bound between Asia and Alaska, flowed streams of human migration, spreading down and across Canada, into the valleys and plains and forests of the United States. Still the migrations continued south and spread to Mexico, Central and South America. When the Eskimos migrated from Asia and "settled" in Alaska and the circumpolar regions, they blocked, so they say, further migratory activity. But the evidence of this great Mongolic invasion to the Americas is clearly verified in many ways. Findings on the campus of the University of Alaska are similar to those excavated in Neolithic community sites in the Gobi desert of Mongolia. This migration gave eventual birth to many outstanding cultures in the Americas, as the Mayans, Incas, Aztecs, etc. Also of deserving importance were the confederacies of the American Indians, like the Iroquois, who may have possibly suggested or emphasized to our early colonial pioneers the principles of democracy in action or of a confederacy consisting of a number of sovereign states.

For the student who desires to follow up the various cultures of mankind and the countries where these cultures grew, changed, faded, etc., the amount of knowledge and courses, books, are almost endless. All we are here concerned with regarding the cultures of mankind is their effect upon the individual and his value of Time. With some review of a few of the world cultures, religions, societies, we will find it worthwhile in the attempt to understand our critique of religion and cultural forms and why the rational approach for analyzing these institutions is so essential to us if we are to reach a philosophy equitable to ourselves and the societies in which we live. With this as our sistrum we may continue with our quest.

China is probably the oldest culture we presently know of, predating other cultures by thousands of years, including India and Egypt. According to Ralph Linton: "A protohominoid, *Sinanthropus,* occupied Northern China during the last interglacial. This form was similar in most respects to the Java Man, *Pithecanthropus,* although somewhat closer to modern man in his evolutionary position. On the basis of physical characteristics, any surviving representative of this group would probably be consigned to a zoo. Nevertheless, the behavior of Sinanthropus was much more human than anthropoid. The species used fire and made stone tools which show considerable skill in the techniques of blade-striking. Some of these are so well-shaped as to suggest that they were deliberately designed for use as specialized tools: knives, scrapers, or even projectile points. The species also followed the human practice of cannibalism, unmistakable evidences of which have been found in the single site of this period so far excavated. Since no indication of burials or other ceremonial treatment of corpses has been found, it seems highly probable that they ate their own dead. This practice has been current among several historic peoples, where it was instigated not by hunger but by an understandable desire to keep the virtues of the deceased in the family." [16]

The religious and social structures of prehistoric China were well organized. The priest-king, a sort of half deity, acted as intercessor between the people and the gods. Slavery was not too important because there were so many (and still are) people that it was more economical and less bothersome to hire "free" men rather than to obtain slaves and be responsible for them. As China came into the historic period we find a well-cultured feudal system. The nobles became an oligarchy of war lords, as the Shogunates of Japan, and grew in power, often having more power than the king. The people had many skills, used bronze primarily for tools and weapons. The country was now and then overrun by the Mongol barbarians of the north and west, but the invaders got "lost" and became absorbed into the tremendous human resources of the country. Their religious societies were not fully organized as other world religions but still sufficiently well organized to be carried on from generation to generation, being a generous assortment of celestial beings but the principal ingredient of the religious porridge was ancestor worship, which was fairly general throughout the Asiatic mainland and in the southeastern islands.

China has given us some very practical philosophies and philosophers. Confucianism stands out as a great social and natural philosophy, having nothing to do with supernaturalism and with gods. K'ung Tzu (Confucius) was a teacher of ethics, of individual and social discipline and decorum, and the importance and value of family ties, all this based on the instinctive desire for people to group and help each other. "Confucius," states Professor Frederick Starr, "taught no religion. He avoided discussion of the supernatural. . . . He did not recognize man's relation to deity. He neither claimed to be divine nor founded a religion." [17] Fung Yu-Lan supports this view in his *History of Chinese Philosophy*: "Confucianism is no more a religion than, say, Platonism or Aristotelianism. It is true that the *Four Books* have been the Bible of the Chinese people, but in the Four Books there is no story of creation, and no mention of heaven or hell. . . . Confucianism cannot be considered a religion." [18] " 'Is there one word,' asked K'ung Tzu, 'which could be adopted as a lifelong rule of conduct?' The master replied: 'Is not sympathy the word? Do not do to others, what you would not like yourself.' " [19]

Taoism, a religion-philosophy, was related to pantheism mixed with ancestor-worship, gods and demons. The priesthood never rose in power in China as the Brahmans have in India, or the priest-kings of Judea and Egypt, or as the popes of Rome. Today the predominant philosophy and religion is Buddhism, and Buddha would not recognize it if he got up and lived again!

Buddhism spread through the Himalayan mountains and valleys enroute to China and paused in Tibet. Here it grew into a real mixed-up pudding of Buddhistic Hinduism with a million demons and ogres enough to scare the quills off a porcupine, enshrining and worshipping beliefs, rituals and catechisms completely in opposition to the principles of Buddha. "As the stronger side of the Buddha's teaching was neglected," writes J. E. Ellam, "the debasing belief in rites and ceremonies, charms and incantations, which had been the especial objects of his scorn, began to spread like the Birana weed warmed by a tropical sun in marsh and muddy soil." [20] . . . "Thus it came to pass that, from a system which recognized no Creator, there has developed a religion with priests who claim to hold the keys of heaven and hell." [21] He explains further: "In the teaching of the Buddha, as set forth in the Pali books, there is no doctrine, dogma or theory concerning any First Cause, or origin, of the physical universe, whether by special creation or otherwise." [22] . . . "The idea of a 'soul' as a self-separate entity, was set aside by the Buddha as a fallacy, which cannot be established by appeal to experience, or by any reasoning . . . The notion of a 'soul' is repudiated as a thing wholly impossible even in thought." [23] And "the Buddha did not relegate the responsibility for existence, with all its sorrow and evil, to any supernatural agency." [24]

It would be unjust to attribute to Buddha any philosophy

that had anything to do with religion, with gods, devils, and supernaturalisms. The capital of Tibet is Lhasa and if the people who visited the place can be believed, it appears to be the cesspool of the world. Ellam describes it well: "Lhasa! It is described as 'the metropolis of filth,' and the Chinese are rude enough to say that it is a 'city of devils who feed upon excrement.' . . . Not even in China, they say, can any city be found that is so indescribably dirty. Not only are the streets full of holes into which the unwary may stumble and fall, but there are in the middle of them open cess-pools, specially constructed, and used both by men and women, without concealment. The filth, the stench, the vile abomination of the streets of Lhasa are, from all accounts, beyond anything of the kind in the world." [25]

Japan was prehistorically inhabited by a race of hairy, short, stocky humans called *Ainu,* although most of the ancestors of the Japanese people today probably came over from the mainland, perhaps by way of the Korean peninsula. Here, too, the religion was and still is, to a great but less serious extent, ancestor-worship and nature/animal worship. Here also a feudal and caste system arose. Shintoism, a complicated and involved ancestor/nature religion, is considered the "state" religion, although Zen, contemplative, and other cults of Buddhism prevail and are very popular. The Sumurai were a warrior class who made fighting and protecting their professional careers until it became a caste, being handed down from father to son. The present emperor is supposed to be descended from the sun-goddess, but this myth exploded when Hirohito disclosed to his people that he is tired of being called a god and wants to regain his human standing so he can begin to enjoy life like anybody else.

India may have been the first country into which the prehuman proto-Australoid hordes poured to occupy this land a long, long time ago. Later, much later, the Aryan people from the north swept over the face of India and also streamed across the southwest towards the Mesopotamian valleys. The social and religious organizations of India were always and still are highly complex, with additives and mixtures of every description from soup to nuts. The Vedic period brought about the various caste systems and sprouted asceticism of all kinds and degrees. The Vedas believed in many gods, too numerous to list here. *Agni* was the fire-god; *Indra,* sky-god; *Vishnu,* the sun. *Varuna* was another sky-god and *Siva* came later as another god of the heavens. While there were thousands of higher, medium, and lesser gods, the Hindus considered all these gods as partial manifestations of a Universal Whole or One Cosmic Deity. This can be recognized as a sort of pantheistic pantheon.

The *Brahmans* were priests who really organized themselves into a first class holy clan, created themselves into a sacred caste of divines who could be expected to intercede with the gods or have direct communion with them. They became a holy segment of Indian society and still are. The Brahmans rate at the lower rung of the godly ladder but above the nobles and the rulers and are so listed on the pagoda levels. The Hindus figure it this way: "The universe is subject to the gods; the gods are subject to the spells (mantras); the spells are subject to the Brahmans; therefore the Brahmans are our gods." [26] So sacred is a Brahman that "he who gives water or shoes to a Brahman will find water to refresh him and shoes to wear on the journey to the next world, while the gift of a present house will secure him a future palace." [27] "In the Brahmanic mysticism the important point is not so much redemption from the misery of existence and liberation from the world as the experience of being exalted above the world in union with the Brahman." [28]

The Hindus are deeply steeped in ancestor worship and this is tied to the belief of *Karma* which is the reincarnation of souls in the form of lower or higher animals, lower or higher human castes according to the merit system, of the experiences and characteristics of each life determining subsequent reincarnations of the spirit as it travels from one life to another, striving to achieve the highest unity with the Universal One. In a way this can be related to the Catholic purgatory. "For faults committed in a previous existence," writes Edward B. Tylor, "men are afflicted with deformities, the stealer of food shall be dyspeptic, the scandal-monger shall have foul breath, the horse-stealer shall go lame, and in consequence of their deeds, men shall be born idiots, blind, deaf and dumb, misshaped, and thus despised of good men. After expiation of their wickedness in the hells of torment, the murderer of a Brahman may pass into a wild beast or pariah; he who adulterously dishonors his *guru* or spiritual father shall be a hundred times re-born as grass, a bush, a creeper, a carrion bird, a beast of prey; the cruel shall become bloodthirsty beasts; stealers of grain and meat shall turn into rats and vultures; the thief who took dyed garments, kitchen herbs, or perfumes, shall become accordingly a red partridge, a peacock, or a muskrat." [29] Thus, each caste is resigned to its own fate, as inescapable and part of the program. To even try to better themselves might confuse the records and set them back even worse. The very low castes, to properly fulfill their mission, will purposely keep away from you so as not to pollute the air you breathe. So low can a human being think of himself that if you were a member of the Rodiya caste in Ceylon, you would know that nothing is lower than you. Ernest Crawley, in his celebrated *Mystic Rose,* writes: "On the approach of a traveler they would shout, to warn him to stop till they could get off the road and allow him to pass without risk of too close proximity to their persons." [30]

Of course, as a result of this thinking, there is nothing in nature without a soul. Every blade of grass, every bug that flies or creeps, every ant, every reptile, is possessed of a soul that is in some degree or stage of punishment or redemption enroute to heaven, maybe. Poisonous snakes roam freely all over the country since no one likes to kill something that might be his paternal grandfather, or swat a Malaria-carrying mosquito because it might be his uncle just trying to make a social visit. "A curious commentary," writes Tylor, "on the Hindu working out of the conception of plant-souls is to be found in a passage in a 17th century work, which describes certain Brahmans of the Coromandel coast as eating fruit, but being careful not to pull the plants up by the roots, lest they should dislodge a soul." [31] Souls are always coming and going and must be constantly watched as they might go off for a walk or stray without knowing how to get back to the owner. Frazer tells us that "when anyone yawns in their presence the Hindus always snap their fingers, believing that this will hinder the soul from issuing through the open mouth." [32] A soul can wander off while one is sleeping and the dream is proof that it did; for this reason it is considered unlucky and dangerous to awaken a sleeper who may be dreaming, for the ectoplasmic thread attached to his meandering soul may be broken and the soul lost. Children, having young and delicate souls, are not beaten for this reason so that their souls may not be hurt while they are still growing. In Bali, Covarrubias tells us, "there is the belief that a child is the reincarnation of an ancestor whose life-giving spirit comes down to earth in the form of dew, which is inadvertently eaten by the parents, the process of gestation taking place after intercourse." [33]

Cremation of the dead is fairly universal throughout India and in those neighboring countries and islands where Hinduism still prevails as the religion of the people. While this is very practical inasmuch as there seems to be more people than land, the significance of burning is to purify the soul and so that it can properly reincarnate or be received by the gods, otherwise it may become a roaming ghost tantalizing its relatives or strangers. For thousands of years the custom of *Suttee* or *Sati,* that is, the tradition of the wife or wives being burned alive in the cremation fire of the deceased husband, was widespread in India until prohibited a little way back by British law. "The Brahmans were especially interested and insistent in advising wealthy widows to perish with their husbands; it was usually they who were the benefactors of the widow's estate." [34]

I observed this cremation ceremony of purification in Bali, where Hinduism and Brahmanism still prevail as the religion of the Balinese. There are here several kinds of priesthoods, the *Pemanakus* who tend to the temples and arrange the daily offerings to the gods; the *Sungguhu,* whose business is to keep the devils busy and appeased; the *Balians* who specialize in magic and fortune-telling; and of course, the *Brahmana* caste of priests who are next to the gods and will be gods themselves when they die. The dead have to be very patient and often wait in their graves for many years before the local village saves enough money to arrange this elaborate and highly decorative ritual, but finally come good fortune their bones are exhumed in white cloth, put into beautifully-painted hollow white cows and cremated, thus releasing the ghosts and purifying their souls. According to the Balinese, a person's soul is traveling from one life to another until it can properly be evolved into a blessed and pure status to ensure bliss and eternal happiness. "Their love of display often goes to extremes," continues Covarrubias, "as in the case of the costly towers, biers, and other accessories for the cremation of their dead, which are destroyed in a few minutes after hundreds of guilders and months of labor are spent to produce them." [35] This reminds me so much of Catholic villages where one can see poverty all around him and surrounding the central Church, the insides of which are sheathed with gold and silver and the place loaded with treasures and wealth. In Bali each little village prides itself on its temples which are walled-in open affairs, open to the skies and the stars, devoted mainly to the feeding, adoration and worship of the family ancestral spirits. The head patriarch of the village is considered as the village god and intercessor for the people in their need for favors and good fortune from the spirits and gods. The women carry things on their heads, the men carry things on their shoulders or in their hands. In eating, no dishes or cutlery are used as these things are considered unclean and an imported bad habit. Here, too, are castes and each lower caste respects the higher caste and will always bend to a lower level. Higher is always better and lower is always worse. The gods live up in the high mountains; the demons and evil gods live in the sea. "A Balinese observes the rank of his head in relation to the rest of his body, and for this reason no one would stand on his head or take any position that would place his feet on a higher level. It is an offense even to pat a small child on the head and there is no worse insult than 'I'll beat your head!' " [36] My time was well-spent on the island of Bali, as I enjoyed the people immensely and found their culture, their crafts, their music and dances, most inspiring and profoundly interesting. To see a boy leading his little phalanx of parading ducks homeward bound from the rice fields is a picture one can never forget, and to see the children do the monkey dance under the stars is a performance that lingers for a lifetime in one's memories. These things have more color, more meaning, and bring greater enjoyment and contemplation than the fast-moving, constantly producing with incredible speeds, technics of the mechanized, dehumanized cultures of the Western world.

Among many of the weird and terrible gods and goddesses of the Indian pantheon is the goddess *Kali* or *Kalee-Ma'ee,* associated with Siva in the general sex symbology of Indian culture. Goats were regularly decapitated and the blood spilled over her figure. Allen Edwardes describes her well: "Kalee Ma'ee the Dark Mother . . . is luminous black. Her four limbs are outstretched and the hands grasp two-edged swords, tools of disembowelment, and human heads. Her hands are blood-red and glaring eyes red-centered, and her blood-red tongue protrudes over huge pointed breasts, reaching down to a rotund little stomach. Her yoni is large and protuberant. Her matted, tangled hair is gore-stained and her fang-like teeth gleam. There is a garland of skulls about her neck; her earrings are the images of dead men and her girdle is a chain of venomous snakes." [37] Nice girl!

Like the cultures all over the world, the sex act and the mystery of creation were closely associated with religious and social rites and practices. No other theme of human experience offered such an intriguing mystery as the act that created life and which was a gift of the gods. "From Egypt to India, the statues featured conspicuous genitals. Shiva in India and Osiris in Egypt were seen testimonially clutching enormous phalli. And from China humorously refined pornography depicted an Oriental Hercules with gigantic member strongly erect, shattering a copper pot with one blow." [38] Many of the temples of India depicted sculpture demonstrating almost every possible method of sexual intercourse, which to them was not shameful or vulgar but a presentation of what the gods ordained and exemplified themselves. Max-Pol Fouchet, in his brilliant work, *The Erotic Sculpture of India,* writes: "And the behavior of these gods! Quite often far from transcendental! What with their battling and knavery, their quarrels about precedence and property, there is much that they could teach the gods of Greek mythology. Siva, for instance, cuts off one of Brahma's four beards with his sword and throws it into the Ganges. And the goddess Parvati deceives Siva, her husband, with Agni, god of fire, whom she hides in her body, just as if it were a wardrobe in vaudeville. Krishna, one of the incarnations of Vishnu, whisks the veils off some cowgirls when they are bathing, gazes upon their nakedness, and bewitches them with his flute-playing. And the Aryan gods might even be charged with racial prejudice, for they refuse to invite the Dravidian Siva in their festivities, yet bow before him in a somewhat cowardly manner when he threatens to wipe out the universe. They fly into a rage, create monsters and set them loose, then wonder how to deal with them when their anger subsides. So they have to join their breaths together and create a goddess to protect them from the buffalo demon . . . there is no division between gods and men, particularly as regards physical love." [39]

Like almost in every nook and corner of the earth, man enslaved woman and held her to be no more than a chattel and a medium of pleasure and very often his cultures and religions reflected this attitude in more ways than one. The Brahmans and other holy men took priorities with the temple prostitutes and with the general female population which considered it an honor and a blessing to be used by these intercessors. "In many parts of India, midday was the signal for certain holy men, *gooroos, saddhoos,* and others, to leave the temples, ringing bells and beating drums and blowing conch shells, summoning barren women to come to fondle their privy members in order to become fertile." [40] Socially, a woman was a slave, mentally and religiously and politically suppressed. Regarding the status of women in the Indian religion, in the *Code of Menu* we find the following: "There is no other god on earth for a woman than her husband. The most excellent of all the good works that she can do is to seek to please him by manifesting perfect obedience to him. Therein should lie her sole rule of life. Be her husband deformed, aged, infirm, offensive in his manners; let him also be choleric, debauched, immoral, a drunkard, a gambler; let him frequent places of ill repute, live in open sin with other women, have no affection whatever for his home; be blind, deaf, dumb, or crippled; in a word—let his defects be what they may, let his wickedness be what it may—a wife should always look upon him as her god, should lavish on him all her attention and care: paying no heed whatsoever to his character and giving him no cause whatsoever for displeasure. A woman is made to obey at every stage of her existence." [41] No wonder James G. Frazer wrote that "if women ever created gods, they would be more likely to give them masculine than feminine features. Men make gods and women worship them." [42]

Another interesting feature of the Indian religion and culture is their methods of purification, absolution, and with it the adoration of the sacredness of the *cow.* This is one country where the cow gets a break, leads a life of leisure and goes where she pleases—until a Moslem gets ahold of her if the Hindus are not looking! Cow urine and excreta are not usually wasted into the ground but considered holy and great purifiers of the body and spirit. "Brahmans and other holy men in the fields or bazaars gathered hallowed cow urine *(moot)* in their brass *lotahs* or, catching and sipping it in cupped hands, smeared it over face and limbs. Thus imbibed, the water of Mother Cow was believed to cure all ills and obliterate bodily sin, unintentional or deliberate. Directly after a Brahman committed worldly acts he sipped of *moot* or purified himself by drinking other liquids consecrated to the gods. Cow dung *(upuleypun)* was also considered a sacred, purifying and healing substance. It was daubed fresh and moist over the countenances and in the mouths of penitent holy men; otherwise, fashioned and dried into chips or cakes *(uplah)*, it was used as fuel, wall plaster, or flooring, or in the making of idols. Ashes of cow dung *(bhuboot)* was a favorite powder of *saddhoos* and other religious mendicants who would roll in it, spray their hair, and rub it on their foreheads. Urine baths were also used for external purifications." [43]

Before we leave India we should note the tremendous amount of scientific discoveries, inventions, mathematical and mechanical knowledge, that the people had acquired in spite of all this Brahmanism. "The Hindu people worked out, among other things," writes Muthu, "the atomic theory, the evolution theory, the theory of motion, of gravity, of sound, of light and heat potential, the presence of ether, the diurnal motion of the earth, the principle of differential calculus, the calculation of lunar periods and eclipses, the humoral theory, and the circulation of the blood, centuries before they became known to Europe. They were the inventors of numerical and decimal symbols, of mathematics and algebra, of chess and fables now used all over the world. They were the first to practice dissection of the human body, first to provide dispensaries and hospitals for men and animals, first to employ minerals and mercury internally, first to introduce skin-grafting and plastic surgery, cataract-crouching, and major operations like amputation, lithotomy, Caesarean section, etc." [44] Even if a fair part of these things are true and historical it is still a formidable achievement.

Ancient Egypt, the archaeologist's delight, was a theocracy of priestcrafts and deified pharaohs. The idea that the welfare, health and good fortune of the state and people are dependent upon the gods or their earthly god-agents, that is, the king, chief or priest, was fanatically accepted by the people and became part of the cultural frame. This same religious theme is very common in many parts of the world, especially in most parts of Africa, from the Barbary Coast to the Cape of Good Hope. Most probably this is how kings started in the first place, gradually getting more and more power until the king was ruler over the people rather than a medium for the safety and welfare of the country. J. H. Denison tells about "a king in Central Africa who cannot be approached by strangers until they have bathed for two successive days and until pepper has been rubbed in their eyes." [45]

The ancient Egyptians worked hard and long to build great monuments for the gods and colossal tombs for the mortal kingly or priestly "deities" because these things would be a way of continuing their own security and progress. "We seem to be justified," writes Frazer, "in inferring that in many parts of the world the king is the lineal successor of the old magician or medicine-man." [46] Magic, a worldwide prerogative of peoples everywhere and the asexual parent of all religions, was really organized into top echelon management by the priesthoods of Egypt who made a speciality of magical procedures and utterances which *they only* were capable of doing or saying and which the gods had no alternative except to obey. The Egyptian *Book of the Dead* is really volumes upon volumes of these magical prayers or "commands" and those who possessed the secrets of magical words could not only "talk" to the gods but make them his servants.

Magic is not only the forerunner of religion but it also "mothered" the whole idea of prayer and other forms of attempting to communicate with the dead, the spirits, with ancestors, with gods, or to be accredited by the gods with good intentions for future judgments. Dr. Maspero, the celebrated Egyptologist, clearly states that "Ancient magic was the very foundation of religion." [47] Frazer adds that "Magic is older than religion in the history of humanity." [48] Besides, "it must always be remembered that every single profession and claim put forward by the magician as such is false; not one of them can be maintained without deception, conscious or unconscious." [49] So powerful was the "sacred" word of magic considered that "even the sun's course in the heavens could be stayed by a word." [50] Man's language, written and spoken, has been his key to pass through the chambers of the gods and goddesses and talk to them "in their own language," that is, the language the man understood himself. As Sir Wallis Budge points out, "There is no doubt that the object of every religious text ever written on tomb, stele, amulet, coffin, papyrus, etc., was to bring the gods under the power of the deceased, so that he might be able to compel them to do his will." [51] There was no shortage of gods and goddesses either in the Egyptian pantheon of celestial beings. "The names of at least 2,000 Egyptian gods are known to us yet there were none of these whom their worshippers regarded with real affection or before whom they felt genuinely powerless. Every deity could be circumvented and controlled if one could only learn the words of power . . . [as a result] there was a complete union of church and state, with a correspondingly complete control of the subjects' bodies and minds." [52] Thus we can see that the Egyptian didn't see his gods with awe or abject fear but merely as beings operating outside the sphere of mortal limitations and which sphere could be brought "within" the mortal field for the protection and continuity of one's life. So, wrote Bronislaw Malinowski, "Between myth and nature two links must be interpolated: man's pragmatic interest in certain aspects of the outer world, and his need of supplementing rational and empirical control of certain phenomena by magic." [53]

The Egyptians had all kinds of animal, insect and reptile gods. Many birds became very important gods. To *Bast,* the cat-goddess, they even created and erected a whole city! "The worship of sacred animals," writes Dr. W. Oldfield Howey, "commenced in Egypt before the dawn of history and survived for many thousands of years in conjunction with other creeds." [54] Of course, the sun and giver of life, like in many other religions, was top banana; *Ra* was the sun-god. Osiris, Isis, and Horus, the regulation trinity so commonly found in so many religions. The *Ba,* in the form of a bird, was the soul of a person, leaving the body at the instant of death and traveling to the world of spirits. The *Ka,* or spiritual double, lived apart from the body during the mortal existence of the person but joined the body after death. It was to this combination of dead body and live spirit that the priests directed their magical commands during the funeral and mummification ceremonies in order to give the body all the functional powers it had during its mortal life. So long as the body existed the *Ka* had a home to roost and was there too. If the body disintegrated the *Ka,* now homeless, was also lost. As a result, mummification of bodies became an important business and the celebrated art of the ancient Egyptians. To elaborate this point, Frazer states: "Man has created gods in his own likeness and being himself mortal he has naturally supposed his creatures (the gods) to be in the same sad predicament. . . . The great gods of Egypt themselves were not exempt from the common lot. They too grew old and died. But when at a later time the discovery of the art of embalming gave a new lease on life to the souls of the dead by preserving their bodies for an indefinite time from corruption, the deities were permitted to share the benefit of an invention which held out to gods as well as to men a reasonable hope of immortality. The mummy of Osiris was to be seen at Mendes; Thinis boasted of the mummy of Anhouri; and Heliopolis rejoiced in the possession of that of Toumou. The high gods of Babylon also, though they appeared to their worshippers only in dreams and visions, were conceived to be human in their bodily shape, human in their passions and human in their fate; for like men they were born into the world, and like men they loved and fought and died." [55]

It is interesting to note that the animal trait of human nature that considers the existence of the body as the existence of life may be strangely tied to the origins of mummification. Robert Briffault tells us that "monkey mothers carry about the putrefying corpses of their dead babies, not being able to comprehend what has happened, and expecting the visible form to wake up from sleep. Savage women in Australia do the same thing. All primitive peoples believe that something of man survives as long as visible portions of his body remain." [56]

The ancient Egyptian State, being a hierarchy of priests with a deified head-priest, the Pharaoh, was as dogmatic and inflexible as a theocracy could be, almost as theocratic as Ireland today. The whole population was in actuality and in their culture an army of religious slaves for the priesthoods. A person's fate or destiny was predetermined before he was born and there was no way of changing it—*Fatalism*. They also believed fanatically in the idea that the gods or divine powers made known their wishes or gave advice or warnings through the medium of dreaming, and whatever they dreamt they painted and many of the finest frescoes of Egyptian art are stories of what they saw in their dreams. Of course, the content of every dream was part of reality and when they dreamt of another world, that, in itself, proved the existence of it. All in all, the Egyptians had an amazing culture and it became a sort of a supply depot for later religions to take as much of it as pleased them to incorporate in their own newer religions. "There is indeed," writes Professor John P. Mahaffy, "hardly a great or fruitful idea in the Jewish or Christian systems, which has not its analogy in the ancient Egyptian faith." [57] The Jews and Christians have to be thankful, too, to the Egyptians for their own written languages because it was out of the Egyptian hieroglyphic characters that the Phoenicians and other Semitic people got their alphabet and from which, in turn, the Greeks, Romans, etc., arrived at theirs.[58] There seems to be no better way to close our short trip to ancient Egypt than to quote a brilliant line from Winwood Reade's *Martyrdom of Man*: "Egypt stood still and Theology turned her into stone." [59]

However, the Egyptian story and culture are two of the greatest in the world and of profound and wide influence upon the surrounding countries, their cultures and religions. They advanced the science of agriculture, many crafts of painting, sculpturing, engraving, metal-working, weaving, to a marked degree. In analyzing and trying to understand the motives and the influences as well as the actual ritualisms of their religion, or any other religion or culture, it is not our purpose or intent to ridicule by sophistry or satire; our intent is to try to obtain some sort of realistic picture, if we can, of the movement and growth of these religions and cultures so that we can so much better understand the meanings of our *own* cultures and religions *today* which, after all, evolved out of the ancient ones and as one unbroken chain, regardless of the many changes, connections and modifications enroute, in the story of man's religious and cultural heritage and experience.

I should repeat that the purpose of this chapter is not to attempt to present an adequate ethnological story of these countries and cultures but, for the purposes of our premise, to give the reader a "smattering" of cultural identities reasonably sufficient for him to get a fair idea of past human history as it went on in different parts of the world, so that this can be used as a base for the general premise of this book, the reevaluation of the Time factor in our lives. For those who are intrigued with this most dramatic and fascinating of all stories, there are hundreds and hundreds of excellent and thrilling books by great scholars and investigators that comprise an education in themselves; there are also many lecture series and courses in the museums, universities and forums of the country, most of which are fully available to the public. To know one's past is to understand one's present and the possibilities and hopes of the future. However, it would take too much space to even briefly survey *all* the countries, religions, and cultures, in the history of man, but we will just have to be satisfied with a few predominant and salient ones.

Somewhere between 2000 and 1500 B.C. hordes of "Indo-Europeans" of mongol origin swept down from the central mountain ranges of Eurasia, from the Balkans to Burma, and overran and conquered the peoples to the south. The invaders were called *Arya,* from which originated the ethnic term Aryan, a far cry from Hitler's Germanic waxenhaired "superman" homosexuals from the Rhine, quite different from the Belgi and Gauls of Central Europe in the pre-Roman days. These invasions and mixing formed a good part of the Turko-Tatar races who settled large parts of the Continent, including eastern and southeastern Europe, Asia Minor, the Near East in general and the countries lying between the Mediterranean and Western India. They invented many ingenious things, among them the loom, the plow (which some scholars claim to have originated in southeastern Asia), and the potter's wheel.

The *Semites,* spread all over from North Africa to eastern Asia Minor, also brought about many outstanding inventions, including the smelting of metal alloys, revolutionizing techniques and innovations which hastened the permutations of the various cultures and sprinkled their influences and seed throughout the Mediterranean world and later spilled over into western and northern Europe. The Alphabet was created here, which gradually evolved our form of writing. The Semitic cultural trait of expecting a bride to be chaste permeated the customs of this entire region and still persists in the countries bordering the Mediterranean, including the Catholic and Greek Orthodox countries of this area, and laid the foundation for the later prudish ideas of Christianity in which the "purity" and "virginity" of women were basic requirements of "proper" and "gracious" living. In South Italy and Sicily to this day there is the custom of the groom's parents stealing into the bridal chamber the next morning to assure themselves by the blood on the bedsheets that their good boy married a virgin!

Circumcision of both boys and girls, men and women, was practiced, in part or whole, by all the Semites. The Jews were not the only Semites; the list is long and includes most of the peoples of North Africa, Sumeria, Assyria, Phoenicia, part of the Hittites, Babylonia, Carthage, later the Moors, the Arabs, among others too numerous to mention here. Circumcision was a cultural practice predating both ancient Judea and Islam; it goes back thousands of years before the legendary time of Christ and is carried on today in the jungles of the Belgian Congo and below the snows of Kilimanjaro. "The Jews adopted circumcision (*meshookim*) from the ancient Egyptian priestly ritual." [60] The patrilineal culture of the Arya was also part of the Semites. Patriarchy was firmly established and institutionalized way back in lithic times, all the way from the family to the priesthood and state. This strong patrilineal custom is evident in the Jewish people right up to the present time. Ralph Linton states: "The family control was rigidly patriarchal. The father had complete control over his wives and sons throughout his life and even beyond. A father's blessing was an important asset, while a father's curse could ruin his son's future. . . . The average Semitic father seems to have taken more pride in his sternness than in his justice. The son's attitude toward him was one of fear and respect. . . . The whole situation was such as might be expected to develop a strong and censorious superego in the individual. The Hebrew picture of an all-powerful deity who could only be placated by complete submission and protestations of devotion, no matter how unjust his acts might appear, was a direct outgrowth of this general

Semitic family situation. . . . The Hebrew Iaveh was a portrait of the Semitic father with his patriarchal authoritarian qualities abstracted and exaggerated. The combination of patriarchal suppression and sexual deprivation has left its mark on the Semitic basic personality. From Moses to Freud, Semites have been preoccupied with sin and sex." [61] This has much to do later on with the Christian attitudes towards woman and sex.

With some modifications the story of Adam and Eve and the story of the Genesis were taken from the Babylonian beliefs. The code of Hammurabi strongly influenced the judicial and social controls of the Jewish ways of life, in the same manner as the dominating influences of Babylonia affected the whole Mesopotamian region and all the Semitic peoples and nations in that area. For myself there are probably thousands of more beautiful and far less cruel "beginning of the world" stories that I would rather like to read and allow children to read. The Hindu genesis of Shiva with Adamyi and Yeva is a sweet, peaceful and enjoyable fairy-tale that is human-like and worldly-wise. From tribes and peoples all over the world, small and large, come stories of how the world began and many of them are really beautiful poetical tales of primitive fantasy and imagination. The idea of Yahweh creating man out of clay was a regional idea widespread at that time and not particularly original to the Hebrews. "The creation of man from clay," writes Professor H. R. Hays, "was an idea common to Babylonia, Egypt, and Greece, as well as Israel. . . . Creation from clay was a myth also found in Tahiti, New Zealand, Australia, and among North American Indians. In many cases the clay was described as red. The Hebrew word for Man, *adam,* was similar to that for the ground, *adamah,* and the word for red was *adom.*" [62] We should also know that the "Ten Commandments," as Clyde Kluckhorn informs us, "are seen to be derived from the earlier code of Hammurabi, a Babylonian king. Some of the *Book of Proverbs* is taken from the wisdom of the Egyptians who lived more than two thousand years before Christ." [63] E. O. James tells us that "the Book of the Covenant in Exodus xxi-xxiii. 19 is now recognized to have been a fragment of a much longer Hebrew analogue to the Code of Hammurabi, comparable to similar Hittite and Assyrian legal codes, and belonging to a more or less common corpus of ancient customary laws current probably in Palestine and elsewhere in Western Asia at the time of the Israelite settlement in Canaan between 1500 and 1000 B.C." [64]

The Jewish religion was not handed down from behind a burning bush but evolved out of the same cultures, social and religious mores, magics, and superstitions, as the rest of the Semitic people in that part of the world. The religion of the early Hebrews was *monolatry,* not monotheism; the existence of other gods besides Yahweh was taken as a matter of common belief and acceptance. In his *Hebrew Origins,* Dr. Theophile Meek writes that "the early Hebrews, in so far as we are able to discover, were no more monotheistic than any other ancient people in the same stage of development." [65] The Hebrews of ancient Judea did not deny the existence of other gods besides their own Iaveh, or Yahweh, later Jehovah, who had a history of his own all the way back from the good old days of Assyria. "In the earlier phases of Hebrew religion, however, there can be little doubt that the attributes of Yahweh were in many respects identical with those of Baal or Hadad elsewhere in the surrounding region." [66] . . . "Primarily Yahweh has every appearance of having been a desert deity who was possibly worshipped among the Kenites before he was encountered by Moses." [67] . . . "In spite of the Palestinian colouring of some of the sources of the Yahwistic narrative, the creation and deluge stories are fundamentally Mesopotamian in their origin and setting. It was, therefore, in the Sumerian and Babylonian cosmology that the Hebrew conception of creation and the cosmic order was rooted and grounded, though in coming to fruition in Palestine as the higher religion of Israel, the ancient stories were expurgated, arranged and adapted by the Yahwist and Priestly editors to serve the purposes of its fundamental doctrine of ethical monotheism." [68]

Even the Ten Commandments do not imply the existence of only one god; it merely "commands" the Jews to worship Yahweh and that other gods should not be worshipped as he is a very "jealous" god. " 'Yahweh' is an abbreviation of an original *Yahweh-'asher-yihweh,* which could be translated 'he causes to be what comes to pass,' i.e., he is the continuous creator of all that daily comes to pass. According to the author of Ex. 3:14, this apparently was the explanation of the divine name and this was the new revelation that came to Moses; and that may possibly be true, because in it we have an old Egyptian liturgic formula that must have been well known to Moses and a conception that is characteristically Egyptian, the idea of a god who continually brings about all that day by day comes to pass, the ever-present creator and guide of all . . . it was the old god that Moses took over, but the old god in a new dress borrowed from Egypt." [69]

History reveals that while the Semitic peoples substantially evolved the idea of monotheism in the evolution of religion, and that while the Jews had their Yahweh, the other Semitic peoples had their own "one" god religions. Even Akhanaton, the Egyptian Pharaoh, tried to enforce monotheism upon his people but he was weak and the priests were strong. This monotheistic theme of "One Supreme Being" is saturated all through the Semitic field and later expressed itself in Islam's Allah. "What is often described as the natural tendency of Semitic religion towards ethical monotheism, is in the main nothing more than a consequence of the alliance of religion with monarchy." [70]

The Arab, like the Jew, believes in circumcision and does not eat pork. And yet Lane reports that "the Jews are regarded as so unclean by the Moslems that their blood would defile a sword, and therefore they are never beheaded." [71] "Mohammed would not eat lizards, because he thought them the offspring of a metamorphosed clan of Israelites." [72] On the other hand, Rudwin states that "the Jews taught that all other races descended from the demons, while the other races believed that all Jews had horns." [73]

Though monotheism is accepted in principle, many pagan and magical forms are accepted in practice, knowingly or unknowingly. The *Torah* (the Law) itself is a sacred and divine creation or object and it is carried about with reverence like the ancients carried around their idols. The *mazoozah* is a pagan vehicle of adoration and magical power to ward off evil spirits from the household, very common in the ancient world. We still spit three times with one eye closed to chase away or "neutralize" the evil eye, and the Jewish mother still fears anyone who looks staringly too long at her child; and before baking bread, she throws into the oven a small piece of dough to bribe and mislead the ogre who might pounce upon her loaf. The making of *matzohs* was not a "patented" idea of the Exodus but was a common method of making bread-cakes in the desert by nomadic people all over Asia Minor and the Mesopotamian region. After a funeral the Jewish people wash their hands and face to "clean" their souls and little stones are put on the gravestones to keep away the evil spirits or defend the living from the haunting ghosts. "The pre-prophetic beliefs of the Hebrews about the dead and their requirements in and beyond the grave were much the same as those that had always prevailed in Palestine and the rest of Western Asia. The importance attached to lamentation, laceration and other mourning rites, and to the disposal of the body and its provision with offerings of food and drink, platters, jars, bowls and lamps —women with beads and other ornaments and men with their weapons —attest to the belief that the afterlife was conceived in relation to an earthly pattern in which these gifts were essential requirements." [74]

When a Jew sneezes his brethren say, "God bless you!" because he just got rid of a small army of little imps or devils, a very common superstition and generally practiced among all Christian peoples. The very orthodox keep their windows tightly closed on the Sabbath so that "bad spirits" may not enter the sanctum of the home on the holy day and so defile it. We can go on and on with a thousand more little superstitions most of them predating the Jewish religion and many additives in their rubbing-shoulders tour as they carried on their faith for thousands of years in so many parts of the world. I met a Jewish Chinaman in Hongkong who never heard of blintzes or borscht and I've met many inferiority-complexed "high class" German Jews who feel they are going to end up in Valhalla on Woden's lap instead of in the *Genáydin* with Yahweh and the other Russian Jews. "Jews have mixed so much with the varying physical types of the different countries in which they have lived that by no single physical or physiological feature nor by any group of such features can they be distinguished as a race." [75]

The *Mogen Dovid,* like the other religious symbols of the Cross, Lingam, Swastika, like the Egyptian sistrum sign of eternal life, is a phallic emblem. Phallicism is very much stressed in Semitic cultures because of the emphasis on sex and the slave role of the female. In most of the Semitic historical religions the women had little or no participation in the ceremonies and rituals of the creeds. In Bali, where Hinduism prevails, a man isn't bound to be faithful to his wife, nor is he expected to, but if any of his wives should be unfaithful it is considered a capital crime; if a Balinese man wants to get rid of his wife and divorce her all he need do is to claim that she is lazy or quarrelsome. "In [ancient] Chaldea a man could divorce his wife by saying, 'Thou art not my wife,' by repaying her dowry, and giving her a letter to her father. If she said to him, 'Thou art not my husband,' she was drowned."[76] Similarly, the Jewish attitude toward women was also on a double-standard basis; if a Jewish husband wanted to divorce his wife he need not give any reason at all, if he decided to, while his wife had no rights to divorce unless it was given *voluntarily* by the husband. "The Mosaic law demanded purity from girls, but not from youths. It was one of the biggest inequalities between the sexes. In this respect man was undoubtedly privileged. He was entitled to return his bride to her father if the bridal night showed that she had not come to his bed a virgin. It was not even necessary to divorce her; there had been a species of fraud, so the marriage was invalid."[77] In Judea the women were technically not even members of the religion or tribe, which is more or less a patriarchal and patrilineal affair. The orthodox Jewish religion allows and accepts the institution of polygamy; where the laws or customs of the land did not allow this, as in Christian Europe and America, the Jew had to abide with these restrictions and he gradually became accustomed to a marriage with only one wife. One of the present problems of the State of Israel is the immigration of Jewish families from North Africa, Yemen and other parts, where the Jewish man usually arrives with more than one wife while the adopted custom of Israel is the European custom of only one wife to one man.

The participation of women in the Jewish temples is unorthodox and simply the result of the general cultural change of modern times in European and Westernized countries, and the emergence of a greater, higher level of the social status of women. There is no ceremony or ritual celebrating the birth of a Jewish girl; the new celebration of *Bas-Mitzvah* for girls is just an added vehicle of revenue for the Reformed and other less orthodox temples, and a way of satisfying the demands of the women for equal suffrage. Yahweh, the Great Father, really never gave his consent but, like the other Old Timers, simply can't buck the style and trend of the march of progress. Fifty years ago the Christian churches wouldn't tolerate and even publicly opposed *any* form of gambling. Today, they have even forced God to play Bingo in Brooklyn, learn the best methods of using little children to panhandle raffle books in Suburbia, and share the profits of the numbers racket in Panama!

The *Kaddish* or prayer for the dead does not mention the dead; it only glorifies the Lord and keeps repeating the glorification with awe and reverence. This "awe" and "reverence" make up the constant and daily schedule of the very religious Jew. To indicate the depth of this "reverence" here is a typical story of a Jewish man who was so desirous of making sure he "awed" and feared his God that he just went quite out of his way to prove it. In one of the *Hasidic Legends* we find the story of Susia:

> "Once Susia prayed to God: 'Lord, I love you so much, but I do not fear you enough! Lord, I love you so much, but I do not fear you enough! Let me stand in awe of you as one of your angels, who are penetrated by your awefilled name!' And God heard his prayer, and his name penetrated the hidden heart of Susia, as it comes to pass with the angels. But at that Susia crawled under the bed like a little dog, and animal fear shook him until he howled: 'Lord, let me love you like Susia again!' And God heard him this time also." [78]

The early Christians "stole" the religious concept of *free will* from the Hebrews, a concept that hardly any neurologist or psychiatrist could possibly accept unless he, too, is a religious fanatic and has waived away all his knowledge of the sciences. Here is the Hebraic idea of free will as expressed by Maimonides, the great Hebrew philosopher: "Man has been given free will: if he wishes to turn toward the good way and to be righteous, the power is in his own hands; if he wishes to turn toward the evil way and be wicked, the power is likewise in his own hands. . . . There is none who forces him, none who determines him, none who draws him toward the one or the other of the two ways. It is he himself, who, out of his own volition, turns to the path he wishes to take." [79] Maimonides was truly a great man, a great teacher and a fine scholar, but when it came to understanding the *will* of human nature it's too bad he couldn't have had Spinoza for his teacher. Our "free will" is composed of our heritage, our acquired experiences, our ability to translate both intelligently into decisions and actions, and these, in turn, are subject to our environment and the circumstances and events which make up the experience of living. It doesn't mean that whatever anyone does, good or bad, is justifiable or punishable; this is a question of judgments which in themselves are questionable and open to inquiry. It does mean that they are *explainable* and if we "know" what is good then we must proceed to *teach* and *cultivate* the explanation of goodness so that this identification of good is understandable and becomes part of one's knowledge and experience. If we "know" what is bad then we should expose it in such manner as will explain and teach people to withdraw from it or not to commit it for their own best interests and gradually it will come to people's minds that thinking and doing good or evil are not questions of choice but of judgments and that these judgments depend upon what we are from what we have been and what we could be from what we understand of ourselves and others.

We *think* we have the choice of picking our own destiny, but it is not so. *Destiny* is in every step we take and every new step is not far from the one we took before—this is the law of *causality* and anyone who really understands it realizes that we are all a series of causes and effects spontaneously generating new causes and effects by the nature of existence and experience. Very often it is not even a question of judgment; the poet, the professor, the shoemaker, the doctor or the child, on the line to be cremated by the Nazi butchers, should they cry out to Maimonides that this is not the road they chose to take! Destiny implies a goal—a destination—but in life goals are illusions regardless how objective they may appear to us in our minds. It is the constancy of *going* there or anywhere which is the real destiny—the process of living and this process knows no goal but only a going or *becoming*. Life is *living* and living is the *process*. If the Chinese believe in luck, as they say, it is because they are very wise philosophers.

It is a strange and ironic paradox that the Jewish people, who may have originated the idea of the *scape-goat* in their own religious practices, should have become the scape-goat themselves for everybody else who persecutes and uses the Jews as a medium of trying to rid themselves of their own guilts, shortcomings and jealousies. In ancient Judea on the Day of Atonement it was customary for the high priest of the Temple to put his hands on the head of a goat, then confess to it all the sins of his people, thus transferring all these sins to the goat, who, now pretty loaded with sins, was chased out of town and into the desert, thus giving absolution to the people.

One of the most dramatic and heartrending stories of all time is the persistence and tenacity of the Jewish people to survive and to preserve, more or less, their religion and culture in the face of the severest and cruelest persecution of people in the history of mankind. This, in itself, is one of the most remarkable stories of a people to endure. Wherever they went and tried to settle, the other peoples, through fears and bigotries of their own religions and cultures, always "threw the book" at them. "Those who are insecure themselves manifest hostility towards others." [80] Anti-Semitism is almost world-wide. Christianity and the Christian people have tried every way to kill them off or make their lives miserable, untenable, and restricted. Christianity somehow seemed to want to eliminate the Jews in order to eradicate forever the subconscious loss of prestige and the absolution of its own innate jealousy that Jesus was born and lived as a Jew. "Christianity," writes Russell, "has been distinguished from other religions by its greater readiness for persecution . . . Anti-Semitism was promoted by Christianity from the moment the Roman Empire became Christian." [81] Actually, I believe the persecution of the Jews throughout thousands of years and the attempt to get rid of them by any and all means, is the conscious sense of inferiority of the persecutors and the use of the Jewish people as a scape-goat to conceal the bigot's own shortcomings and insecurities. Dr. D. T. Atkinson plainly states: "The most cursory study of history will find the foot prints of the Jewish scientist everywhere along the paths of progress. In all departments of human endeavors he has cut deep his niche, and often he has accomplished this in the face of oppression and ostracism. Though he has been despoiled and spurned, he has usually preserved an inimitable optimism. . . . In Medieval times the Jewish doctor of Europe was confined within the limits of a ghetto. This was an expedient to block his progress, instituted by gentile physicians who realized that they could not compete with him." [82]

Yet, nothwithstanding the torrents of cruelty, hate, and every form of bestiality and abuse heaped upon them, here is a people who, for their very limited numbers, have contributed more to the intellectual, scientific, material, artistic, cultural, and even religious advancement of the world than any other people. The creation of the *Talmud,* in itself, is so monumental and so overwhelming an achievement in world influence, stamping for all time the mark of the human being to sustain a code of ethics, to keep to the forefront the importance of humanity in its rise from animalism, to impress upon society the essentiality of a *human* system of conduct, of justice, honesty, integrity, equity, compassion and respect for the truly *civilized* personality of man, that as long as there remains a single human being in this world, he can still live and abide with the philosophy which dedicates itself to the premise that the pen is mightier than the sword, the mind wealthier than money, and the heart far more effective and stronger than arrogance and power. If it took thousands of years of persecution to make the Jew realize, more than any other people or race, the definitive meaning of peace and the preciousness of liberty and the right of privacy, then the world should be forever grateful to this fearless, enduring, courageous people, who, in the maelstroms and vortiginous centuries of slaughter and slavery, have held high, above the furies of peoples and above the heads of tyrants, and across the horizons of nations, their torch of *Sholom!*

The Hebrews, we have just learned, were just one of many Semitic peoples and cultures. Long before the advent of the Jewish people in the historical limelight were the Sumerians in Mesopotamia. It was here at Sumer, that Hammurabi once ruled and spread the famous code that bears his name. They were law-makers and contract-makers; they had laws, rules, agreements and regulations for every kind of transaction, commercial, domestic, social and religious. It was here that *lex taliones* was the law of the land. Here is a sample from the *Code of Hammurabi*: "If a physician operate on a man for a severe wound, with a bronze lancet, and cause the man's death; or open an abscess in the eye of a man with a bronze lancet, and destroy the man's eye, they shall cut off his hands." [83]

Justice in old Babylonia was pretty rugged; perhaps the nature of the people and the experience of the time required it. Regarding the domestic relations the Code of Hammurabi was very severe to protect the men, not the women. "Infidelity on the part of the man entailed no punishment except a possible indemnification to the wife. An unfaithful wife, on the other hand, risked being thrown into the water to drown. . . . Besides the ordeal by water, an unfaithful wife might lose her nose, while her lover could be castrated." [84]

It was here in the Tigris-Euphrates valley that geologists calculate was the scene of a tremendous inundation that flooded the area and destroyed many cities, which they figure might have occurred about 3500 B.C., laying the basis later on for the myth of Noah and his Ark. The ethnologists so far are unable to positively trace or identify the racial origins of the Sumerians but some claim they rose from some neolithic Armenoid type of people; others claim that they were an early mixture of Hittite and Aryan combinations; still others claim that they came in from the Mediterranean islands and regions and mixed with the early prehistoric Semites. However, we should take into account that "Assyrian, Phoenician, Hebrew, Arabic, are sister-languages, pointing back to an earlier parent language which has long disappeared . . . of the original primitive language of mankind, the most patient research has found no trace." [85] The Arab hails the stranger with *salam alaikum!*—"peace upon you!" and the Hebrew would say *shâlôm lâchem!*—"peace to you!" The Arab says *bismallah* and the Hebrew says *be-shêm hâ-Elohim,* both meaning "in the name of God."

Anyway, the Sumerians found and used many metals, especially copper, later bronze, also gold, silver; even made *wire,* all kinds of pottery by the invention of the potter's wheel. They were the first people to *coin* money. They also were one of the earliest creators of the *walled city,* due to the constant danger of attack from other nomadic Semites. To defend themselves they invented the military tactic of the *phalanx* formation, which is often erroneously attributed to the Greeks; they also were the first ones to *drill* and *organize* regular army units.

In Sumeria the principal temple to the gods or god was called the *Ziggurat,* usually centrally located on a higher level for all to see from almost every point of the city, similar to those of the Mayans and Aztecs of Mexico. While each Sumerian town worshipped as a rule many gods there was *one* god or goddess who was *supreme.* Here was worshipped the important goddess Ishtar who later was adopted by the Hebrews as their legendary Esther. "The Book of Esther," writes Sir Frazer, "is non-historical, Esther being the Semite goddess Ishtar; Mordecai, the Babylonian god Marduk, and Haman, a god of the Elamites." [86] Sumeria, like Judea later, like Egypt, was a theocracy pure and simple. Here, too, the priests were top brass and they ruled the people every hour of the day and night.

The *Hittites* gradually overwhelmed them and the Assyrian "commonwealth" began about 4500 years ago. The Hittites advanced the use of metals by smelting iron. The famous city of Troy is said to be one of the Hittite cities and its walls kept the Greeks off many a time, but Troy was taken soon after the empire gradually fell away about five hundred years later. Here again no one seems sure of the origins of the Hittites, probably a combination of Indo-European and Semitic stocks. It lasted, in all, about six or seven hundred years and broke up about 1200 B.C. The capital city of Hattusus was burned to the ground by the new migrating hordes from the west and north who poured over the Hittite lands, probably Phriggians, Mysians and more Indo-Europeans from the Caucasian steppes. The Hittites were not defeated by the Egyptian Rameses II at Kadesh as many historians have reported. The Hittite religion was really a diversification of many religions and their temples boasted of a "thousand gods." [87] Feudalism got its first real start here and became the precedent for the later feudalisms of Europe. Here, too, flourished the judicial code of *lex taliones*—an eye for an eye, a tooth for a tooth. This method of judgment and punishment is still carried on today in Abyssinia, Turkestan, Persia, Arabia, and in too many places of Africa in general.

The *Phillistines* and the *Canaanites,* both semitic, occupied what is now Israel about 1300 B.C., the approximate time of the downfall of the Hittites. It is from the word Phillistine that the word *Palestine* is supposed to be derived. The Hebrew wandered into this territory and subsequently subdued both and thus created the Judean empire.

Another Semitic people, the *Phoenicians,* founded many cities around the Mediterranean, among others the famous Carthage whose Hannibal brought elephants over the Alps to try to conquer Rome. They dominated and practically controlled the Mediterranean. Some people think they might have migrated to Ireland, western France and the Iberian peninsula. *They invented the alphabet,* a Semitic product from which the Greeks and Romans derived their alphabets. They built ships and traded all over the place and left their mark of culture and settlement at most points. Their main god in Carthage, *Baal-Astarte,* demanded human sacrifices. The maypole with ribbons streaming from the top and little children skipping around it and singing to herald the advent of Spring, is supposed to have come originally from Carthage but there it was enacted quite differently. In the Spring the women cut off the genitals of a bull, impaled them on a pole and danced around this phallic symbol in the nude to attract and entice the gods to make fertile their lands and to ensure bountiful harvests.

The greatest contribution of *Greece,* I think, is the idea of Democracy. The Greeks came out of a combination of Cretan, Aegean and Indo-European stocks with a possible "salt and pepper" flavoring from the nearby Semitic cultures and races. The gods and goddesses of the classical period are too numerous to list here but they personified the experiences and natures of human beings and they also represented the forces of the universe and of the immediate environment. As the philosophies of Greece matured more and more, the free expressions and hopes of the individual became more emphasized and as a result, a greater number of schools, cults, and mystic religions evolved to satisfy the personal desires and dreams of the individuals to survive after death and to enjoy a happy and pleasurable eternity. During the Hellenistic period these cults increased and added to themselves the influences and deities of the surrounding cultures, from Egypt, Asia Minor, Crete, Persia, etc. From Egypt the mystery cult of Isis flourished in Greece and from Persia came the worship of Mithra who many centuries later was to provide the Christian church with a birthdate for Jesus. The constant emergence of gods and goddesses in man's and woman's image continued unabated in Greece as well as in other parts of the world. "Primitive man creates his gods in his own image. Xenophanes remarked long ago that the complexion of Negro gods was black and their noses flat; that Thracian gods were ruddy and blue-eyed; and that if horses, oxen, and lions only believed in gods and had hands wherewith to portray them, they would doubtless fashion their deities in the form of horses, oxen, and lions." [88]

The Greeks were great borrowers, imitators, copyists, like the Romans who borrowed from the Greeks; this is an excellent historical anatomy of culture for the student who can easily see here the constant instability and mobilities of culture itself and the nature of almost perpetual change which is in the cyclic ertia of life itself and in the seed and fruition of human experiences. The attitude of the Greeks for analysis, curiosity, the quest for basic motives, explanations, and origins for the actions and beliefs of men, and probably the first attempts to pierce through the "backdoor" of the pantheon of the gods to reveal the secrets of the physical world and some explanation of its nature and the whys and wherefores of life and death, laid the foundation for the Greek contribution of the elements of the scientific approach and method, and the philosophies which tried to give it analysis and explanation in terms of the universe, of man, and of his State. The Greeks must be given ample credit for this "piercing through" the *blind belief barrier* and for making possible man's early attempts to *rationalize* and discover, as a result, the existence and meaning of *individuality.* Not only did this rationalization bring about a higher form of ethics and political philosophy, giving birth to the idea of Democracy, but it also freed the individual to become a *thinker,* without which it is doubtful whether man could have attained the degree of progress up to present times. By *thinking,* man discovered himself as an individual, apart but part of the society in which he found himself. By *thinking* he gathered up his past heritages and examined them, in contra-distinction to the primitive who was a *participant* but not an individual personality factor in the tribe or clan. By *thinking* man freed himself from his "termite" boundaries and became a *person* with a self-reflective and outer-reflective personality. The second primary stage of man's metamorphosis from more animal-than-human to more human-than-animal began; the struggling war of individuals and peoples to pressure the movement of humanity in one direction or the other of these two processes, is still on and whither it will go is far from being determined or foreseen.

Rome continued to tread the highways laid out and firmed down by the Greek and Etruscan cultures, and evolved to a high degree the science of *law* and the science of *conquest,* governmental controls and *taxation.* Rome also laid the groundwork for modern political campaigning by which method the practical politicians cannot concede any degree of intelligence attributable to the voters. Even Constantine pulled a political trick, to save his own skin, when he told the Christians of Rome that he saw a sign in the sky with the words "By this sign I conquer!" The sign was the Cross and it polled all the followers of Jesus to fight for him just when he ran desperately short of soldiers. That's how Christianity became the religion of Rome, by political and military necessity. Another facet of Roman culture was her contribution to the art of conquest. The way to power, wealth, fame, security, was by *war and conquest.* Any reason for invading another country or people was purely irrelevant and not even thought of. They just went on their way, invading, robbing, looting, killing, enslaving. Greece's population very often was more slave than free, and when Rome took over, the Greeks joined the swelling slave ranks to serve the children of Romulus and Remus. Incidentally, the origin of the Fascist or Nazi salute is traced to ancient Rome; freemen or citizens always shook hands but the slaves had to salute like the Nazis who were, in many ways, far worse slaves than the innocent captives of Rome. We may add that Rome advanced the political pattern of dictatorships and oligarchies and the ruthlessness and cruelty that go along with such power leaders and combines. "Marius put to death one-third of the Senate who opposed him, and when Sulla returned and assumed the dictatorship, he put to death two-thirds of the Senate that remained." [89]

The *Etruscans,* also called the Tyrrheni, probably came from

Asia Minor, of some Semitic and mixed stocks, and settled in Italy in the pre-Roman era. The king was also the high-priest and Linton excellently points out how the kingships and priesthoods were so interrelated as to have evolved from a common base of leadership: "Throughout Etruscan history the nobles always retained control of religion, and the priesthood was recruited exclusively from their ranks. Their presumed magical powers no doubt helped to maintain their control over the commoners. The Roman pattern, by which the priests of the higher orders were civil officials elected for secular rather than religious reasons, probably derived from the Etruscans." [90]

Before we reach the outlying outposts of Christianity, it may be wise to get past Islamism with a few remarks. Mohammed was a faker, pure and simple, a clever one at that, and if you should ever stay for even a short time in any densely Moslem-populated city or country you will soon find out quick enough what I mean. The well-experienced traveler knows what I mean. When I was in Djarkarta, Indonesia, not so long ago, my traveling companion and I were "bumped" by the Garuda Airline agents. We were enroute from Bali to Singapore and we had confirmed tickets. We were forced, as a result, to hang around till the next day when we were fortunate to get a B.O.A.C. plane to Singapore. We learned in the interim from responsible sources that it was a common practice for the agents of this government-subsidized airline to "blackmarket" the passages of tourists even though they had confirmed tickets in proper order. Well, one of the B.O.A.C. agents, a native Indonesian chap from one of the smaller islands, and who helped us in the most friendly and sympathetic way, joined us for dinner that evening. I noticed that even though it was pouringly hot and humid he refused to take his jacket off, as we did, and hang it from his chair. He told us plainly he needed his coat and was afraid that some Moslem would steal it. If one wants a taxi, you can walk where you want to go so much faster than the time it would take to come to an agreement with the Moslem cabdriver regarding the fare and whatever money-bill you may offer him he will never have the proper change. Every trick, regardless how small or big, how cruel and indifferent, merciless and mercenary, is pulled on everybody and anybody strange or between themselves.

Born in Mecca, about 570 A.D., Mohammed had a rough, impoverished and struggling youth, being knocked about and among many relatives and from house to house before he grew up; he certainly had a good reason to get even with the world. During his earlier years he was a shepherd, a trader, a caravan helper, and finally wound up marrying a rich widow. When he was about forty years old he went into a cave near Mecca and came out with the incorporation papers for one of the cleverest business organizations in the world. He made a trade agreement with the city of Mecca to allow him to return there peacefully and securely and he in turn would make the city into Allah's headquarters and thus bring trade, people and commerce, to the city from all directions. To this day Moslems from all over the world crawl, if necessary, to reach Mecca to perform their *hadj* pilgrimage. The *Koran* is their bible and most of it was adopted from the Jewish legends and scriptures, and from the usages and customs of the area. Once an old lady who was a tough customer and antagonist, irritated him so much that Mohammed "made the statement that no old woman would be accepted in heaven, since heaven was designed as a peaceful place. He later relented on this dictum, but the question as to whether or not women have souls is still a point of doctrinal disagreement among various sects of Islam." [91] Yet Mohammed managed to get married fourteen times and when he died he left nine widows.

Islamism was never, from its birth to the present moment, either a peaceful or a kindly religion. It's a "good deed" to kill any non-Moslem and to take away, that is, to steal his property is merely to "restore" it to its rightful owner, the Moslem. Soon after Mohammed's death, a Caliphate began and in the whole history of the Caliphs very few of them died a natural death; they killed, assassinated, poisoned, chased each other all over the place and the intrigues and cruelties would make any decent god resign or hang himself off a cloud. Mohammed himself was poisoned and Allah didn't come down to save him in the same manner as Jehovah wasn't able to get away to save the Jews being slaughtered by the Hitlerites. I don't think Allah even tried and Jehovah must have had troubles protecting some other chosen people on Venus.

Since the Arabian lands are mostly arid, hot, sandblown

deserts, the Moslem heaven is a cool, breezy place with forever-fruitful date trees, bowing palms, harems with *young,* tender girls who, regardless how many times they are slept with, never lose their virginity, cups with cool, clear water which never empty—in short, whatever they suffered with here was not in heaven and what they dreamed of getting they found there. "To understand the Moslem's mind," writes Professor Edward B. Tylor, "we must read the two chapters of the Koran where the Prophet describes the faithful in the garden of delights, reclining on their couches of gold and gems, served by children ever young, with bowls of liquor whose fumes will not rise into the drinker's heads, living among the thorn-less lotus-trees and date-palms loaded to the ground, feasting on the fruits they love and the meat of the rarest birds, with the houris near them with beautiful black eyes, like pearls in the shell, where no idle or wicked speech is heard but only the words Peace, Peace!" [92]

The Eskimos's heaven is a hot place where there aren't any winds and where one doesn't have to wear furs. The Saliva Indians believe that their heaven is on the moon where there are no mosquitoes.[93] "The Zulu idea that the blue heaven is a rock encircling the earth, inside of which are the sun, moon, and stars, and outside of which dwell the people of heaven. . . . The Finnish poem which tells how Ilmarinen forged the firmament of the finest steel, and set in it the moon and stars. . . . Some Australians seem to think of going up to the clouds at death to eat and drink, and hunt and fish as here. . . . The modern Iroquois speak of the soul going upward and westward, till it comes out on the beauteous plains of heaven, with peoples and trees and things as on earth." [94] No wonder Islamism prospered and spread.

Catholicism, the parent of all forms of Christianity, in its attempts to *individualize* the relationship between man and his God, accomplished the reverse of it and perpetrated, in the end, a great crime upon humanity, as Islamism accomplished the same end in its own territories. It was also, considering the history and evolutional intent of other religions in the story of mankind, the most opulent, the cruelest, the most injurious, the most mercenary, the most powerful and the most ambiguous. It was, particularly, a complete take-off from paganism, from the Jews, and from plagiarisms from almost every culture in contact with the early Christians. Here we find the supreme copy-cat in religious history. To even enumerate the first words of every plagiarism would, in itself, take a large volume and all I can afford, for our purpose, is to give a few pages to this creed that really put the devil into business like nothing before or after.

To begin with, the monotheistic idea of a supreme being, either completely alone or an incarnate composite of many god-manifestations, was as old as the hills when Catholicism came into establishment and power. The Egyptians believed in one god and their story was copied by the early Christians to produce the legend of Jesus who is not yet established as a historical figure but purely a Biblical one. If he did live he would have been completely lost in trying to understand the Trinity, the Catechisms of the Church, all the rules and regulations, the orders and even castes, all the echelons and phalanxes of saints and more saints, and how they came to allocate all these things to his life and family. If he lived he grew up as a Jew and died as a Jew without any intent or idea of founding any new religion. The idea of a god or divine being dying to save the people, being resurrected, taking over the duties of after-life judgments, the glory of heaven and the absolution of sins of his followers, is widespread all over the world—the Orient, India, the South Pacific Islands, darkest Africa, in the Americas. It was a common belief and practice of most primitive people. Regarding the take-off from ancient Egypt, Sir Wallis Budge, probably the greatest of all Egyptologists, writes regarding the belief in a single god by the Egyptians: "The fact remains that they did believe in one God who was almighty, and eternal, and invisible, who created the heavens, and the earth, and all beings and things therein; and in the resurrection of the body in a changed and glorified form, which would live to all eternity in the company of the spirits and souls of the righteous in a kingdom ruled by a being who was of divine origin, but who had lived upon the earth, and had suffered a cruel death at the hands of his enemies, and had risen from the dead, and had become the God and King of the world which is beyond the grave." [95]

Even the Jews were not the first or the only ones who

believed in *one* supreme being. The Mayans, far from the Garden of Eden, believed in and worshipped *Hunab Ku,* which means in Mayan *Hun*—one, *Ab*—state of being, *Ku*—god. Of course, corn being the most important thing in their lives, Hunab-Ku created the first humans out of corn.[96] Their neighbors, the Aztecs, worshipped *Tloque Nahuaque* as the supreme being. Professor George C. Vaillant states regarding the Aztecs: "The supreme god, Tloque Nahuaque, created the world; and after 1716 years flood and lightning destroyed it," [97] which gives us another "flood" story similar to Noah and his Ark, another tale copied from other religions and a very common fairy-tale in ancient mythology and legends. Charles W. Mead relates that "it is a curious fact that in the folklore of many primitive peoples of South America there are at the present time, or have been in the past, mythological fables that resemble the Biblical accounts of the great flood and of the Virgin birth." [98] The Hindus believed in a supreme being, the *Universal One,* even though it manifested itself in a thousand gods, not any less than the Catholic god and a thousand saints who are deifications or lesser gods of the Catholic hierarchy. Compared with other religions, Catholicism may be more paganistic and substantially less spiritual. Actually, any religion does not stop at the *pure* worship of a single divine being; while many of them recognize one supreme being as the pivotal "top story" epitome of their religious structure, below are the lesser divines, all the way from the Pope to the small parish priest. Even among the Hebrews many of the noted rabbis throughout their history were often considered and approached by the Jews with the respect and reverence given to a divinity. In Catholicism we find saints of every description and for every purpose while the Pope is looked at and believed in by Catholics as being some incarnation of the godly essence or spirit, and even speaks for the god, not even taking into account that cardinals, bishops, priests, etc., are accepted as divinely authorized intercessors between the god and his followers. Winwood Reade reveals to us: "The patriarch of Alexandria was the Abyssinian pope, as he is at the present day . . . by way of blessing he spits upon the congregation, who believe that the episcopal virtue resides in the saliva." [99]

The scape-goat idea of a god dying or being killed to "save" mankind is another idea as old as primitive society and is manifested in a thousand and one forms by various religions, sects, cultures, including the Jews, all over the world prior to the emergence of Christianity. Kings, conquerors, emperors, national heroes, became or were translated into gods. "The Egyptian kings dressed up in the traditional costume of Amen-Ra at certain periods, and the kings of Rome appeared as Jupiter Capitolinus." [100] "In Uganda so holy was the king considered that when he died they preserved his jaw-bone which was supposed to talk, through the mind of the witchdoctor, of course, to the people, keeping them out of trouble and to make them prosper." [101] But the idea of killing a king, priest, agent-prophet, or substitutes, to absolve the people of evil or sin is very old, dating back thousands of years before Christ and a very common religious pattern of most of the cultures in that part of the world in which Christianity had its beginnings. Dr. David Forsyth makes this point clear when he states: "Many gods besides Christ have been supposed to die, be resurrected, and ascend to heaven. The idea has now been traced back to its origin among primitive people in the annual death and resurrection of crops and plant life generally." [102] The scape-goat practice was also in the Aztec religion of Mexico. Tracing the various parallel patterns of dying gods and scape-goats, Sir Frazer states: "The Aztec ritual which prescribed the slaughter, the roasting alive, and the flaying of men and women in order that the gods might remain forever young and strong, conforms to the general theory of deicide. . . . On that theory death is a portal through which gods and men alike must pass to escape the decrepitude of age and to attain the vigour of eternal youth. The conception may be said to culminate in the Brahmanical doctrine that in the daily sacrifice the body of the Creator is broken anew for the salvation of the world." [103] The Aztecs even had the ritual of eating bread to symbolize the eating of the body of the god long before the first Spaniard landed in Mexico.[104] "Kings were revered, in many cases not merely as priests, that is, as intercessors between man and god, but as themselves gods, able to bestow upon their subjects and worshippers those blessings which are commonly supposed to be beyond the reach of mortals, and are sought, if at all, only by prayer and sacrifice offered to superhuman and invisible beings." [105] "The custom of killing men whom their worshippers regard as divine has prevailed in many parts of the world." [106] So much did the makers of the early Christian pact steal from the Egyptian religion that the Egyptians could hardly tell the difference and it was easy to convert them. In the preface of his brilliant work, *The Gods of the Egyptians,* Sir Wallis Budge tells us that "the knowledge of the ancient Egyptian religion which we now possess fully justifies the assertion that the rapid growth and progress of Christianity in Egypt were due mainly to the fact that the new religion, which was preached there by St. Mark and his immediate followers, in all its essentials so closely resembled that which was the outcome of Isis, Osiris and Horus, that popular opposition was entirely disarmed." [107]

Regarding the Catholic beginning of the world itself, Archbishop Ussher of Ireland actually pinpointed the time of creation as Sunday, October 23rd, 4004 B.C. This is a strange deduction indeed, as China, India and other great cultures, were already flourishing with cities, civilizations, and empires at that time and the Mongols were still enroute down the American trail all the way from China to Chile! James Harvey Robinson writes: "Until the middle of the 19th century, practically everyone in Europe believed that the earth had existed for not more than five or six thousand years . . . St. Augustine declared confidently in his *City of God* that not six thousand years had elapsed since the creation of man." [108] "There were more things in heaven and earth, and more particularly under the earth," writes Geoffrey Bibby in *The Testimony of the Spade,* "than were dreamed of in the philosophy of Archbishop Ussher." [109]

While the historical life of Jesus has not been verified or established, it really doesn't matter because any honest student of comparative religion knows that Jesus was a take-off from the similar parallel patterns of Osiris, Tammuz, Adonis, Attis, all gods who were scapegoats *killed* or sacrificed so that they could take away the sins of the people. All were *resurrected;* all *ascended* to take to their thrones of judgment gods and saviors. "In the resurrection of Osiris the Egyptians saw the pledge of a life everlasting for themselves." [110] "If we ask why a dying god should be chosen to take upon himself and carry away the sins and sorrows of the people, it may be suggested that in the practice of using the divinity as a scape-goat we have a combination of two customs which were at one time distinct and independent. On the one hand we have seen that it has been customary to kill the human or animal god in order to save his divine life from being weakened by the inroads of age. On the other hand we have seen that it has been customary to have a general expulsion of evils and sins once a year. Now, if it occurred to people to combine these two customs, the result would be the employment of the dying god as a scape-goat. He was not killed originally to take away sin, but to save the divine life from the degeneracy of old age; but, since he had to be killed at any rate, people may have thought that they might as well seize the opportunity to lay upon him the burden of their sufferings and sins, in order that he might bear it away with him to the unknown world beyond the grave." [111]

A. Powell Davies, one of the fine, great liberals of our century explains the things that man has done to satisfy his ego for immortality: "He invented a Christ to die for his sins, to buy forgiveness from a Father-God for them; he invented blood redemption, and confession and absolution and sacraments and the whole traditional Christian system. And he did this, not alone in Christianity, of course, but in many religious systems. It was something which could stand between him and the reality of his sins; something which could relieve him of the responsibility of lifting his life to the level of his conscience. He was afraid of moral life . . . so he invented a kind of God who would bear his burdens for him, who might relieve him of the necessity for courage, who would stand between him and naked reality. For he could not bear to look upon the truth. He still cannot. . . . He wanted to be his own self for all eternity, and so he invented heaven and refused to die." [112]

Then we have the parallel patterns or take-offs of Mary and Jesus in the combinations of Cybele and Attis, Isis and Osiris, Esther (Ishtar) and Tammuz, Aphrodite and Adonis. "When we reflect how often the Church has skillfully contrived to plant the seeds of the new faith on the old stock of paganism, we may surmise that the Easter celebration of the dead and risen Christ was grafted upon the similar celebration of the dead and risen Adonis, which, as we have seen reason to believe, was celebrated in Syria at the same season. The type, created by Greek artists, of the sorrowful goddess with her dying lover in her arms, resembles and may have been the model of the *Piéta* of Christian art, the Virgin with the dead body of her divine Son in her lap, of which the most celebrated example is the one by Michael Angelo in St. Peters." [113] It is more probable than not that the Virgin Mother Mary brought more wealth to the Church than God Himself. "The worship of Mary," writes Father Emmett McLoughlin, "is one of the largest sources of revenue to parish churches and religious orders. Shrines and grottos dot the country and the world. At all of them, prayers are answered and donations before and after the favors are solicited. Offerings are made for medals, pictures, rosaries, statues, candles, and votive lights. The faithful throughout the United States were solicited to contribute money and precious jewels for a diadem to be placed above a painting of Mary in Rome when the Pope proclaimed her Queen of the Universe in 1954." [114] He states further: "In 1945 Pope Pius XII called [the Virgin Mary] Our Lady of Guadalupe the 'Empress of America.' American Catholics are now trying to get into the act. Mary has been chosen as Our Lady of the Airways, Our Lady of the Runways, and Our Lady of Television." [115]

The Bible (Psalm CXVI, vii) plainly states that "all men are liars." At least the men who wrote the Bible not only had some first hand experience but had good and just reason for making the statement. This is so true of the Trinity, which is another take-off from previous religions and cultures. You will find trinities by the dozens all over the world. Joseph Campbell points this out when he says: "We find that such themes as the Fire-Theft, Deluge, Land of the Dead, Virgin Birth, and Resurrected Hero have a world-wide distribution. . . . No human society has yet been found in which such mythological motifs have not been rehearsed." [116] For us it will be sufficient to examine the parallel of Isis, Osiris and Horus, of the Egyptians. Trinities are ever being manufactured to the present day, such as the Trinity of the Matswa sect of the Congo who believe that the real trinity is Matswa, Jesus and General Charles de Gaulle, and that they all live in a beautiful palace in Paris, that Jesus is a black man and that Matswa, their savior, will return someday and chase all the white crooks out of their country. The Cross, an old phallic symbol, predates the Christian era by so many centuries that it can scarcely be considered an "original" with the Jesus idea. Lee Stone brings to light that "the Christian boasts that the cross is a Christian symbol, when in fact, it is one of the oldest, if not the oldest symbol known to man. For ages the cross has symbolized the phallus and its appendages. The Egyptians used the cross (tau) and it is to be found on hundreds of monuments all over Egypt and India and in some other parts of the world, even among the American Indians, the Mexicans, Aztecs, and the inhabitants of Yucatan and Peru." [117]

So many churches, both Roman as well as the Greek Orthodox, claim to possess a piece of the original cross that if they were all collected together one would be able to build a house besides a cross. Such respect did the Church have for its followers that they claimed to possess quite a number of the foreskins of Jesus and if Jesus really possessed all these things he must have been quite a man. Dr. Goldberg, in his *Sacred Fire,* states: "The old idea of the sanctity of the lingam survives in the Christian attitude of reverence toward the Holy Prepuce, the foreskin of Jesus. Until very recently, there were twelve such prepuces extant in European countries, and many a legend was woven around them. One of these, the pride possession of the Abbey Church of Coulcomb, in the diocese of Chartres, France, was believed to possess the miraculous power of rendering all sterile women fruitful. It had the added virtue of lessening the pains of childbirth." [118] "The number of foreskins of the Saviour reputed to be still extant are said by Davenport to be twelve in number. One was in the possession of the monks of Coulcomb, another at the Abbey of Charroux, a third at Hildersheim, in Germany, a fourth in the Church of St. John de Lateran, at Rome, a fifth at Antwerp, a sixth at Pur-en-Valay, in the Church of Notre Dame, and so on." [119] The Church that has so much to say about "spiritual" matters has used almost every kind of artifice for its followers to worship, besides foreskins, such as wrist bones, hands, pieces of the bodies of saints, which is a curious take-off from the embalmed priests and kings of Egypt. So paganistic were the early Christians that when their prayers to the saints went unheeded and didn't bring the desired results, they used to beat the images of these saints, which after all, were just a continuance take-off from their predecessor pagan gods. [120] Throughout Catholic religious history the followers of the Church were just as pagan as the previous religions from which they borrowed almost everything they could put their hands on. "The medieval Christian was essentially more polytheistic than his pagan predecessors, for he pictured hierarchies of good and evil spirits who were ever aiding him to reach heaven or seducing him into the paths of sin or error. Miracles were of common occurrence and might be attributed to either God or the Devil; the direct intervention of both good and evil spirits played a conspicuous part in the explanation of daily acts and motives." [121]

The very birthday of Jesus, which is known as December 25th, was stolen and adopted from another religion. Edward B. Tylor, considered as the father of anthropology, states: "The Roman winter-solstice festival, as celebrated on December 25th (VIII. Kal. Jan.) in connexion with the worship of the Sun-god Mithra, appears to have been instituted in this special form after the Eastern campaign of Aurelian A.D. 273, and to this festival the day owes its opposite name of Birthday of the Unconquered Sun, 'Dies Natalis Solis invicti.' With full symbolic appropriateness, though not with historical justification, the day was adopted in the Western Church, as the solemn anniversary of the birth of Christ, the Christian Dies Natalis, Christmas Day. Attempts have been made to ratify this date as matter of history, but no valid nor even consistent early Christian tradition vouches for it. The real solar origin of the festival is clear from the writings of the Fathers after its institution. . . . As for modern memory of the sun-rites of mid-winter, Europe recognizes Christmas as a primitive solar festival by bonfires which our 'yule-log,' the '*souche de Noel,*' still keeps in mind; while the adaptation of ancient solar thought to Christian allegory is as plain as ever in the Christmas service chant, '*Sol novus oritur.*' " [122] Such was the wanton plagiarism and unabashed thievery of other religions; no wonder Michelet referred to the decline of the Roman Empire and the Dark Ages as "the thousand years without a bath!" It must have been very filthy, indeed. "Faith seems to flourish better in filth," wrote Duff.[123] Everybody seemed to be waiting to die to get to heaven rather than to bathe and clean the flesh. Dr. Atkinson in his *Magic, Myth and Medicine,* informs us that "in some of the cities of

Spain may still be seen the street paving and other evidences of sanitation put there by the Moors at a time when London was reeking with filth and Paris threw her refuse into the gutters." [124] It didn't matter, so long as they managed to get to the Church, give their last few pennies, if necessary, into the never-filling pot and so cleanse their souls.

Let's now consider the Catholic premises of a devil-chieftain or Satan, purgatory, and the torments of an "everlasting" pain and torture in the realm of Hell. This brought more customers than the belief in a benevolent god. People are not so much concerned with going to heaven as they are with not wanting to go to hell. The idea of a devil was imported from the religion of the Persians. "Christianity borrowed one very important thing from it (Zoroastrianism): the idea of the Devil, which left the outcome in doubt." [125] In Zoroastrianism, which was the Persian religion, life, existence and the cosmos were divided into two camps of the good and "light" forces on one side and the evil and "darkness" forces on the other; one cannot defeat the other and the struggle is a constant balance-counter-balance of equal forces; the life of each person is a battle ground in which these two forces express themselves. Regarding this doctrine, "its influence on Jewish, Christian and Muslem apocalyptic literature was very considerable. As has been considered, this was most apparent in Judeo-Christian eschatology and angelology, and it recurs in relation to dualism and the conception of evil. Thus, the transformation of Satan and the rebel angels into figures not very different from those of Ahriman and his demonic host in the apocalyptic literature has every appearance of having been due to Iranian contacts." [126]

The concept of good gods of the sky and the bad demons in the regions below is one deep-rooted in the prehistoric primitive mind of humans and was expressed in most religions all over the world. In Bali, the people look up to the mountains as the abode of the good gods and the demons are in the sea or below the earth and as a result the Balinese do not take to the sea for fishing. I have seen Balinese adults squatting on the low ebb tide beach at Sanur sprinkling the sea water upon themselves from very shallow tidal pools as their kind of bathing limitation. The Rangdas are more feared and respected than the gods on the tops of their pagodas, and in their dances the good gods are never finally overtaken but the bad ones are never defeated. Rudwin states: "The legend of Lucifer has no Biblical basis. The ancient Hebrews knew of no devil whatever." [127] Yet Justin Martyr, in his *Second Apology for the Christians,* said "If there was no hell, there was no God." And Chrysostom, in his *Epistle to Philemon,* said, "It is because God is good that he has prepared a hell." Bunyan, from the *World to Come,* wrote that "the saints shall rejoice that we are damned, and God is glorified in our destruction." St. Thomas Aquinas, Catholicism's master copy-cat and twister of Aristotle, wrote in his *Summa:* "That the saints may enjoy their beatitude and the grace of God more richly, a perfect sight of the punishment of the damned is granted to them." [128] What we are really contending with here is a lot of sadists in tremors extremis. "Peter the Hermit would now be sent to a lunatic asylum," wrote William James in his *Will to Believe.*[129] Martin Luther, centuries later, bellowed: "Behold a matter on which there is no room to doubt and that is that the plague, fevers, and other diseases are the work of the devil," and the Catholics countered by preaching that Luther himself was a witch and a tool of the devil.[130]

The idea of purgatory was also of primitive origin and a very common concept and religious belief in many parts of the world. The Aztecs used to make fires on the grave so that the soul could keep warming itself when the weather was cold, and the Winnebagos made fires on the grave each night so that the soul could find its way on its journey to heaven,[131] and the Hindus' system of *Karma* was a slow reincarnating, purgatorial trial of the soul in its efforts to reach purity and heaven. The Egyptians had it and the Greeks had it. It was nothing new. The Catholic version "was an invention of Pope Gregory the Great and it proved to be extremely lucrative by encouraging payment for masses for the dead." [132] To give you an idea of what *masses* meant to these fear-stricken unfortunate human beings, I will quote some passages from the Will of Pedro de Cieza de Leon, the "Chronicler" of the Incas, who died in 1554 in Seville:

> "I direct that there be said on my behalf five hundred masses . . . and the usual alms should be given for each mass.
>
> "I request that on each Friday, when two years have passed from the day of my death, there be said a Passion Mass for my soul

in the Colegio de Santo Tomas of this city, by the monks . . . who are to be paid the usual alms.

"Futhermore, I request that there be said another fifty masses for the souls in purgatory where my executors think best.

"Furthermore, I request that there be said fifty masses for the soul of my wife; for the soul of my mother, twenty masses; and for the soul of a saintly sister of my mother, ten masses; and for another sister of mine who died, ten masses; and for an Indian woman called Ana, eight masses. . . .

"Furthermore, I request another ten masses to be said for the souls of the Indian men and women in purgatory who came from the lands and places where I traveled. . . .

"Furthermore, I direct that at the end of a year another three hundred masses be said for the souls of the persons. . . .

"Furthermore, I direct that every Sunday and on the Feast Days of the Apostles and of Our Lady and on the Feast Day of St. Catherine and St. Elizabeth and St. Joseph falling during the year after my death, masses be said on each of these days. . . .

"Furthermore, I direct that one ducat be given to each of the monks who reside and are in the said Colegio de Santo Tomas, who are ordained priests, so that they remember to pray to God for my soul." [133]

I took the liberty of using up so much space on masses because I feel it is important to realize to what ends people would go on account of their groundless fears and to what extent the wealth and power of the Church have grown because of it. It reminds me of the yak-hide prayer wheels or cylinders which the Tibetan monks push around, each push giving themselves credit for many thousands of prayers with a swing of the hand. The idea of sin and purgatory was a great idea to increase the power of the Church and place it in an unalterable position as sole intercessor between God, his mercy and prerogative of forgiveness, on one hand, and the mandatory torment of purgatory for those whose sins have not been absolved and forgiven. "The more sins you commit," Bertrand Russell so aptly tells us, "the more the Church profits by the steps you have to take to mitigate the punishment. The system is convenient both for priests and for sinners, but it is preposterous to pretend that it promotes virtue. What it does promote is mental docility and abject fear." [134]

The idea and evolution of the sin-concept go back and deeper than the beginnings of Christianity, but it was the Church that was first really to put it on a successful pay-as-you-go basis. Sir Frazer states that "It seems probable that originally the violations of Taboo, in other words, sin, was conceived as something almost physical, a sort of morbid substance lurking in the sinner's body, from which it could be expelled by confession as by a sort of spiritual purge or emetic. This is confirmed by the form of auricular confession which is practiced by the Akikuyu of Kenya in East Africa. Amongst them, we are told, sin is essentially remissible; it suffices to confess it. Usually this is done to the sorcerer, who expels the sin by a ceremony of which the principal rite is a pretended emetic: *kotahikio,* derived from *tahika,* to vomit. Thus among these savages the confession and absolution of sins is, so to say, a purely physical process of relieving the sufferer of a burden which sits heavy on his stomach rather than on his conscience." [135]

However, in no religion in the world up to that time was the concept of sin incorporated with such a dreadful fear of inescapable punishment at the hands of the Devil who was following out, it would follow, the judgments of a terribly angry and unforgiving God. "Nor did Christianity scorn to borrow details," Tylor writes, "from the religions it abolished. The narrative of a medieval visit to the other world would be incomplete without its description of the awful Bridge of Death; Acheron and Charon's bark were restored to their places in Tartarus by the visionary and the poet; the wailing of sinful souls might be heard as they were hammered white-hot in Vulcan's smithies; and the weighing of good and wicked souls, as we may see it figured on every Egyptian mummy-case, now passed into the charge of St. Paul and the Devil." [136] To satirize about sin, Lin Yutang says: "The idea of a God who desired the smell of roast meat and could not forgive for nothing! . . . You can't make a man a Christian unless you first make him believe he is a sinner." [137]

To further implant fear and resignation into its slave-followers the Church had to discourage any resurgence of democratic ideas that still flickered from the ruins of the Greek pantheons and schools. "The

Church contended that it was Satan who inspired the opposition against priestcraft and kingcraft, and that it was the Devil who filled man with the love of liberty, equality and fraternity. Diabolus represented discontent with existing conditions in matters social, political, and ecclesiastical. He was identified with the spirit of progress so disturbing to those who are satisfied with the existing order of things. Every democratic institution, every social reform, was attributed by the reactionaries to the machinations of the spirits of hell." [138] "Samuel Butler has remarked in his *Note-Books* published posthumously in 1912, that we have never heard the Devil's side of the case because God has written all the books." [139]

To interrupt briefly, the reader may come to pause and ask himself: Why do we spend so much time and space and detail in trying to expose these things or in building up a relentless criticism against religions? Is this necessary in order to bring about a constructive philosophy of Time? The answer is self-evident. The most important task of this essay is to reveal true aspects of these things and once we understand them as creations of men, minds, and groups, we will no longer fear the products of their nightmarish ideologies, liquidate the myth of hell and begin using our precious time to live better, to be happy, to breathe freely every moment of our lives and not to fear death. As Philip Wylie so nicely puts it: "Death is not the worst thing—only the last." [140] We must face the happy fact that *hell is a myth,* a fraud, and the expression of crazed sadists who would rather die an agonizing death rather than to spend one hour of love or sleep one night of true peace and tranquility. John Dewey, America's greatest philosopher, plainly exposed this fraud when he wrote that "theories paint a world with a God and a Devil in Hell." [141] Even Pope Gregory VII admitted that "there is hardly a bishop who has obtained his office honestly." [142] Joseph Wheless, one of the most outstanding of Bible critics, exposes the fraudulent base upon which the main pillars of the Church were imbedded. "The Church," he writes, "was founded upon, and through the Dark Ages of Faith, has battened on monumental and petty forgeries and pious frauds, possible only because of its own shameless mendacity and through the crass ignorance and superstition of the sodden masses of its deluded votaries, purposely kept in that base condition for purposes of ecclesiastical graft and aggrandizement through conscious and most unconscionable imposture." [143] Father Benavente, the rebel Mexican priest, summarized this well when he told the peones that "while preaching of heaven, they appropriated the earth!" Alas, no religious institution has valued life so little, or for that matter, all the lives in the world so long as a fictitious, outdated, outmoded compendium of stupidities should still be considered, in spite of all truths, as a holy institution. "Only the arrogant steal the intent of life and meaning of Nature," writes Philip Wylie, "to furnish heavens for their vanity and hells for those whom they cannot bring to subjugate agreement in that vanity." [144]

The "good" Fathers of the early Christian centuries knew every chamber in hell and described its equipment in every detail. Imagination to instill fear and horror became of age. It wasn't enough that every hour of the day and night were filled with ogres, imps, monsters, all kinds of creatures and macabre beings, in greater supply and variety than Hieronimus Bosch and the Pieter Brueghels could paint if they lived five hundred years, but they described in sadistic detail every agony and every device to agonize. Let's look at just a few samples of the "blessed" and "most gracious" work of a few of these lunatics:

St. Fulgentius, in his *De Fide,* writes: "Be assured, and doubt not, that not only men who have obtained the use of their reason, but also little children who have begun to live in their mother's womb and have there died, or who, having been just born, have passed away from the world without the sacrament of holy baptism administered in the name of the Father, Son, and Holy Ghost, must be punished by the eternal torture of undying fire; for, although they have committed no sin by their own will, they have, nevertheless, drawn with them the condemnation of original sin by their carnal conception and nativity." [145]

Hoenus, an Irishman, in 1153, visited hell and reported that he "saw people of every age and of both sexes nailed to the ground with red-hot nails, and being whipped by devils; men, women, and children boiled in cauldrons; people lying on their backs with fiery dragons, flaming serpents, and toads dining upon their bowels; people hung on iron hooks driven through their feet, hands, eyes, nostrils, ears, or navels, over flames of burning sulphur." [146]

But this is nothing. To reprint the monkish monologues on hell is to invite nausea and disgust. No one in his right mind could possibly imagine Jesus to have such thoughts or harbor such hate against the unborn, babies, and children. Father Furniss, a most "benevolent and paternal" soul, specialized in writing about the punishments to children in his *Books for Children.* He describes the punishment handed out to a child *for wearing a nice dress:* "Punished in hell by wearing a bonnet of fire. It pressed down close over her head; it burns into her skin; it scorches the bone of the skull and makes it smoke. . . . Think what a headache that girl must have. . . . She is wrapped in flames, for her frock is on fire. . . . There she stands burning and scorched; there she will stand forever burning and scorched. . . . When that girl was alive she cared only for one thing, and that was dress! And now her dress is her punishment!" [147]

This priest-paranoic describes further the punishment of a child in hell: "The little child is in this red-hot oven. Hear how it screams to get out! See how it turns and twists itself about in the fire. It beats its head against the roof of the oven. It stamps its little feet on the floor. You can see on the face of this little child what you see on the faces of all in hell—despair, desperate and horrible! . . . *God was very good to this child.* Very likely God saw that this child would get worse and worse, and would never repent, and so it would have to be punished much more in hell. So God *in His mercy* called it out of the world in its early childhood." [148] This crazed sadist would have made a wonderful adjutant at the Five Chimneys of Auschwitz. It is sad, but this is the religion which claims to represent Jesus who, according to the Bible, taught his followers that they should not fight or express cruelty, that they should not punish adultery, that they should not hate, that they should not go to the temples, that only *as children can they enter the kingdom of heaven!* A former priest writes thusly about his childhood Catholic education: "To the Catholic child, even at the age of seven, eight or nine, hell is a vividly real thing. By that time already it had been depicted to me so often in word and picture that I could see and almost breathe its leaping, lunging flames." [149]

There was once another lunatic by the name of Tundale and he had visions (as described in Delepierre's *L'Enfer décrit par ceux qui l'ont vu)* and in these visions Tundale and others did not overlook any detail, small or large, which entailed the torture of people or the suffering being witnessed by the blessed looking down from the heavenly balconies. "The Devil is represented bound by red-hot chains on a burning gridiron in the center of Hell. The screams of his never-ending agony make its rafters to resound; but his hands are free and with these he seizes the lost souls, crushes them like grapes against his teeth and then draws them by his breath down the fiery cavern of his throat. Demons with hooks of red-hot iron plunge souls alternately into fire and ice. Some of the lost are hung up by their tongues, others sawn asunder, others gnawed by serpents, others beaten together on an anvil and welded into a single mass, others boiled and then strained through a cloth, others twined in the embrace of devils whose limbs were afire. The fire of earth, it is said, is but a picture of that of Hell; the latter is so immeasurably more intense that it alone could be called real. Sulphur is mixed with it, partly to increase its heat and partly too in order that an insufferable stench may be added to the misery of the lost; while, unlike other flames, it emits according to some visions, no light, that the horror of darkness may be added to the horror of pain. A narrow bridge spans the abyss, and from it the souls of sinners are plunged into the darkness that is below." [150]

How many devils are there really in Hell and roaming around the earth tantalizing people? "Johannes Wierus, a pupil of the famous Cornelius Agrippa and author of the learned treatise *De praestigiis daemonum,* 1563, went to the considerable trouble of counting the devils and found that the number was seven and odd millions. According to this German demonologist, the hierarchy of hell commands an army of 1,111 legions, each composed of 6,666 devils, which brings the total of evil spirits to 7,405,926, 'without any possibility of error in calculation.' A professor of theology in Bastle, Martinus Barrhaus is, as far as is known, the last man to take the census of the population in hell. According to this infernal statistician, the devils number exactly 2,665,866,746,664. If we are to believe Richalmus, an abbot of the first part of the thirteenth century, the number of devils exceeds all calculations, being equal to the grains of the sands of the sea. Three friars, so runs the legend which confirms the view of this monk, hid themselves one night near a Witches' Sabbath, which happened to be held in the valley in the Alps, in order that they might count the devils present at the affair. But the master of ceremonies, upon discovering the friars and divining their intention, said to them: 'Reverend brothers, our army is so great that if all the Alps, their rocks and glaciers, were equally divided among us, none would have a pound's weight!' The fecund imagination of our ancestors peopled the air, the earth, and the flood with devils. Paracelsus tells us in the 16th century that the air is not so full of flies in the summer as it is at all times of invisible devils; while another philosopher maintains that the air is so full of devils that there is 'not so much as a hair's breath empty in earth or in waters above or under the earth!' "[151] Dr. A. D. White, first president of Cornell University, relates in his classic work, *The History of the Warfare of Science with Theology:* "Tertullian insisted that a malevolent angel is in constant attendance upon every person. Gregory of Nazianzus declared that bodily pains are provoked by demons, and that medicines are useless, but that they are often cured by the laying on of consecrated hands. St. Nilus and St. Gregory of Tours, echoing St. Ambrose, gave examples to show the sinfulness of resorting to medicine instead of trusting to the intercession of saints."[152] . . . "Thus the water in which a single hair of a saint had been dipped was used as a purgative; water in which St. Remy's ring had been dipped cured fevers; wine in which the bones of a saint had been dipped cured lunacy; oil from a lamp burning before the tomb of St. Gall cured tumours; St. Valentine cured epilepsy; St. Christopher, throat diseases; St. Eutropius, dropsy; Ovid, deafness; St. Gervase, rheumatism; St. Apollonia, toothache; St. Vitus, St. Anthony, and a multitude of other saints, the maladies which bear their name. Even as late as 1784 we find certain authorities in Bavaria ordering that any one bitten by a mad dog shall at once put up prayers at the shrine of St. Hubert, and not waste his time in any attempts at medical or surgical care and cure. In the twelfth century we find a noted cure attemped by causing the invalid to drink water in which St. Bernard had washed his hands. Flowers which had rested on the tomb of a saint, when steeped in water, were supposed to be especially efficacious in various diseases. The pulpit everywhere dwelt with unction on the reality of fetish cures."[153]

The saints were very busy fighting the demons for the sake of mankind. Every person had devils entering or leaving his body day and night. "For hundreds of years this idea of diabolic possession was steadily developed. It was believed that devils entered into animals, and animals were accordingly exorcised, tried, tortured, convicted, and executed. The great St. Ambrose tells us that a priest, while saying mass, was troubled by the croaking of frogs in a neighboring marsh; that he exorcised them, and so stopped the noise. St. Bernard, as the monkish chroniclers tell us, mounting the pulpit to preach in his abbey, was interrupted by a cloud of flies; straightway the saint uttered the sacred formula of excommunication, when the flies fell dead upon the pavement in heaps, and were cast out with shovels! A formula of exorcism attributed to a saint of the ninth century, which remained in use down to a recent period, especially declares insects injurious to crops to be possessed of evil spirits, and names, among the animals to be excommunicated or exorcised, mice, moles, and serpents. The use of exorcism against caterpillars and grasshoppers was also common. In the thirteenth century a Bishop of Lausanne, finding that the eels in Lake Leman troubled the fisherman, attempted to remove the difficulty by exorcism, and two centuries later one of his successors excommunicated all the May-bugs in the diocese." [154] . . . "Satan could be taken into the mouth with one's food—perhaps in the form of an insect swallowed on a leaf of salad, and this was sanctioned as we have seen, by no less infallible an authority than Gregory the Great, Pope and Saint." [155]

One can imagine the constant battle of the theologians in trying to keep the devilish population from increasing in numbers, and no doubt many solemn conferences took place by these learned lunatics to resolve procedures and stratagems to combat the ever-pressing invasion of these tempters. "During the witchcraft epidemic, from the fourteenth century onward, many persons were put to death for lycanthropy or other forms of shape-lifting, whether the change was believed to be actual or illusory, the result of demoniac influence. The phenomenon was never proved from the first-hand evidence of educated people. Those who were accused, like the mass of medieval folk, believed the change to be real. Evidence of it, or of the demoniac power which seemed to cause it, was wrung from victims by torture, and in addition to this there were a few stock cases which kept recurring at trials as evidence of the possibility of change. Those who condemned witches to a cruel death were much given to accepting from others, as proof of objective facts, statements which were evidence only of subjective illusions." [156] Anyway, during the Dark Ages, "men in general believed that the majority of mankind would find their way to hell." [157] Once in a while some bishop or monk would become the mouthpiece of the Devil and give him a chance to express himself through them. Jacobus de Voragine, Archbishop of Genoa (about 1275) has the Devil say: "The Christians are cheats; they make all sorts of promises so long as they want me, and then leave me in the lurch, and reconcile themselves with Christ as soon as, by my help, they have got what they want." [158] Incidentally, the origin of the "printer's devil" is traced to a black slave employed by Aldus Manutius, the Venetian printer; regarding the evil of printing itself, as late as 1851 it was claimed that the invention of printing money was the work of the devil.[159]

It would take tomes of paper and print to go through all the strange and bizarre capsules which the Catholics are supposed to swallow in order to avoid eternal damnation and get to heaven. I may repeat for emphasis: Everybody seems to fear or love God or think they do both, but *nobody seems to be in any hurry to meet him.* Everyone adores the glory and ecstasy of heaven, yet everyone seems to desire to keep missing the bus that is supposed to go there every day! Actually, no one wants to die but keep living and religion is that department which sells the idea of living after one dies. In order to accomplish this, religion ends up by demanding that people start dying once they are born; and to be happy as normal and natural people is a disgrace and can only be "purified" by torment and suffering.

The Catholic Church even goes further: It tells its caterpillars that they are endowed by the Almighty with a mind that can judge right from wrong in accord with the rules in the Almighty's mind, and if they should "choose" right he yells Bingo! and if they should commit what the Almighty already knows in *His* mind to be wrong, then it is torment and the fires of hell. For any intelligent person to assume that everyone has the same intuitive, introspective, rational potential and competency to pick right from wrong from the

Lord's book of rules is to reduce any form of intelligence to an absurdity. And on a basis of faith to assume that any person knows what's in the Lord's mind, as a guide or direction, at any time, is to reduce the behavior of a human being to a caterpillar without a mind or direction. Edmond N. Cahn, in his *Sense of Injustice,* writes: "A wholly free will in a finite world is a fair definition of insanity." [160] To insist, in spite of all knowledge and common sense, that man *has* the "knowledge" of what God has ordained and *has* the ability or power to accept and live that "knowledge" or deny it and do "wrong," can only come from people who have put blindfolds on their minds, eyes and hearts, and have attempted to be "completely" immune from any contact with the world, experience, and with life itself. This is a form of hypocrisy that comes from the introversion of the monk to the extroversion of the Pope with his tri-storied tiara, bejeweled and ornated like a Cellini masterpiece, being carried about on the shoulders of human beings like a Chinese porcelain doll-god, both a complete antithesis of the equality-and-fraternity philosophy of Jesus and from whose Biblical legend we can only picture a man of simplicity and the only jewel he could have possibly possessed was the sunshine from the heart of a man who condemned no one, loved everyone, had compassion and understanding for every creature and circumstance, taught that no one has the right to throw "stones" at anyone as we all live in glass houses, more or less; taught that to err is human, to forgive divine. Jesus did not mention the devil, hell, or purgatory. He couldn't have imagined hell or purgatory because he was a Jew, lived as one and was brought up in the Jewish culture and religion of the times, and the Jewish culture contained no devil, hell, or purgatory.

During the Dark and Middle Ages every cat was killed on sight because she was considered an accomplice of the witch and a symbol of Satan. Listen to what Jesus had to say about cats: "As Jesus entered a certain village He saw a young cat which had none to care for her, and she was hungry and cried unto Him; and He took her up, and put her inside His garment, and she lay in His bosom. And when He came into the village He set food and drink before the cat, and she ate and drank, and showed thanks unto Him. And He gave her unto one of His disciples who was a widow, whose name was Lorenza, and she took care of her. And some of the people said, This man careth for all creatures . . . are they His brothers and sisters that He should love them? And He said unto them, Verily these are your fellow-creatures, of the great Household of God; yes, they are your brothers and sisters, having the same breath of life in the Eternal. And whosoever careth for one of the least of these, and giveth it to eat and drink in its need, the same doeth it unto Me; and whoso willingly suffereth one of these to be in want, and defendeth it not when evilly entreated, suffereth the evil as done unto Me." [161]

As stated, the whole structure of sin seems to be predicated on the idea of free will, that all persons have the ability to pick right from wrong, and when they pick wrong they are going to be punished for it. Considering that the Catholic population is a minority percentage of the total world population, one can imagine how busy the Devil must be checking in new arrivals. Business must be kind of slow up in heaven, in comparison, with almost everyone except the forgiven Catholics going to hell. On a basis of statistics and numerals, I would say that the Devil has always been, by far, the most successful and hell the most occupied. The saints sitting in their partierre boxes according to rank surely must be having one hell of a time looking down upon all these trillions of people cooking in brimstone!

No human is possessed with free will any more than a human can "will" to fly off a cliff like a bird or "will" to live in the sea like a fish. What he can actually will is determined by his heredity, the way he has been brought up and all his experiences and relationships of living. The ability of a person to evaluate right from wrong, or this from that, is a *process* of judgment due to his being what he is; it is not a "substance" given to him on a platter with his baptism. The things a moron can will is most definitely limited to a moron; what a woman or man can will is limited to being a woman or a man. Schopenhauer put his finger on it when he said that "a man can surely do what he wills to do but he cannot determine what he wills." Fridtjof Hansen, the famous explorer, explains his version of free will more fully: "Will is not free and independent. It is itself a quality which has been inherited and may have been strengthened or weakened by education and environment; its functioning at any moment is determined by previous causes. In reality, therefore, an individual can no more shape his own character than a tree can shape its branches. Whether a man becomes what is called good or bad, moral or immoral, and what views he holds, depends entirely on his inherited qualities and how they have been influenced by education and environment." [162] Spinoza explains it from his philosophical point of view: "The decisions of the mind are nothing save desires, which vary according to various dispositions . . . there is in the mind no absolute or free will; but the mind is determined in will in this or that by a cause which is determined in its turn by another cause, and this by another, and so on to infinity." [163] Dr. L. Guy Brown, of the University of Missouri, gives his view from the psychological approach: "The individual does not possess human nature at birth nor does he have a world in which to live. Both are achieved through experience in social situations where meaningless objects are defined and attitudes created concerning them. There is nothing in the biological processes that predetermines the definitions the individual will give objects and situations." [164] The Catholic Church, by predicating its rules and punishments on the idea of free will, has imposed upon its followers a false philosophy without any foundation or verification in the experience of living, an unnatural and antinatural philosophy without any regard to the acquired knowledge and realities of the world and the humans in it, and without any attempt to translate these realities into rational and verifiable meanings and values. If any "free will" could exist it must be based on some free way of choosing by thinking about anything and yet the Church *always opposed free thought.* A former priest and a teacher for many years in Catholic parochial schools, emphatically and clearly states that "although Catholic students do not know it, freedom of thought is formally condemned by the highest authorities of the Roman Catholic Church." [165] "In the eyes of the Church, thinking was equivalent to blaspheming and so it imprisoned, tortured, hanged or burned every person who dared to think for himself." [166] Dr. Kaufmann put it clearly when he wrote: "Most of the Christian dogmas as well as most propositions about God are essentially ambiguous. . . . The history of the development of Christian dogma is a continual fight for the abundance of mystery and not for rationalistic clarification. . . . Or to put it more simply, they are through and through ambiguous." [167]

The Catholic religion was imposed upon the people of Europe by force, intimidation, intrigue, massacre and the constant play of fears. They replaced every pagan god with a saint and replaced every pagan myth with a Catholic myth. Those pagan gods, goddesses, mythologies, which they could not possibly incorporate into their own hierarchy they condemned as devils and witches. When the pagans followed their own religious ways in secrecy in fear of being murdered, the Church condemned them as witches and people possessed by demons. It reminds me of the "paternal" and "compassionate" attitude of Diego de Landa, the archbishop of Yucatan, who burned at Merida every book of Mayan culture he could put his hands on and who ordered the slaughter of any Indian who would not accept conversion to Catholicism.

Many of the leading scholastic philosophers, besides most of the popes and hierarchy of the Church, believed in witches and demonology. Thomas Aquinas and Albertus Magnus both believed in the existence of witches and witchcraft and favored the persecution of "witches," and they were two of the leading theologians of the 13th century. "The witch-madness and witch-trials were not confined to the Catholic world; the Reformation provided them with fresh fuel from both sides. In England witch-hunts reached their climax under Queen Elizabeth; in the next century they spread to the North American colonies, but even in Europe witch-mania was still rampant in the century of Galileo and Descartes. It was not a case of individual instances, but of mass annihilation of human beings. A Saxon Magistrate boasted that he had read the Bible through fifty-three times and had condemned twenty thousand witches." [168] Henry Charles Lea, the great historian of the *Inquisition,* brings out that "a bishop of Geneva is said to have burned five hundred within three months, a bishop of Bamburg six hundred, a bishop of Wurzburg nine hundred. Eight hundred were condemned, apparently in one body, by the Senate of Savoy. . . . The spring of 1586 was tardy in the Rhinelands and the cold was prolonged. This could only be the result of witchcraft, and the Archbishop of Treves burned at Pfalz one hundred and eighteen women and two men, from whom confession had been extorted that their encantations had prolonged the winter. . . . Paramo boasts that in a century and a half from the beginning of the sect, in 1404, the Holy Office (the Inquisition) had burned at least thirty thousand witches who, if they had been left unpunished, would easily have brought the whole world to destruction." [169]

A whole "science" of precautions and preventatives gradually acculturated itself into the Dark Ages to protect people from the witches and demons. The mistletoe, an old pagan idea, continued into the Catholic era, like so many other things. On All Hallow's Day it became a charm against witches. In parts of France the farmer hung mistletoe in his stable to protect the cattle from Satanic contamination. In the pre-Catholic Norse religion mistletoe was used to find gold or precious stones and in one of the myths it was used as an arrow to kill the beautiful god Balder, showing that this item was also copied, like thousands of others, from pagan folklore and religions.

Almost everyone is fairly familiar with the periods of witchcraft which dominated the cultures of Europe and early America. In his *Origins of Culture,* Tylor writes: "In the 13th century, when the spirit of religious persecution had begun to possess all Europe with a dark and cruel madness, the doctrine of Witchcraft revived with all its barbaric vigour. That the guilt of thus bringing down Europe intellectually and morally to the level of negro Africa lies in the main upon the Roman Church, the records of Popes Gregory IX and Innocent VIII, and the history of the Holy Inquisition, are conclusive evidence to prove." [170]

How could there be any humane or sacred value to life under such a dark pall of superstition and demonology? During these many centuries, for almost two thousand years, Science stopped breathing, Philosophy had to use the word God as innuendo to meanings and interpretations out of fear of being quartered, or burned; the Arts reflected, though with splendor and skill, the vast insane asylums of these cultures and peoples. Professor Elie Metchnikoff, once the Director of the Pasteur Institute in Paris, comments that "the intimate connection between the depreciation of human nature due to Christian dogma and doctrine and the inferiority of the art of the Middle Ages cannot be denied." [171] Regarding any disagreement in rules, forms, or catechisms, the good Fathers wrangled and killed each other. At the council of Nice several monks got killed for disagreeing as to how many angels can stand at the top of a needle! As Professor Burnham P. Beckwith so aptly puts it: "There is no logical and peaceful method of determining which revelations are true and which false. Therefore, from the beginning of history, religious leaders have frequently resorted to forceful measures—torture, enslavement, crucifixion, burning at the stake, religious crusades, holy wars, etc.—in order to enforce their decision that certain professed revelations should be accepted. During the Inquisition heretics were tortured or killed by millions. . . . Religious men have no non-coercive method of settling disagreements on religion." [172] To "enlighten" any dissenter, Dr. D. T. Atkinson reports regarding how they "persuaded" people in Scotland: "In Scotland, the iron boot was used for the purposes of torture. In this receptacle the foot was crowded and iron wedges were driven in at the sides until the foot was crushed. This method of inflicting punishment was in vogue chiefly to settle differences in ecclesiastical opinion and was used to convince the victim of religious truths." [173] To even scratch the surface of the history of the brutalities, the slaughters, the torture and murder of innocent people by the millions, the intrigues and ambitions of charlatans and fakers, would take volumes. We have no choice but to abbreviate. "From the age of Constantine to the end of the 17th century, Christians were far more fiercely persecuted by other Christians than they ever were by the Roman emperors." [174] The story of the Crusades will be forever a blot and a clot upon the history of mankind. An ex-priest confesses: "Peter the Hermit didn't lure the vast Children's Army of the early Crusades to their death through reason. None of the Crusades was based on reason. The frenzy to conquer the Holy Land and recover the Holy Grail was nothing but emotion." [175] Everett Dean Martin remarks, regarding the Crusaders' capture of Jerusalem: "The capture of the holy city, Jerusalem, was followed by one of the most bloody massacres in history. Mohammedans at prayer in their houses of worship were slain without mercy. The crusaders rode their horses into the great mosques in blood up to their bellies." [176] The Papacy was not any better in the dark and bloody centuries behind us. "In the tenth Century the papacy descended to the lowest depths. With no protector and assailed by Saracen and robber barons, popes were insulted, chained and murdered, while their murderers assumed the tiara. The fingers were cut from the body of Formosus and his corpse was flung into the Tiber with insults. Stephen IV was strangled. Two corrupt women, Theodora and Marozia, disposed of the papacy to their favorites for many years during an incredible orgy of bloodshed and vice. Alberic, the supposed son of Marozia by Sergius, controlled the city and made and unmade four popes." [177] To go into the history of the Borgias is too vile and polluting, but in general if one would study closely the history of the papacy he could not come to any conclusion that the judgments of the popes are infallible and the word of God.

We should not leave out the recurring idea by Christians that the world is about to end, the periodic flushes of *Dies Irae*. They expected to see the world crumble in the 17th Century and hundreds of thousands of people, including whole villages, committed suicide, so they could be dead when the Christ came! People paid monks to kill them and regular establishments were put up by the religious orders to kill on a "fatherly" basis for which the poor victims left everything they possessed to the Church. The priests waited somehow and lived. Bertrand Russell tells a witty story in this regard. "I knew a parson," he says "who frightened his congregation terribly by telling them that the second coming was very imminent indeed, but they were much consoled when they found that he was planting trees in his garden." [178]

One wonders why the Catholic stress on right and wrong and heaven and hell when it has hardly, in all its history, ever lifted its head to support and encourage humane, equitable and *just* forms of ethics and morals. St. Augustine, after a life of dissipation and disease, emasculated by his excesses, suddenly saw the "light" and condemned women as the cause of the fall of man. "The story of the Fall of Man in the third chapter of Genesis appears to be an abridged version of savage myth. Little is wanted to complete its resemblance to the similar myths still told by savages in many parts of the world." [179] "Almost every man whom Charon ferries across complains that his wife was responsible for his downfall." [180] To touch a woman is pollution and of course, priests in order to remain "pure" must not get married or have any sexual intercourse with the opposite sex, although

I do not know of any priests including all the popes, who haven't come from some mother's womb; fathered by some man who just got "polluted." This Catholic sexual facet is both hypocritical as well as condemnatory of the very life which their god is supposed to have created. Even Jesus protected Magdalena!

St. Paul himself said that "it is good for a man not to touch a woman." [181] "And Pope Gregory the Great proclaimed that he himself was corrupted because of his own parents' sexual intercourse and that everyone is born of sin." [182] "Copulation was the Evil, whether committed in the brothel or the marriage bed. There is no material difference, wrote St. Augustine, between the *copula carnalis,* between man and wife, and the *copula fornicatoria,* or physical union with a whore. Both are sinful. And Pope Gregory, two hundred years later, endorsed St. Augustine's doctrine: even marital intercourse was never free from sin . . . Clement of Alexandria goes so far as to say: 'Every woman should blush at the thought that she is a woman.' . . . Origen, one of the greatest and most original thinkers of the Alexandrine School, chose to give mankind an example by emasculating himself." [183]

The disrespect toward womanhood is one of the main pillars of Catholicism. Anselm of Canterbury wrote as follows: "Woman has a clear face and a lovely form, she pleases you not a little, this milk-white creature! But, oh! if her bowels were opened and all the other regions of her flesh, what foul tissues would this white skin be shown to contain!" [184] This is a typical example of the general class of woman-haters who constructed and pushed Catholicism all over Europe. Gibbons, referring to the Dark Ages, wrote: "It would not be easy within the same historical space to find more vice and less virtue." [185]

It was not since the beginning that Catholic priests were forbidden to marry or to have concubines; the early centuries openly allowed this and it was not until the 10th and 11th centuries that the Church first actually decreed that a priest must not marry or have sex relations in any way. This problem brought conflict with the Eastern churches and finally led to the break which separated the Orthodox Church from the Church of Rome. But strangely, even though the Church of Rome riled against womanhood and sex, this did not interfere with the Church arranging and accepting the profits of brothels in order to build for itself its own churches. "The curia had partly financed the building of St. Peter's by a tax on prostitution, on the classic model: this brought the papal exchequer 22,000 ducats, a vast sum for that age—four times as much as Leo X expected from the sale of indulgences in Germany. Had it been possible to squeeze a little more out of the prostitutes, the whole transaction in indulgences might have been unnecessary." [186]

Also, I cannot find in history any genuine move on the part of the Church to stick up for the rights of the poor, of the serfs, of the common man in his fight against the tyrants, the feudal lords. They never supported the trade guilds or any form of progressive and just rebellion on the part of the workers, the farmers, or the craftsmen. They opposed the Masons, and still do, and considered them as witches and coming out of the bowels of hell. When Luther, a former monk, became a power, he sided with the German lords against the peasantry, and hordes of defenseless, starving farmers and their families were slaughtered and the rest brought into submission. For humaneness the Church never reared its head against the abominable evil of slavery; neither did her offspring, the Protestant church, ever oppose slavery with very few exceptions. Let me quote just two items regarding the relationship between the Christian god and the poor kidnapped slaves from Africa, as written by Denison in his book, *This Human Nature*: "Jack Hawkins, in 1562, shipped 300 blacks from Sierra Leone to Santo Domingo. Ship was named *Jesus*. When the wind was calm and the ship couldn't go, threatening the precious cargo, they prayed to God. The wind soon blew up and in the log we find the following: 'Almighty God, Who never suffereth His Elect to perish, send Us a Breeze.' " [187] And here is the second item, which is indicative of the culture and the mind of Christian society in those days: "British laws upheld slavery—quite rightly from the point of view of the new business morality. Here is the actual wording of a bill of lading used by the good ship Thomas just over 150 years ago: 'Shipped by the Grace of God, in good order, and well conditioned, by James Dodd, in and upon the good ship Thomas, master under God for this present voyage, Captain Peter Roberts, and now at anchor at Calabar, and God's grace bound for Jamaica, with 630 slaves, men and women, branded

DD, and numbered in the margin 31 DD, and are to be delivered in good state and well conditioned, at the port of Kingston (the dangers of the sea and mortality alone excepted) unto Messrs. Broughton and Smith. In witness thereof the master and purser of the ship Thomas hath affirmed to this Bill of Lading, and God send the good ship to her destined port in safety. Amen. October 31, 1767.' "[188]

Another criminal assault upon the integrity and value of life is the Catholic damnation of unbaptized infants and children. In the Balkans and the Grecian islands the peasantry are still obsessed with beliefs that stillborn infants or children who die unbaptized are apt to become vampires, were-wolves, or giant ghostly insect-like ogres who prey upon the people in the dead of the night. When the Conquistadores came to Mexico and Peru they baptized the Indian children and then clubbed them to death.

When the missionaries came to the South Sea Islands and gradually bribed the natives with nails, hammers, mirrors, etc., they also gradually coerced the Hawaiians to wear clothes. Apparently, it was too much for these Boston prudes to view such beautiful, tempting bodies, so they persuaded them to wear "nightgowns" covering their bodies from their necks to their toes. As a result of this unnatural imposition upon their hygiene the Hawaiian people were decimated by disease. Believe it or not, there is more morality and less hypocrisy in the Congo or in Bali than in Paris, Rome or London. "I must confess," writes Hendrik De Leeuw, "that Western cities are indeed more immoral in many senses of the word than the Eastern ones."[189] "Shame is merely coquetry or a fear of causing disgust to those of a different sex; it is largely a product of civilization and a characteristic only of the clothed human being. In women it is actually an attraction to males, and well they know it."[190] "Among ourselves, the people who are regarded as moral luminaries are those who forego ordinary pleasures themselves and find compensation in interfering with the pleasures of others."[191] "Customs in any case constitute moral standards."[192] Mark Twain was so wise when he said that "there are many humorous things in the world; among them is the white man's notion that he is less savage than the other savages." Napoleon knew what missionaries are and what they could do, and therefore he proclaimed that "It is my wish to reestablish foreign Christian missions, for missionaries may be very useful to me as spies in all the lands they visit, in Asia, Africa, and America. They will be protected by the sanctity of their dress, which will serve to conceal their political and commercial investigations." Politics and priestcraft came from the same potion; they are inseparable unless there is the attempt of law to separate them. The hierarchies of religion know well that this is the truth. An ex-priest reveals this bluntly: "Lay Catholics and priests frequently try to tell non-Catholics that the hierarchy does not enter into politics and does not try to tell the faithful how to vote in either local or national elections. This is simply not true. Whenever it is to the advantage of the Church, it is in politics up to its ears." [193]

One of the worst depreciators of life, and consequently of its value of Time, a great deterrent to happiness and normal living, the cause of untold misery, frustration, neuroses, and psychosomatic ailments, is the Catholic-Christian concept of "morality" regarding sex, courtship, and marriage. This they didn't copy from the Greeks, Romans, Gauls, or the Norse; they took part of this dreadful doctrine from the Near East where the women had no voice, choice, or religious rights of any kind. The Catholic prohibition of divorce is stupid, criminal, and anti-natural. Even the savages and primitives all over the world, and many of them were or are monogamous in marriage custom,[194] allowed divorce without difficulties. The Christian influences and mores have created a sense of shame, prudery, and frigidity, throughout their history in Europe and then exported this plague into America and this sense of shame and prudery is unfortunately deep-rooted in the American culture. Albert Schweitzer states that "Christianity also brought European thought into relationship with world and life negation." [195] Actually we can trace the origin of this process of self-humiliation to primitive Sumerian and subsequent cultures wherein the man was master and "owner" of his wives, and the primitive value placed on a chaste (and young) girl as more "purchaseable" and physically desirable. This life of "moral" social custom has a natural animalistic basis and its expression is common in the social and sexual life of animals.[196] Louise Marie Spaeth, the brilliant social anthropologist, writes regarding the origin of the chastity idea: "A woman who has not been bought has neither social standing nor worth in the tribe, and the amount paid for a woman determines the family's social position. Only chaste women are bought—a great inducement to morality in the younger women of the tribe. When women quarrel, they taunt each other with the low price for which they were purchased, and even the children boast of the prices that were paid for their respective mothers." [197] The wedding band given to the bride originated in the custom of "making public" that this woman is the property of the husband and the band was usually inscribed with a symbol or signature of the household or master. It is only human to make mistakes and forgivable, according to Jesus, but according to the Catholic Church it cannot be undone and if the marriage is broken then it is an unforgivable and carnal sin punishable in hell. Of course, there are annulments and usually for a price. The poor Catholics must usually resign themselves to misery unless they have the courage to break loose; the rich Catholics are in a much better position to "buy" their way out via annulment, not in every case but the examples are far too numerous.

Concerning shame and prudery, the Puritans really did a good job of it when they came to settle here in America. Shame and prudery bring about the very reverse of their intentions; in the 19th century there was more venereal disease in America and Europe at the height of the Victorian period, when propriety and modesty were more emphasized than at other times. In the Orient you may find cheese-cake and strip performances mostly in the larger cities where the infiltration of Western culture has penetrated. When I was in Japan I attended their various theatrical performances and they were beautiful, interpretive, classical. In Kyoto I attended the annual Geisha Cherry Blossom Festival; I shall always remember it as long as I live, it was so beautiful and wholesome. Even the floor show of the Benibashi night club in Tokyo was a performance that could be held in Carnegie Hall or the Metropolitan Opera House, it was so beautiful; the only things revealed were the finely sensitive and esthetic emotional plays of the Japanese. Yet these people have no sense of shame or any concept of prudery regarding the human body. People of all ages and both sexes bathe together in a common pool and find no self-humiliation or sense of shame in so doing. Yet courtship, marriage, and love-making, are on a fine and high and very clean plane, more so I think than is usual in Western culture. In Bali I found the sense of shame even less existent. In the more rural villages, the Balinese women and men only wear a sarong from their hips down and in such a climate it is comfortable as well as understandable. Here, too, the infiltration of Western culture, especially in the larger towns as Den Pesar, has caused the enactment of laws preventing the people in exposing themselves as they have done for many centuries. I witnessed the Christian culture of shame at work when I visited the Elephant Cave located in one of the smaller villages in Bali, a kind of temple carved out of the living rock. I saw several European Christian men entice with money a group of Balinese girls to denude themselves for picture-taking. The Balinese men wouldn't think of doing this in a thousand years! Yet, as Dr. B. Z. Goldberg writes, there is no sense of shame when "even today in some parts of Java, when the season of the blossom on the rice is at hand, the husbandman and his wife visit the fields after dark and unite for the purpose of promoting the growth of the crop. It is a form of *mana,* a magical way of getting the generative divinities to do likewise and to bless the world with fertility." [198]

Even prostitution was unknown among savages and primitives until the "moral" and "sense of shame" Europeans came into the picture and not only demoralized and degraded the women wherever they went but also "gifted" them with veneral disease. Felice Belloti tells us in the book, *The Fabulous Congo*: "Prostitution does not exist among savages; it only started in its venal form after the arrival of Europeans and with the birth of a sense of shame—a sentiment which up to a few decades ago was completely unknown in the Congo." [199] The same occurred all through the Pacific Islands, the American Indians, the Mexicans, the Caribbean, and all the way through South America. Most of the brothels in the British West Indies, it is said, are or were owned by English gentlemen who sip tea or drink whiskey and soda in their clubrooms far removed from their houses of prostitution. Brothels in Latin America are usually owned or controlled by the chiefs of police, and the prostitutes usually give a percentage of their earnings to the local church as an offering to their patroness saint who "protects" them. While prostitution existed in India, Babylonia, Greece, and other parts (the sign of the prostitute's quarters was very unmistakably self-evident when I visited the ruins of Pompei), it was mostly as temple vestals or "prostitutes of the gods," being part of the religion and the receipts went completely to the priesthood, the ancient pimps; but it took European culture to really put prostitution on a public and paying basis throughout the world. Between the missionaries and the traders, the Christian idea of "virtue" was thoroughly peddled around wherever they set foot and their ideas of shame and purity vulgarized and degraded, and wherever they settled the area became libidinous, lascivious, lewd, and licentious.

In the Pacific Polynesian islands the lovelife of the natives was beautiful, natural, normal, and uninhibited until the Boston Puritans came to despoil their cultures and their islands. While the women were considered inferior to men, which concept is almost world-wide with rare exceptions, the low position of women was religious and social; so low was woman indeed considered in the religious light that they were not even accepted as sacrifices as the gods would be polluted with them, a parallelism to the Hebrew custom (until a short time ago) of forbidding women to enter the temple. But the sexual life among the people of Polynesia before the arrival of the New England ascetics, was so beautiful and wholesome and the people were so happy that even the Europeans and Americans considered and still consider these islands as the paradises of the world. Fortunately, not all of their own culture has been despoiled and when I visited the islands it made me feel good to know that the missionaries were not able to do a complete job. Sex is considered as a normal, natural habit like eating, drinking, sleeping or breathing. The more experience a man has before marriage and the more he is desired by women makes him so much more desirable to his bride, and the more pre-marital sexual experience a girl has makes her more desirable to her bridegroom. Bengt Danielsson did an excellent job in his book, *Love in the South Seas.* He tells us: "On all the islands the children's genitals are massaged . . . In the Marquesas Islands, which were the America of the South Seas, where people always wanted to be ahead in everything, they tattooed even the tongue and the genital organs. The Maori women for their part discovered that red lips were repulsive and therefore tattooed them blue." [200] In comparing our attitudes toward these things, he writes further: "It cannot be denied that compared with us Westerners the Polynesians seem abnormally oversexed. But this is primarily because we ourselves are so repressed and puritanical; in reality it is not the Polynesians but our own attitude towards sex and love which is peculiar and abnormal. . . . The cause of our unfortunately adopting such a unique position in this respect is, of course, the fact that our sexual morality is based on Jewish and Christian religious doctrines. No such relationship between religion and sexuality as is seen in our Western culture has ever existed in Polynesia and neither gods nor men were able to understand that sex life could be sinful." [201]

To me it seems that every living thing has the right, by its own nature and life, to express its emotions and perform its natural functions in any way it wants to express it so long as it doesn't encroach upon the similar rights and attitudes of others and no one else is actually harmed by it. Sex belongs to the individual, not to custom, religion or state, and any imposition or influence which tends to inhibit it or put fear into the person so that his or her free exercise and satisfaction of the basic physical and emotional needs are suppressed, is a violation of the natural rights of the individual, a trespass upon the sanctity of life and a bar to the natural flow of happiness in living experience. Philip Wylie was so right when he said: "For it is the 'righteous,' alas, who push more people around than the 'wicked.' " [202] In the study of human culture we find the "righteous" and the "pure" to be most often the wrongful and the dirtiest. This is the sickness and vulgarity that Christian Puritanism and Catholic dogma have heaped upon humanity.

This is just a small, fractional part of the dreadful story of the Church and its culture, and it is another reason to realize that if people ever want to know the meaning of Time, the value of each hour, the preciousness of life itself and the respect for other lives as well as for their own, then they must liberate themselves from this fearful dragon which has festered and feasted itself upon the lives of millions and millions of misled and innocent people. "It is possible," writes Bertrand Russell, "that mankind is on the threshold of a golden age; but, if so, it will be necessary first to slay the dragon that guards the door, and this dragon is religion." [203] People cannot regain life and Time once spent is never retrieved. Life and Time are not retraced, and absolutions, confessions, and dispensations will not bring back one single moment. "Recklessness and deceit do not automatically excuse themselves by notice of repentance." [204] We have tried to expose the basis of the power of Catholicism—as *a myth built into a fear and a threat.* "Without a mythology a religion could hardly operate. . . . The Biblical stories of the creation and fall of man, of the Mosaic legislation, of the divine government of the world and the day of judgment; the doctrine of the Trinity, the lives of the saints, and the love of God—all this religious material forms the very heart of the practical devices by which the Christian religion operates. Take away the use of the mythology and you take away the religion. Take away the *belief* in the truth of mythology, and you begin to understand the religion for what it really is." [205] Whether the Christian people will free themselves from their chains I do not know, nor when, but I do feel that the "hold" is not as strong as in the past. Yet I like to hope, for their sake, that perhaps someday this could come to pass. George Santayana feels more confident and writes: "The shell of Christendom is broken. The unconquerable mind of the East, the pagan past, the industrial socialistic future confront it with their equal authority. Our whole life and mind is saturated with the slow upward filtration of a new spirit—that of an emancipated, atheistic, international democracy." [206] Bernard Shaw says: "The Bible . . . is hopelessly pre-evolutionary; its descriptions of the origin of life and morals are obviously fairy tales; its astronomy is terracentric; its notions of the starry universe are childish; its history is epical and legendary; in short, people whose education in these departments is derived from the Bible are so absurdly misinformed as to be unfit for public employment, parental responsibility, or the franchise. As an encyclopedia, therefore, the Bible must be shelved with the first edition of the *Encyclopedia Britannica* as a record of what men once believed, and a measure of how far they have left their obsolete beliefs behind." [207]

We have seen very plainly and historically that one of the prime and basic foundations of the religious background of all cultures is the adoration or denial, or both, more or less, of the creative or sex principle. Gods were deified as the *natural,* possessing or were the actual phallus or yoni itself, or else were deified as the *anti-natural,* "purified" of it as a sexless, ascetic ideology of withdrawal from the world and from procreating further human beings as part of the general world pollution of sin and the cravings of the flesh.

As a result, the Western world and its cultures, fathered by Christianity, is a hypocritical, insincere and anti-natural world of cultures because its ascetic and self-denying parent laid down the principles of its structures and processes—anti-natural, anti-human, and anti-life and anti-reality make-believes which fouled and froze the natural and normal requirements of living and life happiness. Instead of teaching a person how to keep his body clean, his mind clean, and his habits ethical, honest and truthful, for his own enjoyment and preservation, it ordained that soap and water were not needed at all so long as he believed in the Lord and in the coming Judgment.

Fortunately people everywhere are gradually realizing that religion and its history is a history of mythology, of metaphors, and of symbologies; not something to hate or adore, but to understand its meaning in the past, in the present, and its portent in our lives and for those to come. Samuel Miller, dean of the Harvard Divinity School, reveals the increasing admission: "The whole imaginative structure of Christian truth, elaborated in myth and symbol, for the most part has crumbled under the impact of the last three centuries of revolutionary thought, scientific methods, and historical studies. The vision of reality articulated in this great Biblical formulary has evaporated and no longer serves as a frame of reference for elucidating the mysteries of being human." [208]

There are many other religions and cultures to explore, in Europe, Asia, Africa and the Americas, but we will just have to bypass them and proceed briefly to look into the cultures of Mexico and the lands to her south. The main purpose of this essay is surely neither a history of religions nor an ethnological inquiry into world cultures. Although we have barely scrutinized limited selections of these cultures in a few pages I hope it may be sufficient at least to alert us and to serve in a way with a general idea so that we may more properly be able to appraise our natural assets, our historical heritage and the evolutional trek from the jungle to our living room, and thus attempt to better be able to understand ourselves somewhat more clearly and as a result, I hope, this may give us a better way of further reflection upon our philosophy of Time.

The Aztecan people lived in the area which is now Mexico City and its environs. They were a warrior society headed by a priesthood and the top priest, a demigod, was also the chief of state. One might call it a military theocracy because the nature of its religion required so many victims for sacrifice constantly, and as the number of victims from its own population became more limited, the Aztecs soon were compelled to invade other tribes and bring home the captives for sacrifice. Their main source of food was corn and corn depended on rain and sun, so human sacrifices were made to satisfy the gods of the elements to ensure good crops and fertility. Women and children were killed in these fertility rites to "sympathetically" bring about good harvests. During the reign of Ahutzotl about twenty thousand people were killed as a sacrifice to the Serpent-god, having their hearts torn out of their breasts while alive. Cannibalism was also practiced as part of the religious and magical beliefs that the one who partakes of the deceased will absorb the virtues and the powers of the one who is eaten. This is a common practice in savage and primitive societies and part of the *homeopathic* magic generally permeating human cultures up to present times in a thousand different ways.

To give you an idea of what a religion can do to a people let me quote from Professor Vaillant's authoritative work on the *Aztecs of Mexico*: "The Aztecs performed a hideous ceremony in honor of the Fire God, *Huehueteotl*. Prisoners of war and their captors took part in a dance in honor of the god, and the next day the captives ascended to the top of a platform, where a powder, *yauhtli* (Indian hemp), was cast in their faces to anesthetize them against their ghastly fate. After preparing a great fire, each priest seized a captive and, binding him hand and foot, lifted him onto his back. A macabre dance took place around the burning coals, and one by one they dumped their burdens into the flames. Before death could intervene to put an end to their suffering the priests fished out the captives with large hooks and wrenched the hearts from their blistered bodies. . . . In contrast to the callous brutality of the fire sacrifice, the ceremony in honor of the god *Tezcatlipoca* was strikingly dramatic, tinged with the pathos with which we view the taking of a life. The handsomest and bravest prisoner of war was selected a year before his execution. Priests taught him the manners of a ruler, and as he walked about, playing divine melodies upon his flute, he received the homage due *Tezcatlipoca* himself. A month before the day of sacrifice four lovely girls, dressed as goddesses, became his companions and attended to his every want. On the day of his death he took leave of his weeping consorts to lead a procession in his honor, marked by jubilation and feasting. Then he bade farewell to the glittering cortege and left for a small temple, accompanied by the eight priests who had attended him throughout the year. The priests preceded him up the steps of the temple, and he followed, breaking at each step a flute which he had played in the happy hours of his incarnation. At the top of the platform the priests turned him over the sacrificial block and wrenched out his heart. In deference to his former godhood his body was carried, not ignominously flung, down the steps; but his head joined the other skulls spitted on the rack beside the temple." [209]

One day Cortez and his fellow-plunderers came. The Aztecan king, having dreamed of the end of his reign by the invasion of a white god, saw in the Spaniards the realization of this prophecy. Besides, the Aztecs, who never before had seen a horse, thought these people to be monsters of man-animal gods, were frightened and gave little resistance to their conquerors. The Spaniards slaughtered and robbed them and, of course, the Spanish monks who accompanied them, had a Roman holiday in conversion or death. The story of the part that Christianity played in these countries is pitch black with the blood of millions of innocent people, a story of extermination and robbery and despoilation of graves and temples for gold and treasures unequaled in the annals of human history. Of course, the Catholic Church has remained in religious power since then and the minds of these Indians have hardly progressed, as a result, right up to the present day. "It seems hard to be always attacking the Roman clergy," writes Tylor, "but of one thing we cannot remain in doubt—that their influence has had more to do than anything else with the doleful ignorance which remains supreme in Mexico." [210] Fortunately, since the reforms of the 1920's the Mexican people are now exposed to secular educational methods, illiteracy has somewhat lessened, and the power of the Church has been restricted to some extent.

The Mayan people lay to the south of the Aztecan and Toltec empires. Mayan cities and settlements were scattered over what is now Guatemala and worked their way up to the lands north of it and south of the Toltec domain, and to the north-east to occupy what is today the Yucatan peninsula. They never had or saw cattle, had no metals, only stone and wood. The older lintels were made of wood, the later ones of stone. Here, too, corn was their staff of life and their religious life, their study of the stars, the sun and the moon, the calendric mathematics of the priests, the rituals and the beautiful palaces and temples were erected by the Mayan peasantry so they could continue to grow corn and harvest it. As far as can be revealed and pieced together by archaeologists and ethnographers, the pre-classic archaic period of Mayan history indicates that the head of each family or kin was also the priest who carried out primitive rites and forms of sympathetic, *white* magic to induce rain, sunshine, and good weather to make the corn grow without destruction or injury until husking time. While the Guatemalan well-moistured plains were better suited for growing corn, the spreading arms of the Mayan migrations into Yucatan found much poorer soil, far less alluvial for agriculture but they still persisted in growing corn as this was what they knew and grew up with. The archaic Mayan deified everything around him and about him, everything in his environment, including the dust beneath his calloused feet. There is no telling just when the pre-classic Mayan primitives first formed their cultures in this part of the world; the "good" monks and their fellow-despoilers made sure of that—they destroyed and burned everything and anything they found that had anything to do with Mayan religion, and it was the religious leaders, the priesthoods, who had the records and the keys to all the secrets of the Mayan civilizations. Because of this only a meagre handful of hieroglyphics can be translated or explained; most of the secrets probably went up in smoke before proud Catholic missionary zeal. Fortunately a few extant records were discovered centuries later and from these much has been revealed regarding the life and times of the Mayans. Here was indeed an amazing people and an amazing civilization. While the main premises of their religion and culture had to do with fertility, rain, and their harvests upon which their very life and survival depended, they did not make human sacrifices to the gods until the Mayans were invaded and overwhelmed by the Toltecs from the west and north about a thousand years ago and the new Mayan culture was a combination of both cultures.

There are no remains of the homes of the Mayan peasant or farmer; they were made of wood and grass and faded away without a trace. The only ruins we have are of the temples and the palaces where lived the priesthoods and the royal families, and considering that they only worked with stone and had no metal tools of any kind, it is amazing how they built these beautiful structures that in many ways are more magnificent than the Egyptian pyramids and tombs. Though they used a round doughnut-shaped stone disc for ball games somehow they never thought of using it for a wheel. As their culture rose the priesthood rose until the society became as pure a theocracy as a country could become. As Sylvanus G. Morley wrote: "From the cradle to the grave, the life of the common people was dominated by their religious beliefs as interpreted by the priesthood."[211] Morley continues to give due credit to this primitive but astounding people: "As early as the fourth century after Christ, Maya culture was firmly established; Maya religion had become a highly developed cult based upon a complete fusion of a more primitive personification of nature with a more sophisticated philosophy, built around a deification of the heavenly bodies, a worship of time in its various manifestations never equaled anywhere in the world before or, for that matter, since."[212]

I know what Morley meant because I stood before many of these temples and reflected upon a culture which, in many ways, was more wondrous and more inspiring than the diseased and polluted cultures of Europe and Christianity. "The Mayas," writes Ralph Linton, "at the time of the Conquest were vastly better astronomers than any in Europe and as competent mathematicians. . . . The Mayas, as is shown by the inscriptions on the monuments, were aware of the exact length of the solar year." [213] I often wondered what would have happened if only these people had metals, wheels, cultivated cattle and knew more about agronomy and irrigation. While their main and probably more magnificent centers were located in Guatemala at Tikal, Piedras Negras, Peten, Uaxactun, and Palenque in northern Chiapas, I had the opportunity of visiting the Mayan sites in Yucatan, like Chichen Itza, Mayapan, Uxmal, Kabah, Sayil, Tulum; at Labna we camped outside the portals of the old ruined palace and roasted iguana over the fire which was the only light in the jungle darkness except for the moon which revealed the blue whiteness of the other ruins in the distant clearing. We slept in slung hammocks in one of the chambers and as I half-slumbered there was a constant parade of large bats swishing over my head in almost invisible flight as they went in and out of some recessed pitch-black room to my left, like a procession of Mayan priestly spirits annoyed at the intrusion and conferring about the strangers who came to sleep in so hallowed a sanctuary. By small plane we landed at Tulum, enroute to the Isla de las Mujeres. Tulum! I'll never forget it. A small settlement of ruins hugging the high rim of the coast facing the livid blue Caribbean, hemmed in by jungle; the fascinating, mysterious frescoes painted in the little stone temples, the upside-down sea-god carved in stone, their silent, empty sentry posts on high points overlooking the sea and the jungle plateau. I still have a few obsidian arrow heads given to me by one of the Mayas living there. Before I left the land of the Mayas I visited the excavation work being done at *Dzibilchaltun,* located a few miles out of Merida and which portends to be one of the greatest centers of Mayan civilization yet discovered and which may bring us a knowledge of the Mayas as far back as two or three thousand years B.C.

Long before the Toltec influences brought a sanguinary shade to their culture, the Mayas had acquired a society, theocratic in character, but which was astounding in what they know of astronomy, physics, and the other sciences. However, as in other lands and cultures, so did it happen here that the priesthood which was responsible for its rise was also responsible for its decline and decay. The priesthood controlled the entire society. Charles Gallenkamp, in his book, *Maya,* stresses this point quite clearly: "The keystone of Mayan civilization—its initial vigor and ironically its eventual failure—the insuperable power of the guardian priests who directed its rise along inflexible lines of orthodoxy. Everything the Maya achieved had upon it the stamp of religious striving; it had evolved from a commonly shared and widespread source of inspiration. Leading the way before masses of humble, illiterate peasants, the priests established the tenets on which their existence was to rest throughout their history; the subjugation of life and property to religious mandates, the endless building of ceremonial centers, the adoration of the pantheon—especially the gods associated with fertility, crops, and rain—through rituals and offerings." [214] He writes further: "The priesthoods—whose power was now unimpeachable—sought to apply their crafts to every facet of daily existence. They searched out the symbology and supernatural manifestations of the gods, their favors and displeasures, and the means by which they could be lured in living essence into the circle of human experience. By this endless pursuit of aesthetics Mayan civilization became a convincing testament to the power of religious absorption! Priests stood at the helm of Mayan society. The cities were theirs alone; the populace continued to dwell, as they had from archaic times, in thatched huts on the edge of outlying fields. Only the priests and segments of the ruling nobility entered the temples and shrines which were the labor of the awe-inspired multitudes. For the ordinary man life remained a relentless cycle of tending his fields, manual labor, and participation in ceremonial rituals. There were never-ending tributes and sacrifices required by the priesthoods to support their sequestered realms of astronomy, mathematics, and philosophy. Yet of such things, their subjects were permitted no knowledge whatsoever. Learning also was the sole dominion of the hierarchy." [215]

When the Toltecs overwhelmed the Mayas they brought with them the human sacrifices which were such a dreadful custom and ritual with the Mexican tribes. Most of the settlements or cities of the Mayas had to be located near water and most of this supply came from *cenotes,* subterranean rivers or wells, and many of these were used as receptacles for human and other sacrifices and offerings to appease the various gods, especially the rain-god, *Yum Chac,* for without rain they would have no corn and they would starve. Mayapan was one of the great cities built by the new Mayan-Toltec combination. At Chichen Itza was a large *cenote* into which humans were thrown regularly as part of the religious rites. So important was the ritual of human sacrifice that it became an honor and a blessing to have one's heart torn out and his living skin flayed and worn by the priests attending the killings. To commit suicide by hanging oneself as an offering to the gods was considered especially noble and highly merited to ensure personal immortality and a dignified place among the gods.

As they didn't raise cattle, or knew of any, the women of the tribe took the place of cattle in social status value. They had no voice in the household or in the social order. A husband had the right, by the custom and law of the land, to beat his wife if she didn't have his bath ready when he returned from the fields, or if she didn't do her work properly, or if she was considered lazy *by him.* She was nothing but a slave and if one could consider her from the viewpoint of caste, she was in the lowest order. The only exceptions were the women born into the priestly or royalty families and kinships.

Of course, here, like almost every other place in the world, the belief in souls and continuity of life after death was thoroughly accepted and fanatically believed in. Everything had a soul including every bird, animal, bush and tree. Death and life were not considered seriously as two distinct worlds; death was a phase of life and every death came about because of some spirit, bad magic, evil ghost, the anger of the gods or the slackening or absence of proper offerings and sacrifices. This togetherness of life and death is made more clear by Gann and Thompson in their outstanding work, *The History of the Maya,* in which they state: "The connection between this world and the next is regarded as rather close, and an Indian will whisper a message for a dead friend into the ear of a newly deceased corpse, in the confident belief that it will be delivered with promptness and dispatch." [216]

So far as human sacrifices are concerned, this was not just a provincial affair peculiar to the Aztecs and related peoples; it was common in most parts of the Americas. When the Aztecs and post-classic Toltec-influenced Mayas ate so very often the bodies of human sacrifices, it was not because they were just hungry but because they felt that by absorbing part of the body they also absorbed part of its powers, its life which they could be able to add to their own and so strengthen and lengthen their own lives. "The Indians of Quayaquil, in Ecuador, used to sacrifice human blood and the hearts of men when they sowed their fields. The people of Canar used to sacrifice a hundred children annually at harvest. . . . The ancient Mexicans also sacrificed human beings at all the various stages in the growth of the maize, the age of the victims corresponding to the age of the corn; for they sacrificed new-born babes at sowing, older children when the grain sprouted, and so on till it was fully ripe, when they sacrificed old men. . . . The Pawnees annually sacrificed a human victim in Spring when they sowed their fields." [217]

Of course, pure cannibalism, that is, the idea of eating human beings as food without any additives of superstition is well-known and commonly practiced in different parts of the world, especially in Africa. "The Ikonda," writes Felice Belloti, "and in general the tribes of the Ubangi valleys were considered the most refined in the matter of cannibalism, because they fattened their slaves before killing them and then steeped the corpses in brine to make them as tender as possible." [218] Yet, on the whole, cannibalism and superstition are more closely related than just the mere need for food. "In any case it has been found that the greater part of those who are killed and eaten are victims of superstition. A mentally deficient or hump-backed child is sacrificed so that no others will be born like it. A girl is killed and eaten so that a powerful spirit of evil may not prevent the brave hunters from finding game (though the truth may be quite different: the girl is in love with some youth of the tribe and will not submit to the advances of the witchdoctor). Sometimes an old man is eaten because he would soon have to die and be left as a meal for the wild beasts, and it is better that he should finish in the pot rather than in the belly of a lion." [219] . . . "If for us Europeans cannibalism is one of the gravest crimes, for the Azande and the Mangbetus it is the most normal and legitimate in the world. If burning someone alive who has been denounced by the local witchdoctor for casting the evil eye is a murder in our eyes, to any of the Bantu races it is the work of supreme justice." [220]

Returning to the Mayas, the Spaniards came, as we all know, early in the 16th century and began their deadly work of decimation, despoilation, and destruction, for the glory of the Church and their own pockets. But it didn't go entirely easily to wipe out a culture in short order. It was not difficult to massacre the Mayan priests and burn their temples, but the Mayan peasants, although converted in principle, could do nothing but maintain in their hearts and minds the general flow of their own religious culture and folklore. Gallenkamp enlightens us as to what had transpired: "But the ancient beliefs were not easily uprooted. What manner of men were these Spaniards who wrought unspeakable cruelty while spreading the teachings of a god they extolled as benevolence? And the Maya fled deep into the sheltering jungle by night, gathering in deserted temples to make offerings before smuggled idols and invoke the wrath of their gods on the invaders." [221] Quoting from the Books of *Chilam Balam,* regarding the fate which had overtaken the Mayas, we read: "It was the beginning of tribute, the beginning of church dues, the beginning of strife with purse snatching, the beginning of strife with guns, the beginning of strife by trampling of people, the beginning of robbery with violence, the beginning of debts enforced by false testimony, the beginnings of individual strife, the beginning of vexation." [222] . . . "Lust for gold wore the mask of piety, fanatic personal ambition was heralded as devotion to a national cause, cruelty became the means by which heathens were reclaimed into the family of mankind, enslavement was made their hope of salvation. Against the nations of the American Indian was once again to be pitted the monstrous oppression of overgrown regimes and dogma." [223] This is what Christianity brought to the Americas.

This is what the Spaniards brought to Yucatan, and this is the shadow of the Cross upon these unfortunate people who were discovered, degraded, despoiled and destroyed ruthlessly and recklessly. "The very few survivals of the ancient faith," writes Sylvanus G. Morley, "that have remained are not those of the priestly class, the esoteric cult of the astronomical gods and the involved philosophy behind them, but rather those of the simple gods of nature—the *Chacs,* or rain-gods of fertility; the *alux* or little folk of the cornfields like elves, who though they may be mischievous, are on the whole well-disposed toward mankind; the *xtabai,* or evil sirens, who by day are the *yaxche* trees of the forest but at night become beautiful maidens who lure men to destruction. The homely everyday beliefs of the common people about nature have outlived and outlasted the more formalized gods of priestly invention." [224]

And so, over many centuries the Christian paganism mixed with and acculterated itself with the Mayan religious culture and today the Mayan natives merely have different names for their ancient gods and retain their old-time festivals decorated and modified with Christian ornamentation. Some of the festivals I witnessed in Merida could have been easily recognized as processionals of the old days, the Mayas smearing themselves with a sort of royal blue paste and chalk, just like their ancient forefathers did. *Hunab-Ku* became Jesus; their mother-goddess became Mary. "A motley mixture of Catholic saints and pagan deities. In Yucatan, the Archangel Gabriel and other Christian saints became the *Pauahtuns* of ancient Maya mythology, the guardians of the four cardinal points; the Archangel Michael leads the *Chacs,* the former rain-gods. In British Hondorus, it is St. Vincent who is the patron of rain and St. Joseph the guiding spirit of the cornfields." [225]

The Incas lived in what is now Peru and Chile, from the Andes to the Pacific. They believed in a supreme being also, called *Tici-Viracocha,* maker of heaven and earth, and the other gods like the sun, moon, etc. The *Inca* or head of the state, was also a defied offspring of the sun. The sun-god, *Mocha,* was an all important god and many ceremonies and rituals were celebrated in its honor. The story of the Spanish conquest of Peru, of the Inca peoples, and the desecration of their tombs and graves by the greedy Spanish emissaries, is a tragic story of the New World. To this day the archaeologists are amazed and puzzled regarding the methods used by the Incas to build their superhighways, their temples made of immense boulders of squared and rectangled stone put together without cementing material. Regarding their religion and culture, like the rest of the world, they too believed in a soul called *xongo,* which was the same word for *heart.* When a chieftain or Inca died the whole tribe went into great lamentation and mourning and many sacrifices were performed in honor of the soul of the Inca. When the Inca Huayna Capac died the people "killed more than four thousand souls, what with wives, pages, and many other servants, to lay in the tomb with him, and treasures, jewels, and fine raiment." [226] Pedro de Cieza de Leon, the chronicler of the Inca story, tells us that "the native chieftains of these people were never laid in their graves alone, but accompanied by living women, the most beautiful of them, as was the case with the others. And when these chieftains were dead and their souls had departed their body, these women buried with them in those great vaults which are their tombs await the fearful hour of death to go and join the dead man. They consider it a great good fortune and blessing to leave this world together with their husband or lord, thinking that afterwards they will serve him as they did on earth. And for this reason they believe that the woman who died quickly would the sooner find herself in the other life with her lord or husband." [227]

So much work was done for the building of tombs and vaults for the dead that little was done, in comparison, to better the daily life or dwellings of the people. The Chronicler of the Incas continues to tell us from his travels: "And truly it amazes me to think how little store the living set by having large, fine houses, and the care with which they adorned the graves where they were to be buried, as though this constituted their entire happiness. Thus, all through the meadows and plains around the settlements were the tombs of these Indians, built like little four-sided towers, some of stone only, some of stone and earth, some wide, others narrow, according to their means and taste. Some of the roofs were covered with straw, others with large stone slabs, and it seems to me that the door of these tombs faced the rising sun . . . When the natives died they were mourned with great lamentations for many days, the women carrying a staff in their hands and clasping one another, and all the relatives of the deceased brought whatever they could, llamas, corn, as well as other things; and before they buried the dead, they killed the llamas and laid the entrails in the square before their dwellings. During the days they mourned the dead before burying them, from the corn they owned or the relatives brought they made a great quantity of wine which they drank, and the more of this there was the more they felt they were honoring the dead. After they had prepared their liquor and killed the llamas, they tell that they took the deceased to the fields where they had the grave. If it was a headman, the body was accompanied by most of the people of the village, and beside it they burned ten llamas, or twenty, or more or less, depending on who the deceased was. And they killed the women, children, and servants who were to be sent with him to serve him, in keeping with their superstitions. And these, together with the llamas and certain household articles, were buried in the grave with his body, putting in also as is their custom, several living persons." [228]

The Incas specialized so much in preparing to die and decorate their souls with all the gold and silver and precious jewels they could find that they, hundreds of thousands of them, couldn't beat down a mere handful of European criminals who came to rob and murder them. They were so loaded with fears of the gods that they even believed the Spaniards to be gods and so, fearing them, went down in defeat and submission. So deeprooted was their belief in the eternal life of the "other" world that the friends and relatives of the deceased would only think of going along "for the ride" and thus join the departed to the land where they could all be happy together "forever," so similar to the people in Europe who killed themselves or had the monks kill them in anticipation of the *Dies Irae*. Pedro de Cieza continues to narrate: "They buried with the dead their best-loved wives, and their closest vassals and servants, and their most prized possessions and arms and feathers, and other ornaments of their person. And many of their kinfolk, for whom there was no room in the tomb, dug holes in the fields and lands of the dead lord, or in those spots where he was most wont to sport and pleasure himself, and laid themselves in them, thinking that his soul would pass by those spots and take them along to serve him. There were even women who, to have more of a claim on him and so he would value their services more, fearing there would not be room for them in the tomb, hanged themselves by their own hair and killed themselves in this way." [229]

Time and life meant nothing to these people in their delusion of a "forever" world in lieu of the brief one they knew and lived. The only philosophy of Time these people had was the timeless, measureless, meaningless "eternity" they believed in and hoped for. I quoted these details especially so that we can try to sense the impact of the delusion in the minds of these people, and other peoples all over the world who have forfeited and continue to forfeit this life for something they actually knew nothing about but believed in it because they were taught and cultured to believe in it, to kill others for it and to die for it themselves.

Thus the story of the part religion plays and played in the lives of people everywhere continues and exacts its toll of time and tribute, and how long it may continue no one knows. One thing is certain: the new world of the sciences, carrying with it all the advance of communication, interchange of people and ideas, and the widening and weaving pattern of individuals and societies being culturally knitted more closely together, should eventually cause a gradual but definite fading away of these ancestral ghosts and bring the twilight of the gods so much sooner into the night of a historical but forgotten phase to the evolutional story of mankind. Perhaps the gods themselves, out of pity and sympathy, may pray for our freedom—if we promise to make scientists out of them!

Of course, in this present slow but definite transitional stage of modern ideas and trends, it is important to understand, as a result of these studies, that the further we go back the more we find that man did not distinguish himself as an individual apart from his environment and the other animals it contained and as the later religions rose and widened the events brought a rise of individuality in the religious perspective and one's relationship to the deity. Instead of being part of a collectivized entity tied to a pantheon of divinities, he gradually found himself confronted with individual responsibility and with it a new intimate relationship between himself and his God. The more the pantheons dissolved into monotheistic supreme beings, the more individualized became the human being and this brought, in turn, a widening separation of himself from the world about him and increased his stature and meaning to his God and to himself; he became "human-ized," so to speak, and left behind his cultural and religious ties to the animals, trees and stones that once were part of his spiritual world. He now became a theologian and a philosopher; the witchdoctor became the metaphysican. Instead of the group or tribe, or collectivized society as a unit, forfeiting its life for eternity, now the individual forfeited his life for the same delusion. The results did not change.

One needn't be an expert in sociology or in human behavior to acknowledge that there is a vast difference between the *history* of human beings and the other animals on this earth, and the cerebral evolution which marks the contrast in intelligence and accomplishment between man and his relatives. This is common knowledge as well as public and private reality. It is also common knowledge that while human beings are naturally cruel, psychotic in degree, and which is more or less normal with a million variations for the better or for the worse, the evolution of the human intellect, and with it the rise of religions, philosophies, and various systems of ethical, civic and social conduct, has more and more distinguished the human being for what he has been and what he is today.

But we must also be realistic in analyzing the experience and content of the few and the many. There are many wise philosophers and students of human history who strive to maintain the hope of mankind's realizing its *human* elements over and above the behavioral elements of other animals and feel that the exposition and extension of this human element must be not only preserved but activated to a greater extent if this world is to be hopeful of any peaceful and equitable solution of its many problems; and that the elevation and maintenance of the *human spirit* is the prime possible savior for the individuals constituting a more truly civilized, peaceful, and healthful society.

This would be a great achievement and I agree with Kahler and others that unless this *is* achieved there can be little hope for any such solutions. However, we must also realize that this attempt to *humanize* the human animal is also the expression of the human ego that consciously or unconsciously *wants* to enthrone itself at the top with everything else looking up. While man was willing to enthrone God as above him, he still inwardly felt in word, at least, that this made him *part* of that enthronement, of God itself, in action and content. That man created his gods so he can look up as a child looks up to its parental guardians for affection and security, was a concession in which man revealed his feelings that God and he were *one* person, that faith in his God meant communion or an interflow of spirit with this over-all Godliness. As I have emphasized before in these pages, if the human spirit is to persevere and finally establish its supremacy in the world order, it will come about because of the *pressures of necessity* for sheer physical survival if it ever comes at all. Should this be accomplished then the *truly civilized* individuals and its truly civilized communities will conquer the truly uncivilized individuals and their social, economic, political and cultural forms which are now strangling the peoples by the pressing grip of economic waste and the despoilation of world natural resources, by the regimentation and collectivization of individuals into cultural robots and caterpillars, and by the paranoical ambitions for power and false values of sick and sinister individuals and their sick and sinister philosophies.

The earliest forms of religion appear to be an attempt to preserve the *body* of man; his spirit, mind or mental make-up, were merely the means at his disposal to create the effort and the dream to realize this hope during his life and after death. And it will be by the sheer necessity of preserving his body that the spirit of man may yet be able to preserve the body at least for the life it actually knows and feels to be living. The *human* elements in human nature must be turned to reevaluate the value of himself, as a life, in terms of mortality, in terms of the life he lives rather than to consider his "human" superiority over other life simply because he thinks he is going to live forever and the non-human animals not. The great forces of humanism must be engaged to make this life, the life we live, a happier, more peaceful, more valuable and meaningful reality and upon this new evaluation and its penetration into the minds of people everywhere will rest the firm opposition to the forms of religious, social, political and cultural obscurantisms and the possible freedom that could possibly make the universe into a *living* thing instead of a cosmological speculation on which these obscurantisms have thrived, festered, and multiplied.

One thing we should not overlook, in deference to the religions of the past, and this is the tremendous influence, impact, and dominance they had in the shaping and furtherance of the creative, expressive, esthetic, and imaginative arts in all their attempts in form, sound, color, and substance. In Europe, in particular, I visited many of the more famous cathedrals and stood in almost reverential appreciation at the beautiful paintings, sculpture and carvings, of the many centuries of spiritual search by artistic expression and imagery. Also, I visited the many art galleries and museums of the European countries and many of the great paintings and figures could easily become altars of worship for anyone seeking esthetic appreciation, enjoyment, and philosophy. In every generation and probably in every country are born creative geniuses who, nurtured by the culture they grow up into, give individual expression which have become part of the great heritage of the arts of mankind. For it is the arts of man that primarily differentiate him from the other animals; certainly not his eating habits or his natural cruelties. Whether it be the colorful carved panels of Nikko, the Buddha of Kamakura, the Temple of the Thousand Goddesses of Mercy in Kyoto, the ancient wooden carving of Nara, the Jades of Aw Boon Haw in Singapore, the palaces and temples of Angkor Vat, the temples of Bali or of Java or of Laos, or the Pagodas of Bangkok, the religious art treasures of India too numerous to mention, the stone heads of Easter Island, the beautiful, picturesque men's clubhouses of Melanesia, or the massive window panels of Notre Dame Cathedral, the classic temples of Greece and Rome, the Pyramids of Egypt or the Columns of its Karnak, too numerous to mention arts of almost every description, the carved walls of the Hittites and the Assyrians with their endless parade of gods, the wondrous edifices of the Mayas, the Aztecs and the Incas, they are all beautiful, exciting and exhilarating to see and feel.

I remember well the village cathedral in Taxco, Mexico, built by the local Indians, every part of it hewn and carved by hand including the pews, chairs, and benches. And I remember very vividly a statue of man with his hands outstretched pleadingly to the heavens, sculptured in Barlach style, set in the center of the little churchyard of the town church of Chiavenna, Italy, with the beautiful snow-capped Alps rising up behind, and I stood there for some time enjoying its beauty, its message, appreciating and trying to understand the profundity of man's philosophies seeking guidance, paternal shelter, and answers to the riddles of life.

I think it is just and proper to give credit to the expressive mind of man rather than to the tenets of religion in judging the panoramic procession of the arts of man, because here we have the natural tools with which man was able to actually create them. Religious expression in art emanates from primordial and earthly desires and quests based on very much natural and physical manifestations of the human being. "There are no aesthetic laws which are valid exclusively for works of art and inapplicable to human beings of flesh and blood." [230] There is no question that the spiritual and emotional needs of man "fired" and inspired his mind and body with spiritedness, and animated his creativeness, individually and collectively, to produce forms and expressions that gave him satisfaction, relief, adoration, and a visual translation of his concepts and of his pursuits to overtake or resolve things and ideas he could not physically reach. Also, there is no question that people grow up and into, and are firmly affected by, their cultures whether they be geniuses or ordinary people; and whatever their art forms we will find these art forms more or less attached to, or explainable in terms of their cultures. In Christianity, where the religious nucleus lay in a relationship between the individual and his god, including his fears and loves, and his "hopes of paradise and threats of hell," here we can fairly observe a status of an individual's personality and its *personal* communion or intimacy with his god for his own sake. As a result, his expressions in art forms take on a greater individualistic approach and potential. On the other hand, while each person is basically and naturally interested in his own security and salvation, yet in many other religions, like the Incas or the Mayas, or of many tribes of the African veldt, we find the collective or group approach to art forms because here the relationship was between the people, as a unit, village or nation, represented by the priesthoods on one hand and the gods on the other. Whether it be the crude cave paintings of early man, the totem carvings of the primitives, or the massive sculptures of Michael Angelo, they are all beautiful and expressive, but to take the position that the existence of art depended or depends upon the existence of religion seems to me to be untenable because man's art emanates from his cultural habit, and regardless what forms this cultural habit takes, the potential of art expression in man will continue, more or less, according to the seeding of inspiration and the fruition of concepts in any generation or age. So long as man exists he will continue to bear ideas, hopes, dreams, imagery, fears and hates, loves and affections, poetical fancy and creative fantasy, and these will continually "force" him to art expression.

We should keep in mind that the earliest art expressions of the prehistoric hominoids were to a great extent sex-fertility symbolisms, earth-bound, nature-bound, and mortality-bound, and according to many anthropologists and archaeologists, a very substantial part of art expressions and creativeness were inspired, directly or indirectly, consciously or subconsciously, by the adoration and "worship" of the sex and creative art, and sex most certainly preceded all forms of religion. Practically every emblematic standard bearer of almost every religion is a phallic emblem in itself, whether it is the cross, the Hindu lingam and yoni, the actual image of Shiva himself as a male organ, the Jewish star and the aureate Star of El-Islam—both representations of sexual conjunction, or the golden crescent of the Moslems which symbolizes the vulva (*hheshoom*) "from the mons veneris to the anus," and a thousand and one other symbols of humanities throughout the world and going back as far as we can retrace. "Nearly all of nature from the palm tree to the golden mango fruit supplied phallic or womb symbols." [231] . . . "The extensive history of religion is based upon phallic symbolism, from the Christian Cross (*lingam-yoni*) to the Moslem Crescent. Without it, no form of religion could possibly have survived; for man, desiring partial understanding, had as the demand of vanity and human nature to cling to materialistic as well as spiritual principles in his adoration of one or many gods." [232] To enter into the "womb" of the Church and to get "reborn" through Christ are sexually symbolic metamorphoses, just two among thousands of sex symbolisms expressed in visual art forms but which have lost their original meanings as the cultural changes throughout the centuries brought different perspectives and interpretations concerning them. Also we should keep in mind that the translation of art thought into physical and visual art expression was made possible only because of the prehensilities of man's body and without which no art could ever have been created, and those prehensilities existed, as part of his *animal* nature, long before any human or human savage ever felt the touch of the gods or the need to look for them.

In the study of the arts of the various cultures and their attached religions throughout the human world and its histories we find that where the religion was the dominant power and the most influential agent in the daily life of the people, the creation of art expression most often took the form of religious expression because only the church or temple, and the nobilities or ruling classes co-existing as power factors, controlled the creativity of the arts, with few exceptions, as these had the means of subsidizing the various commissions for individualistic or collective work, or had the means of paying for them, or forcing the people to create or construct them. While the sculptors and painters worked to make masterpieces for the Sforzas, Medici, Colonnas, Orsinis, Savellis, and the Borgias, the people lived like filthy mice and worse than the animals in the woods. While Michael Angelo painted the ceiling of the Sistine Chapel the Roman people couldn't afford to sanitize or paint their house walls. While Peter the Great of Russia amassed together great collections of art for the Russian palaces and churches, his people were suffering under tyranny and oppression. While the Spanish kings filled the rooms of the Prado with their gems, their people were poverty-stricken, ignorant, and enslaved. And still are. Hitler enjoyed, no doubt, admiring his art collections while the crematories were burning innocent people by the millions. Alas, the palaces, castles and cathedrals of Europe most certainly represent a distinctive culture and a great form of art expression, but we should remember that each palace and each castle and each cathedral were surrounded by slaves and their hovels, and the tragic shadows that these cast upon the magnificent edifices of grandeur can never be removed.

The great masses of the ordinary people could neither afford, in those times, such luxuries or were unable themselves due to the constant penury in which they existed and the continual struggle to exist as serfs and slaves to their lords and masters, to either appreciate fine art or possess them. Point your finger at any period of religious density, dominance and control in any country and there you will find art expression constricted to the adoration and maintenance of that religion. Thus no one can tell how much the free, uninhibited, uninfluenced expression of art has been retarded by such unilateral and narrowed channels of suggestion and direction, and no one can tell, as a result, how extensive, appreciative or ingenious could have been the art expression of people if they had not been so influenced and controlled and emotionally overwhelmed by this factor. If Art is to really find its greatest and fullest expression it can only do so when the creators of Art are free spirits, unshackled, uninhibited and unfrustrated people and, though attached to the habit of their cultures as all life is attached to its nature and its experiences, can still be able to *personalize* and *individualize* their expressions and imaginations to achieve, according to their abilities and inspirations, whatever esthetic experience they are free to accomplish.

Without any desire to detract any artistic or cultural values from the *objets d'art,* which once created become both historical and transcendental, we should take into account that there are other expressions of art which are not so fixed, solidified, physically formed, substantive and substantial as the material art forms created by man, but which are, in many ways, even greater than these things in the esthetic life of people. I refer to the "lively arts," the dances and the music of the world, the poetry of language and the songs they contain. These are the arts that *breathe* and *move,* that *live* and give off *sound,* that pulsate and pant, leap and gesticulate, rhyme and articulate, sigh and sing, for they exist in the very life, motion and sound of mankind. These are the arts of the *continual present.* Man may identify his history from the physical forms of art he has left behind but he identifies his presence by his dance, his utterance, and his movement. They belong always in the present, not in the past, because they are constantly relived, regenerated and reborn in our daily lives. The Dance generates beauty, and Beauty, like Art, must breathe freely; it should not be standardized, congealed, metered or judged from rules, beliefs, or customs. Art, like Beauty, is a sense of ecstasy which is translated into an exhilarating satisfaction, a requirement for the restlessness of life, a sensual success which arouses in us a focus of introspection. Beauty is the symmetry of Art which appeals to our admiration. It is the silent language Art speaks with, to mirror our own dreams of perfections and to resolve our imperfections into states of satisfying experiences and acceptances. Beauty is the art which pleases and refreshes; it is the pause which allows the transfixion of our ideas and desires into things about us, to reflect again what we seek and enjoy. If there is any universality in Art, we will find it more than ever in the dances, songs, and music of peoples. The whole world enjoys these. It brings strangers and strange peoples together more than anything else because Art is the sovereignty of *all* of mankind, not merely any of its parts alone. They are happy and thoughtful things and the human being basically wants to be happy and satisfied. They soothe the scores of our mundane miseries. The Japanese lullaby and the Jewish *Raisins and Almonds* sound almost alike; either one would lull to sleep the babies of either country. The syncopation of the Balinese monkey-dance can easily stir any Westerner to reverential heights and the Moslem in New York can enjoy the *Cha Cha* as much as the Latins who created it. Americans love to dress as Indians and do their dances and Japanese girls love to come to New York to show us that they can wiggle their assets just as good as the American girls, with a little assist from our choreographers. Perhaps between the threats of science and the love of the arts, mankind may yet be finally liberated and united into one sovereignty!

Yet the march of the sadists, anti-natural depressors, masochists and ascetics, continues unabated to despoil mankind of whatever joy, comfort, and happiness it could possibly find. Only recently the Cardinal of Toledo, Spain, spoke out very sternly against the "sin" of dancing. He said: "Among diversions, probably none constitutes a graver and more frequent danger than dancing. Modern dances, among which we can classify all those involving an embrace, are a serious danger for Christian morals because they are very close to a state of sin." [233] A good boy, but for what I do not know!

Now let us pause from our cultural trek into the past and proceed to the second half of our venture in *Cultures* by taking a peek or two into what is going on around us today and to see whether people have really changed for the better or the worse. The past does not exist save in the memories we are conscious of as part of our present and the legacy we have inherited in the physical world we live in. Let us try to see what the *present* consists of, whether it belongs more to the past, or to the future, or to us. Let us see what we can learn about them and ourselves so that we can strive for some lucid, happier and more sensible philosophy of Time.

Life is no brief candle to me.
It is a sort of splendid torch
Which I have got hold of for the moment,
And I want to make it burn as brightly
As possible before handing it on to
Future generations.

GEORGE BERNARD SHAW

TIME AND THE *CULTURE FACTOR*: PART II

IN briefly transversing the historical panoramic view of mankind's cultures, one generally finds that primitive or archaic period human beings were already using fire and making crude stone and wooden tools, weapons, calendrical stelaes, idols, dwellings, etc. People, sometimes in different places unaware of each other, contrived to invent similar tools, but this unawareness is merely on the surface; actually the stream of history is a constant fusion and emission of one continuous chain of events and the concept of "isolated" similarities of ideologies and ingenuities is merely the contemporary or apparent periodic identities of situations. Very often migrating and infiltrating nomads or the results of conquest and invasion, brought about the widening use of tools, arts, wares, crafts, ideas, etc., and the interchange of modes and experiences brought newer and more improved creativeness. Also, while in one part of the world the people passed the bronze age into the iron age, there were still empires, races, and civilizations (that is, societies that have built cities or city-states) in the neolithic stages of culture, still using stone and wood, without the wheel or the potter's wheel, without any kind of metal, etc., even without any reasonable way of cultivating or raising cattle or farming. Yet, in other respects, these cultures were far more advanced than and exceeded in many ways, the empires of Greece, Rome, or of Charlemagne. Therefore, the rise of civilization and the emergence of cultures and newer cultures were not a *consistent, one-directional* advance or parade of humanities, but, considering it on the whole, a broken and irregular drama of many phases, peoples, and permutations. Only by visualizing this on a grand scale yet understanding the experiences and histories of provincial and segregated areas can we obtain a rounded and fairly realistic picture of the history of culture itself.

We find, also, that cultures, the building of settlements and later cities, grew as the inventive arts and crafts, social and trade relationships, religions and powerful leaderships, grew. If we carefully backtrack over the histories of the various peoples and their cultures, we will generally find that either the ruling priestcraft or the ruling leaders in the forms of kingship, nobilities, royal castes, etc., or combinations of both growing into and out of each other, left the principal heritage that archaeologists usually find—the ruins of religious and regal structures, tombs, and palaces which the peasantry and slaves built for them and to maintain them, even to sacrifice their lives so that these priests and rulers may have their services in the after-life beyond the grave. What the ordinary people lived in perished like dead leaves, into the earth. Considering the general march of humanities throughout thousands of years of prehistory and history, one thing seems to stand out for a certainty: the individuals of the masses, the multitudes called the populace, never had a chance! They still don't, in the real sense of the word. The "little man," the unit of many groups of units composing the population of any country, was always being made use of. They had to keep struggling to find some compromise in reaching for peace and security and in yielding to the greater pressures of force and religious zeal. No god came down to hand them anything or to protect them in their constant problem of survival. The angels never invented anything or provided the next meal. Mark Graubard, in his excellent treatise, *Man the Slave and Master,* writes: "The harpoon, fire, bow and arrow, irrigation, plow, house, clothes, furs, were devised in the course of man's bitter struggle for survival over a long period of time. Not one of these implements or conveniences, without which life would seem impossible, was donated by supernatural beings. Generations struggled and perished. No supreme kindness stopped the cold from freezing man's tissues, prevented floodtides from drowning entire tribes, communities, nor issued orders prohibiting the starvation of entire tribes by droughts. If man overcame these forces and made his life somewhat less hazardous, it was through his own efforts, along a course paved with an abundance of victims." [1]

The whole science of taxation, tribute and service, is a carry-over of thousands of years of *cultured slavery* of peoples during which periods the poor slobs had to build, work, fight for, die for, the few who empowered themselves by cunning or by force, or both, into the priestcrafts and the ruling, military leaderships of any culture. Whether the taxation or tribute, or any part of them, of any country existing today, is just or unjust, needed or wasteful, is not the purpose or prerogative of this book, but the point is that we should realize that the governments and cultures presently existing are not separate or alien to the previous cultures from which they arose and evolved from. The history of human cultures is not a series of separate and isolated "supernatural" creations, each unaffected or uninfluenced by others, but a continual arterial and veining flow of mixtures, admixtures, and mutations. Human culture is the stew of applied and habituated custom conditioned to keep cooking without ever seeming to get done and in it the young tender years of every child are flavored so that they can grow up and be patternized into the present culture and as they grow up they are forever self-odoriferous with the hash their forebears and parents have covered them with. "What is most humiliating about custom and convention," wrote Dr. Walter Kaufmann, "is that they appear inseparable from ignorance, misinformation, and hypocrisy." [2] The science of psychiatry is the attempt to identify and rationalize this hash for the benefit of the patient or the society, depending on the philosophy of the doctor.

This does not mean, in itself, that either the psychiatrist or the patient have any just recriminations against their cultures and the society which is a part of it, nor does it preconclude that all in any culture is bad or good, that if the individual were to free himself completely (as if he could) of this culture, then his worldly Nirvana or Utopia would be attained. But we can gain some light by realizing that when a person is born into the world he has the right, the inalienable right, to accept that part of his culture that he finds conducive to his happiness, health, intellectual development and the general satisfaction of his life, and to refuse to absorb or be absorbed, or accept any part or all of the culture which hinders his free intellectual expression and the use of his knowledge and intelligence, that operates, directly or indirectly, to make his life unhappy, unreasonable, irrational, which tends to keep him ignorant and which suppresses the innate desire of a human mind to think and to build some reasonable confidence in its ability and right to think, which uses the misguided duress of elders upon the young to acculterate them to believe in things only because their ancestors believed them and regardless of the impact of these beliefs upon a free, reasonably uninhibited, rational and naturally normal way of living, which tends to compromise *real* values of happiness for *unreal* values of prestige, conformance, and approval. Art Young may have exaggerated but he was on the right track when he said that "every child is a genius until it is forced to surrender to civilization." [3] "Liberty," writes Everett Dean Martin, "is a cultural achievement; it cannot be preserved by a populace which is moved by passion and sentiment and has no knowledge of the principles upon which life in any free society must always be based." [4] But it is a cultural achievement which developed from the primeval and deep-rooted trait of the *individual* to be free by nature and because of his nature to be free to express himself. "Individuality," writes Erich Kahler, "does not mean personal interests and their pursuit, but an inner, personal way of thinking and feeling that suffuses the whole of a man's existence and permeates all of its manifestations and forms. In other words: to have individuality is not so much to be a single one as to be a whole one, it is not so much to have a mind of one's own, as to have one's own mind. . . . As man loses his wholeness, his indivisibility, he loses his individuality. To have individuality is to be oneself, or rather it is to have a self to be." [5] "Man today is in danger not only through his lack of freedom, of the power of mental concentration, and of the opportunity for all-round development: he is in danger of losing his humanity." [6]

Just because a child is born into this world is no reason for him to be obligated to carry the burden of a thousand ancestors, their beliefs and equipment, on his shoulders, add to it that some of those piggy-back ancestors insist on his ears being stuffed so that he should not hear, his mouth gagged so that he should not speak freely, and his mind stopped so that he should not think, his body bound up in a million frustrating customs so that he should not live but serve, slave, and prepare to die. Like the custom of the Chinese to bind up the feet of their women so that they shouldn't be able to walk or run away. Thomas Jefferson wrote that "the tree of liberty must be watered with blood once every generation." He should have added that the tree of culture, which necessitates this periodical blood-bath, apparently must be fertilized with the bodies of people who have been and still are victimized by it.

We know today that we lived yesterday and while we are today the product of what we have been, it doesn't mean that we are contracted to carry the past as a blindfold or as a shield to funnel our thoughts and hopes always toward the past, rather than toward today and tomorrow. Yesterday's life has already been spent and is presently non-existent; we don't have to materialize it into some ecto-plasmic ghost and carry it with us today. "One thing alone life does not appear to do; it never brings back the past . . . That life was ever a fixed chain without movement was a human illusion." [7]

Life is the constant surge to rise, to get above, to grow, to sense the light and move towards the sun; like the plant or tree in the jungle, so likewise did the temples and hopes of man try to reach towards the sky and not to the darkness below. "The desire for change," wrote August Bebel, "in all human relations is deeply rooted in human nature." [8] So is it with the Time we live. We must use it to look up towards the sky, towards the light and enshrine each hour with light, with color, with meaning and value, and even in the blackness of the night to allow our minds to shed light so that we can see and understand not only the substance of the darkness but to en-joy its own values, its concomitant vicissitudes and the realizable ap-preciations of the day that may follow.

The difference between the individual and the society is that the life of the individual is brief but the life of the society continues so long as there are other individuals around. Therefore, the society is only of value to the individual if it serves him to make *his* life happier, longer, and more secure.

The great tragic repetitive error of society is that the highest prestige values are usually given to the oldest, who have had time to make the most mistakes and are already too old and dogmatic to change their opinions or their minds; and the lowest prestige values are given to the young whose flexibilities can bring about changes and a rejuvenation of spirit in any social organization but who are kept down by the elders who are supposed to be the keepers of the owl's eye of wisdom because they have whiskers. When the aged are wise to learn from the young they become younger and when the young observe the aged they become wiser.

Of course, there is always a certain quanta of inherited or acquired culture that may be basically "good" for the preservation and welfare of the individual and group, just as there may be a certain quanta of obsolete or "bad" culture which hinders the full, finer, and freer expression of both. We should keep in mind, therefore, as Dr. Bain does, that "the effects of different environment on like heredity is equivalent to the effects of like environment on different heredity." [9] This is the source pot of culture and the producer of "custom" dishes. "Custom," Dr. James Harvey Robinson tells us, "is the god all of us revere except a few who have come to see the casual way in which habit gets formed and the pertinacity with which they are transmitted from generation to generation." [10] What is important to bear in mind is that the cultural base should be constantly weeded, raked and the ground hoed so that the intellect can grow up freely, set new rootlets, mature unrestrained and unhindered by the dead and hardened vegetational rigidity of the past. Realistically this is most often the prerogative of the young. So much pressure is there of culture upon the growing child that he has to constantly fight his way through the ancestral growth of entangled acceptances and so-called inhibited decorum before he has a chance to see the light of day and a chance to think for himself and to cultivate a constructive and critical sense towards himself and society.

Our country, the United States, was created out of the struggle of individuals who fled from rigidity, dogmatism, and the persecutions of Europe where they were not allowed to think or worship freely, where the inalienable rights of liberty were suppressed or denied in order to maintain the power of the then existing cultures. They came here so that they could "breathe" freely, in peace, and feel like human beings should feel instead of being conforming, non-thinking robots of compliance and self-imprisonment. Yet, as our country formed, solidified, settled and grew, our history lengthened, what happened? The rigidities of Europe with which we could not happily and peacefully live, were established here. The persecutions and the inhumanities we left behind in Europe we now aimed at the Indians or at ourselves. Class cultures and castes sprouted all over the place, especially during the present century, and today we find in America an ailment and a situation which is gradually becoming tragically pathetic—the emergence of a *composite,* the *collectivized* individual, the standard bearer of mediocrity, non-thinking, easy-way-out, nonresisting, why-fight-City Hall?, you can't win so join them, non-intellectual, conforming, pecuniary peculiarity type—the *Average Man*—that is translated into the group or mass expression called *public opinion,* which you can usually and more often than not consider to be wrong, and which any alert advertising agency must concede in its practical approach to be of the intelligence of an eight-year-old child!

It is both strange and peculiar, in the study of human nature and its social and political systems in history, that whenever the individual rebelled to gain his freedom as an individual instead of a termite, it didn't take long before he got "termited" again and collectivized. Again and again this has happened and it seems that in the short history of our own country, where the start began with the attempt of the individual to assert his life rights as a person, as an entity in himself, it didn't take long before he again fell into the quicksands of the masses and lost himself into the group or congregational mind, and thus the prime purpose of freedom for which he has fought so hard, was forgotten and lost. Erich Kahler, in his excellent book, *Man the Measure,* elucidates on this factor: "The modern development of man, then, shows one thing with indisputable clarity: the growing tendency toward collectivity and the invalidation of the individual. . . . Since the Renaissance man's development has moved more and more *in the direction of collectivity.* At the very moment the individual was completed, the attempt to establish a collective order began." [11]

The principal approach of many of the modern purveyors of literature, of art, of commercial and producing industries, of the theatre, music, etc., is the "practical" principle, in order to insure success, of "sensing" that people do not desire to think, that the whole idea is to get them to accept or buy more and think less, or to make them feel like great thinkers without the trouble of actual thinking. It's a great make-believe in spite of the fortunate fact that there are honest, sincere, and competent people in the various fields of the arts and industry. It is the new idolization by greed and short-sightedness of a new god—the *Average Man,* the public idiot that public relations men adore and plan by, the criterion of policy-making and subsidy-planning. In this new land of make-believe, rigging, faking and phoniness, talent and ability are calculated on the premise of how "realistic" people are to "put it over." The new retinue of priests who keep the candles burning brightly in this new temple can be found in almost every type of profession, business, services, politics, etc., in which these priests have something to sell, get, or "enlighten" the doldrumated and opiated mass-mind, fragmentized and segmented into the *Average Man.* Very often this unfortunate new product of the American *Symphonie Fantastique* is called, to please the egocentricities of this non-thinking mass-mind—the *Plain Man.*

Modern man is now using his intelligence to defraud himself, to dissociate himself from his own personality; his mind no longer functions in a relaxed way so that he can properly and honestly analyze his own value and the values he places on things outside of himself. He doesn't seem to realize that he is being trapped in a vortex of false meanings and values; he inflates his own balance sheet so that he can, without vigilance or investigation, give more credit to an expanding bubble. He is becoming lost in an increasingly speedier machine in which human values, ethical principles, honest intentions, have gone with the wind. Erich Kahler enlightens us further: "Goodness never ruled the world, but never has goodness been made so hard for man, never has he been weaned away from it as methodically as through the tendency of our age. Out of man's obsession with goods, his entanglements in a net of technical functions and relations, his mental and moral confusion—out of the subservience of man to things have come the barbarism, the state of chaotic participation, that characterize the world of today." [12] Justice William O. Douglas adds his comment: "Madison Avenue experts have dinned into everyone's ears the bright and happy future that will exist if each of us will only drink, eat, sleep, ride, exercise, and think in the prescribed way. It is a uniform society that Madison Avenue promotes, a society engaged in rather primitive conduct, with little or no spiritual or intellectual glow to it. The great financial rewards go to those who can train people in understanding and manipulating response and behavior patterns." [13]

The priests of this anti-intellectual, beguiling and insincere culture in which the so-called public becomes an expedient and an expendable group-slave, operates on the basis that the less you "annoy" the *Plain Man* with "complex analyses," details, the need for reading, learning, figuring or criticism, or even the need of responsibility for decision or judgment, the better and more successful should be this new priest in getting and maintaining a steady supply of "disciples" or "contributors" to his temple. Even in the field of politics, the political machines try to find candidates, not so much for competency, statesmanship, honesty, a free intellect and a constructive, critical and analytical mind, but mainly on the basis whether he can appeal to the Plain Man. If its candidate is in the West it may be good if one can play a guitar, wear a ten-gallon hat, boots and a Western air in his speech and stature. The reason that the parties do not nominate some television two-gun toting marshall for President is because they are too sure he would be elected! At present there are political petty tyrants and crooks in too many provincial areas of our country who actually feel that they are out of the jurisdiction of the United States and are running their own little "empire" of fear-stricken citizens, drunk with their own local power, completely oblivious of every decent and honest obligation of a public servant. So far as the Plain Man is concerned, he wouldn't care if the Capitol were transferred to a Hollywood lot; perhaps he would even like it better and it would ensure a national devotion to more gun-toting television programs. John Dewey saw the coming of the Lords of Public Relations when he wrote: "We are beginning to realize that emotions and imagination are more potent in shaping public sentiment and opinion than information and reason." [14] Douglas also sees the coming of these "Lords": "We move more and more with the crowd and are infected with its mediocrity. The mediocrity of the crowd threatens indeed to condition our management of internal problems and our approach to the world. . . . We have drifted or been propelled by similar fixed patterns of thought. . . . The collective patterns of economic life have been on the increase." [15] Dr. Albert Schweitzer also sees this tragic trend when he wrote: "The bankruptcy of the civilized State, which becomes more manifest every decade, is ruining the man of today. The demoralization of the individual by the mass is in full swing. . . . It is only an ethical movement which can rescue us from the slough of barbarism, and the ethical comes into existence only in individuals. . . . Where the collective body works more strongly on the individual than the latter does upon it, the result is deterioration. . . . That is the condition in which we are now, and that is why it is the duty of individuals to rise to a higher conception of their capabilities and undertake again the function which only the individual can perform, that of producing new spiritual-ethical ideas. . . . A new public opinion must be created privately and unobtrusively. The existing one is maintained by the Press, by progaganda, by organization, and by financial and other influences which are at its disposal. This unnatural way of spreading ideas must be opposed by the natural one, which goes from man to man and relies solely on the truth of the thoughts and the hearer's receptiveness for new truth. Unarmed, and following the human spirit's primitive and natural fighting method, it must attack the other which faces it." [16]

Many newspapers, magazines, book publishers, broadcasting and television studios, advertising agencies and public relation experts, have more or less joined the priesthood of this emergent culture in America and have dedicated themselves to serve the anti-intellectual and anti-individualistic passivities of the Average Man. Before a script is passed, before an idea accepted, before a product released, before a program is ready, the priests consult their oracles in their Temple of Robotry, and if the oracle assures them that their aim will not result in any cerebral activity (God perish the thought!) on the part of the "peasantry," then it's "O.K." and on with the show! It seems quite ensurable that a bank will more readily finance a motion picture of delinquent idiots rock-n-rolling about in a stupid narrative than it would finance a picture of the Life of Thomas Jefferson or the History of the Emasculation of the American Indian. "This is a growing aspect of modern society," writes Loren Eiseley, "that runs from teen-age gangs to the corporation boards of amusement industries that deliberately plan the further debauchment of public taste." [17]

This Average Man must not be alerted to any idea that he can possibly think about. His sense of security and his complacent *modus vivendi* must not be irritated or disturbed. He must be flattered, considered a mental genius without mentality (i.e.: "A thinking man's filter and a thinking man's taste!"), considered a wise judge without wisdom or judgment, considered a humble, regular guy without a brain or an opinion, considered anything but literate, to such an extent that it would always assure the success of the priesthood of the temple. To the extent that man becomes an insipid, non-thinking, non-intellectual imbecile in such a culture, to this extent he can never judge or recognize the state of real values of his own and other lives and things and experiences about him, but in addition he cannot judge or recognize the value of the Time he lives and that he is being depreciated and misled into an illiterate stupor of mere acceptances and conformance to superficial, mindless and injurious impositions upon his person, his hard-earned money, his life, and his Time. "We have entered on a new medieval period. The general determination of society has put *freedom of thought* out of fashion, because the majority renounce the privilege of thinking as free personalities, and let themselves be guided in everything by those who belong to the various groups and cliques." [18]

With very few exceptions (who were the really great), most of the rank-and-file leaders in the political, military, religious, social and professional fields of mankind always operated on the acceptance and maintenance of their own established order and power, and on the assumption that the average or plain man cannot, or doesn't want to, think for himself. To make certain that he doesn't think too much they evolved the cultures, standards, and codes, to which we are more or less tied today and to which the people were tied in any period of man's history anywhere in the world. If he rebelled and smashed them it was usually out of the urgency of sheer pain, threat of extinction or the fear of being overwhelmed by cruelty and taxes. He hardly ever rebelled because he began to think that his life is just as important as the next guy's, whether he be king, pope or president, and that as a human being he has inherited what culture so often has taken away or forgot to give him—*a chance to live* happily, tranquilly, properly, freely, and to *think freely* so that life can possibly be generous, beautiful and enjoyable, and his Time may be possibly beneficent and wisely used.

Blindness to reality, blindness to natural limitations, blindness to the stupidity and non-essentiality of waste, blindness to rational exposure of the lessons of experience extrapolated for present and future, blindness merely to stay fixed, to maintain the *status quo* of authority for its own sake, blindness in belief and principle disregarding the rising realities and problems confronting us, these cannot forsee good for the people of the world. If the Catholic Church, the Karmists of India or the ancestor worshippers of Japan or Bali, had their way and hindered or prevented the people from using birth control methods, the gradual but inevitable explosion of population can only result in world bankruptcy of the essential materials to live and bring on disastrous conflict rising out of the sheer problems of desperate survival, reduce living conditions to jungle levels and spread poverty, disease, and dissension all over the earth. The probability is that widespread war on small and large scales may precede it. The supply of materials needed for the existence of humanity is not unlimited and the material needs of the Church have not come from the supernatural but from the natural, the pockets of the poor and rich alike. So has the material need of the people come from the natural and the natural resources of this world are not limitless. Unless we can control our appetites for waste; unless we can realistically and reasonably control the exploding birth rate, so long as human happiness is not impaired, retarded, or interfered with; unless we can feel the impelling impact of the urgent need to produce in order to live and live better rather than to maintain a supply of sales to the Molochian furnace of waste and overproduction and forced consumption; unless we can restrain and expose the deceptions for greed and profit; unless we can switch our energies from producing weapons for killing to tools for peaceful living, then the human race can only reflect the folly of committing eventual suicide by self-elimination.

This has been often the pattern of individual lives; it has almost always been the pattern of empires and civilizations; it may readily be the pattern of the world humanity itself. Profit for today disregarding reasonable preservation for today and tomorrow, is a poor foundation, an ill-guided investment. Building our deluxe, overgadgeted buildings, based on the principle of predetermined obsolescence and waste besides the monstrosity of almost perpetual debt, is building on quicksand and is a poor basis for a peacefully enjoyable and a reasonably secure and healthy economy. Sooner or later all this opiated pageantry of display, despoilation, depletion, and disintegration, accelerated by the weight of fictitious values and needless waste and antiquated dogmatisms, may sink our civilization and our culture into decline, disillusion and perhaps disaster. Overextension is the dilution of protective resources and the gradual dissolution within. It happened to Rome; it could happen to us.

Our culture is gradually becoming a football, thrown between fraud and folly; a game destined to be won by no one and lost by all. One doesn't have to be a complexed methodological mathematician to see the handwriting on the wall for the American people unless they awaken to the fact that a free individual is hardly of value unless he understands that this freedom, to enjoy and to continue, must be based on individual-social responsibilities, on ethical values, and wholesome intents, on peaceful sympathies and appreciation of the meaning and value of life itself for oneself and for others, on an educated, enlightened and intellectualized plane of honesty that can be reasonably competent to evaluate the real from the false, to distinguish enjoyment from intoxication, ethical intent from phraseological excuses and subterfuges, salubrious sanity from social chaos by indulgence, indifference and a thousand forms of cheating, chiseling, thieving, conniving and misleading. In short, we must stop ourselves from *kidding* others and others from *kidding* us. The evaluation has to proceed, not to justify, even naively, the processes which are slowly engulfing us but, on the premise that we can realize this clearly and therefore return or proceed, or both, to sound and constructive factors that can preserve our freedom by preserving the values that can possibly maintain it or give it real meaning.

There is no sense in being just critical for its own sake or neurotic enjoyment. There is much to be thankful for in our America. But this doesn't mean that we should, by careless believing or thinking, misconcepts and shortsightedness, jeopardize this security, this wonderful land, this peacefulness and joyfulness, by taking for granted that regardless how rampantly we waste our resources and how indifferently we feel for each other, how blind we are to the real and basic needs of sound, sane and sensible living without it being less joyful, less comfortable or less secure, this doesn't mean that we should allow ourselves to "get lost" in a whirlpool of caterpillar submergence to a gang of admen and stylists whose only object is to beguile us with revolving and repeating spectres of desires which are basically and objectively needless, superficial, phony, and deceitful. It is for the people of America to regain their senses, their freedom-loving culture and philosophy in which the economic and esthetic tastes of meaningful arts, crafts, industries, and ways of thinking provided us less with installment plans and more with a rational and physical enjoyment of quality, of good and wholesome ways of living. We have to realize that industry and its machine exist for us, not that we exist for the machine. We have to recover and rediscover ourselves, our individualities, our own integrities as thinking human beings. We must try to reduce our increasingly complexed trends in which we fathom our years outside and beyond ourselves, to the more simple and sound ground upon which individual and social peace and happiness are grown and cultured. Erich Kahler again splendidly exposes this danger: "The condition of the modern world prevents the millionaire as well as the average man from developing his sense of the enjoyment of life, for neither is a free individual. Both lead the same restless, exhausting life of work in the service of a collective function. The big executives rush from conference to conference and usually have an even smaller margin of individual life than their employees. They gulp down their life like a hastily eaten meal, until one day they die of a heart attack. . . . The ruling principle of our world is production, and the true problem is not the welfare of the many against the rule of the few, but the *welfare of man* against the *rule of things*." [19]

Let us hope, then, that the point of least marginal utility

and the point of diminishing returns may ultimately force this awakening upon us, the enlightenment that quality is more necessary than quantity, that change of itself is meaningless unless it is a concomitant factor to need, enjoyment and improvement and not merely another prop for a cumulative process of money power and economic influence. If these new priests of the Temple of Public Relations and Gadgetry, whose sacred idols are the billboards splattered over the countryside, are allowed to have their way too long, they will eventually undermine the American way of life beyond repair and return, and lay the foundation for the disintegration and collapse of our culture. These connivers who, like mice, gnaw away at every decent principle of representation in the various channels of democratic society, gnaw away at every ethical root by which this country was inaugurated and built, they must be exposed and chased out of their alluring temples dedicated to the program that production and sales must go up and up, for its own sake and theirs, and that the people are to be manipulated and handled to make this process continually possible.

People are not born, live and die, in order to make possible the maintenance of any economic factor. Economics is a science of the methods of subsistence and finance is a historical by-product of the means by which these are carried on. They are creations by man, not the creation of man. Unless the individuals who comprise society can accomplish this realistic factor, the machine which man invented to better himself may ultimately become the unconscious master and destroyer of the very life for which it was originally dedicated to serve. Automation can never take the place of human lives, hopes, dreams, nor should it. It should be controlled and channeled for the betterment of the human race. It should not become the new god before whom we dissolve the precious years of Time that are so uncertainly alloted to each living thing.

The machine and the conflicts it causes, its tendency to de-individualize and in-collectivize the people, are phases and revelations of the more dangerous trap into which these things are leading us—the trap wherein man becomes a caterpillar, becomes "termited" and thus loses his individuality, and with it, his humanity. Only by freedom can individualism be maintained; when freedom falls, the individual falls, and with it all the human-ness of which only individualism is potentially capable. Humanity thus can only return to animism and the jungle. "It is neither the machine nor war that is destroying modern civilization," writes Lin Yutang, "but the tendency to surrender the rights of the individual to the state . . . only by recapturing that dream of human freedom and restoring the value and importance of the common man's rights and liberties of living can that undermining threat to modern civilization be averted." [20] The individual must always be the *square root* from which all the problems of society should be weighed and resolved. The individual, himself, must realize the meanings, values, as well as the necessary coordinating socializations which, by the nature of himself and his society, require a realistic and understanding perspective of his own rights and the necessities of social order which can make the fulfillment of those rights possible. Woodrow Wilson so wisely stated: "The history of liberty is the history of limitations on the power of government." [21] John Dewey, in his paper titled *Authority and Social Change,* read at the Harvard Tercentenary Conference of Arts and Sciences, in 1936, said: "The genuine problem is the relation between authority and freedom. And this problem is masked, and its solution begged, when the idea is introduced that the fields in which they respectively operate are separate. In effect, authority stands for stability of social organization by means of which direction and support are given to individuals; while individual freedom stands for the forces by which change is intentionally brought about. The issue that requires constant attention is the intimate and organic union of the two things: of authority and freedom, of stability and change. The idea of attaining a solution by separation instead of by union misleads and thwarts endeavor whenever it is acted upon. The widespread adoption of this false and misleading idea is a strong contributing factor to the present state of world confusion." [22] Any such authority must be one which has the power to serve, not the power to enslave; the power to direct but not to dominate; the power to lead, not to oppress. The *Rights of Man* states clearly: "Men are born free and equal in rights, and so they remain. The natural and imperscriptable rights of men are Liberty, Property, Security, and the withstanding of oppression." [23]

So long as the individual is free and unafraid and courageous to search for the truths and verifications of living experiences and the values that can intellectually and intelligently be derived from it, so long will the individual, knowingly or unknowingly, be the consistent heretic, even against his own former convictions if need be. "The history of philosophy," writes Dr. Walter Kaufmann, "is a history of heresy. . . . Philosophy is a way of life, and, as the Greek word suggests, a kind of love and devotion. It is the life of reflective passion—penetrating experience, unimpeded by accepted formulas, thought about. That was what philosophy meant to Socrates, and if we want to bring philosophy down to earth again, it can mean nothing less than that to us." [24] Again and again Albert Schweitzer and his *Weltanschauung* keep pleading the case for the freedom of the individual: "A man's ability to be a pioneer of progress, that is, to understand what civilization is and to work for it, depends, therefore, on his being a thinker and on his being free. He must be the former if he is capable of comprehending his ideals and putting them into shape. He must be free in order to be in a position to launch his ideals out into the general life." [25]

Such a freedom of individuals and such a free field of societal structure can only be possible in an atmosphere of constant openness and flexibilities of mind movement and expression disregarding barriers, customs, and established orders that may hinder such a free movement of ideas and the translation of such ideas into individual and social judgments. In the past many specialized branches of learning and knowledge were considered as "non-scientific" or "mental" or "interpretative," etc., in contradistinction to the accepted sciences. It was professionally and academically rude, unfair, unethical and improper to cross over into another's territory and take even temporary or partial possession. Today, and increasingly more so in the future, we seem to feel by what we find and uncover, that there is no fine line where one study ends and another begins, where organics end and inorganics begin; in this way, also, there will be no lines where philosophy begins or ends or where the sciences begin or end. If maturity is to come to knowledge and education in general, and if mankind will continue to search for solutions to old and new problems, then there must be a constant fusion, freedom, and interchangeability of ideas and effort in the freest and most coordinative spirit. Mankind will come of age when its mind can become so free as to really be considered open and receptive to new knowledge and new situations no matter from what direction it may come and no matter how shattering it may be to old, accepted ideas and usages. This is the seed of tolerance and the fruition of tolerance brings peace.

This does not mean that the brain or heart, or both, may be competent to achieve a perfect Utopia. That is impossible as the entirety of nature, and all the parts within it, is a process of adaptational change which may be a perfecting or defecting process as optimist, pessimist, or cynic may interpret it. No intelligent person even dreams of an ideal state which, in itself, could be an intellectual prison. Intellectualism and intelligence require, by their recognition and continuity, a "conditional reflex" sensitive to change and the reception of ideas from constant experiences; this is hardly possible from an "order" or "state" of society in which perfection could be even achieved. As it is, it is most difficult to get even an imperfect society to listen to new ideas without labor pains. According to John Dewey, "A state which will organize to manufacture and disseminate new ideas and new ways of thinking may come into existence some time, but such a state is a matter of faith, not sight." [26]

Now we come to a little examination of the caste systems within our own country. We have briefly noted that in every culture and country in the history of the human race there have been caste systems, more in some countries and less in others. While in India there have been and still are hundreds of religious, social, and economic castes, in the Mayoid peoples of Central America and Yucatan there were principally two castes, the people and the priests. In Japan there were and still are a fair number of them, from the peasants up through the business bourgeoisie, nobilities, royalties, military, etc., all the way to the deified rulers. In ancient Egypt there were also different castes, the slaves, the free people, the priests and magicians, the pharaohs and their royalties, and the soldier or military castes. In Mesopotamia we also find castes, the slaves, the agriculturists and domestic animal raisers, the warriors, the merchant class, the priests, rulers, etc. Among the ancient Jewish people the priesthood or rabbinical caste was very powerful and dominant. In Greece and Rome we find the usual array of castes and more slaves, and the freemen ranked into different castes according to the trade or polis. As far as woman was concerned as a caste, she was really an outcast and in many countries today still is; she "was nothing but a means of procuring a supply of citizens." [27] And on and on throughout history we find no classless society; even in Russia or China today the Communistic oligarchic leaders most certainly do not feel they are in the same caste as the peasant or the coolie and are very unwilling to give either peasant or coolie the same authority and power as they possess over them. Karl Marx and his dream of a classless society made the mistake of calculating and appraising human nature on the same basis as production, consumption, and wages; the Hegelian approach to history, which he tried to apply, should have taught him better. The proponents and dreamers of a classless society are dealing with a myth. It is utterly impossible. Humans are not born equals, neither do they grow up that way, and the kind and degree of heritage and experience vary with each individual, even with groups and cultures. "No abstract doctrine is more false and mischievous than that of the natural equality of men." [28] Even the idea of a pure race or of a purely segregated racial nationality is pure fiction. "The idea of racial or national purity is a ludicrous perversion of reality." [29] Socially, Clyde Kluckhorn explains it more to the "skin": "Wild animals ordinarily breed only with others of the same type. The lines of most domestic animals are kept pure by human control of breeding. There are exceptions, such as mongrel dogs. But virtually all human beings are mongrels! For countless thousands of years human beings have been wandering over the surface of the globe, mating with whomsoever opportunity afforded or fancy dictated." [30] Regarding the Germanic fancy of recent times, sociologists Thomas and Hamm bring out that "the theory of either Nordic or Aryan supremacy is nonsense. There is an Aryan language but there is no Aryan race and the term Aryan culture is most indefinite." [31]

As long as human nature exists people will group or "caste" together where their economic, social, business, religious, communal, artistic, esthetic, professional, political, scientific, philosophical, or other activities or patternizing norms cause an infusion or "ranking together" by the coadunating processes of experienced similarities. Whether this is good or bad, wholely or partly, is aside from the point that these things exist because of the nature of the human constitution and its experiences.

Of course, the purpose, hope and intention of democratic theory and attempted practices, are to avoid such coagulations within the society. However, the sciences and philosophies of politics, esthetics, ethics, law and equity, did not pre-date the evolution of the human being; on the contrary, these sciences and philosophies have been trying, giving them the benefit of wisdom, to restrain and to acculterate, if possible, some tranquilizer for the constant, restless and gnawing insecurities of people and to translate and convert this restraint into some form of democratic action that would further our intellectual advancement and educate us so that an equitable yet orderly social order can be achieved. It is the innate fear of insecurity that grips people together by common rank and purposes, standards, wealth, religion or profession, and fuses them into a clinging stratification or status, and makes them helpless segments of an impersonalized caste system which subsequently not only controls and rules them but furthers the continuance of its solidifying and dogmatic status by opposing change, modification, infiltration, interrelationship, open-mindedness and a changeable, flexible attitude which are the generating premises of acquiring knowledge and of a truly democratic society. Unfortunately in our own country the formation and stratification of various castes are taking place very realistically and definitely. Let's place our lens a little closer and try to identify the processes and the castes which are now part of our culture and which may be the forerunner of a decline of our culture and the weakening of our country, which had happened so many times before with other countries and cultures. It is important, however, while we attempt to detect and understand these processes, that we are doing this with the criterion of a free and happy individual in mind as our pivotal center of cultural gravity and the value of the Time factor to him as a gauge of his relationships within the arena of his society.

As I stressed in a previous chapter, there is the unfortunate barometer of wealth that is the measure by which people are calculated and categorized in the caste systems of our country. Although money is the measure of men and classes in most parts of the world, the other cultures are much older and the power of money has not yet superseded to any marked degree the powers and prestiges of religions, nobilities, royalties, intellectual and academic societies, and even the enjoyment of leaderships and statesmanships in other countries as much as it has in the United States. And again I must emphasize, to avoid any accusation of cynical criticism of the rich or of the necessity of economic security: This is not the issue. The issue is the misled and misguided acculteration of our society that considers the power of money and the prestige of wealth as greater values than the basic human values that are really the most important, not only in the process of a person's maturation, not only most important to the general satisfaction and enjoyment of life, but of equal paramount importance to the sustenance and maintenance of democratic principles and ideals. The aristocracies of Greece and Rome, in aspiring toward their powerful influences, also laid the groundwork not only for their own downfall but also for the downfall of their countries. I can repeat again: Money is important, very important, but when it becomes *all* important then something has gone wrong with our sense of values.

By the caste system I do not refer to the grouping or clanning together of specialized fields of human crafts and schools, professions, ideologies, although many of these, too, become too often eventually dogmatic, reactionary, less flexible and more resistant to anything progressive if the progress may imply any threat to or a reduction of their established power, prestige or existence. By the caste system I mean the gradual acceptance of superficial, *veneer* appearance and *status* value. Take the average town, for example. If you look closely you will begin to see various classes categorized according to how wealthy they are, where their home is located, what the father does for a living, and who their grandfather was and from what country he came. Usually the richer people live naturally on the better, higher or upper sections of the town and the classes get lower until you get to the poor unfortunates who live on the other side of the tracks.

Poverty is surely no disgrace; it is unfortunate that the poor weren't so lucky. But to consider poverty and wealth as a measure of a person's dignity, goodness, character, peacefulness, honesty, intellectualism, refinement, is a *fraud* upon the misguided and a criminal assault upon those against whom these standards are directed. Whether a poor person, wearing blue jeans, gets stinking drunk in a cheap bar and grill or whether a rich sartorialized scion of the bank president or company chairman of the board, dressed in tails, gets stinking drunk in his "library," doesn't matter. They are both stinking drunks. If a carpenter who lives in Levittown or a millionaire who lives in Kings Point, if they should be both kind and merciful enough to pick up some stray animal and feed it or give it a home, or both seem to like Hamlet or go to the same show but in differently priced seats, to this extent they have both similar senses of value and it is this sense of value that counts. I have often been to the homes of poor people who had a few scattered books here and there and a few shelves of worn literary accumulation, but they were read and reread and often loaned out to other poor people to read, and they didn't seem to care if they never got the books back, so long as the friends were reading them. And I've been to rich homes and seen walls and walls loaded with fine editions, leather bindings, classics beautifully bound and engraved, and I very often found, in examining a volume from the shelf, that the pages remained uncut and the volume undisturbed. It reminds me of what Bertrand Russell said: "I visited book shops in every part of the country and found everywhere the same best-sellers prominently displayed. So far as I could judge, the cultured ladies of America buy every year about a dozen books, the same dozen everywhere." [32] The stupidity of making false and phony impressions is a curse upon the sacredness of Time and its value and usefulness in the life of a person. In a letter written home from the Galapagos, Charles Darwin wrote: "A man who dares to waste one hour of time, has not discovered the value of life." [33] We really haven't got the time to fool ourselves and others. If only people would stop trying to appear what they are not, and begin to try to live as they really are or hope to be, both they and their friends would surely be happier and more relaxed—and the police would be assisted greatly in knowing whom to lock up!

In a previous chapter I also noted that it is a human trait

to try to overcome, overwhelm something or other and that it is part of the restlessness, the constant mobility of life. Did it ever occur to you that this natural, emotional compulsion is also translated and expressed in the desire of one to appear taller, bigger, stronger, wiser, finer, more beautiful, more powerful, of a higher class with a more expensive house than his neighbor? Did it ever occur to you that this desire to overwhelm and climb up is translated in the desire for a bigger, flashier car, rubbing shoulders with the people of the "next, upper class," that it is one's inner satisfaction to see *his* wife wear a more expensive fur coat, hoping at the same time that his friend's wife wears a shabbier and an older and a cheaper one? This is the kind of culture that puts the people on stage without rehearsal and the only audience that could be around us is the inner guilt of our own stupidities, taking for granted that we do possess a certain quantum of activated brains.

This same compulsion for growth and bigness is becoming more expressed in business, industry, and the professions. Even the scientists want to become bigger, to reach the moon and the stars with space rockets, which is wonderful, but there are still millions of our people who haven't seen their first bathtub! In industry the generals of high finance have found out that it is not successful to grow piecemeal; they have found the secret of mergers and the quick enlargement of assets, companies, and stock fluctuations. High finance today is the art of making stock prices rise perpendicularly straight up without adding a brick to the factory or a cake of soap to any of its toilets. Even our government finds it most difficult to avoid being friendly to those nice gentry in our neighboring countries who facilitate the exploitation of the peoples of those countries by equally nice gentry here. And the strangest thing occurs when someone starts a revolution to free the people of these nice gentry. The first cries of complaint come from these gentry, their pockets bulging with the profits taken from these people, cries of improper, illegal confiscation, uncivilized procedures. It is so usual that the thief, when caught, should hide behind the Constitution! The "fifth amendment" may yet turn out to be the worst enemy of the democracy it was originally enacted to serve and protect. The societies that protect the rights of the Communists to the freedoms we enjoy would cease to be here if these very Communists would take over and destroy the very freedoms by which these societies exist. Even the idealist has to be practical if he desires to *remain* an idealist. The French nobles enroute to the guillotine also couldn't understand, either, why these crazy peasants wanted to cut their heads off. The walls of the Bastille and the minds of these nobles were of the same content and understanding. That's what culture does to people if they don't watch out!

The Bastille is now reduced to a single boulder in the center of a small patch of grass in Paris; I stood gazing at it for a long time, thinking of what went on here and the people who suffered here, of the people who always had too little and the people who unnecessarily had too much! Those in authority and power to enact laws, change style and innovate social moods, very often realize the penetrating force of "conditioning" the individual and the public to any particular ideology sustained by those in power. Professor Burnham P. Beckwith candidly stresses this method: "Even a dog can be conditioned to like a bible. It is only necessary to present him with food and the bible at the same time until he associates them with each other. Christian missionaries in Asia and Africa have found this method of conditioning very effective and use it widely. They provide the natives with food, education and medical care as well as bibles. Prospective converts thus learn to associate Christian dogma with services which they need and enjoy." [34] This same very often treacherous, misleading and hypocritical way of thinking and policymaking must be watched most carefully by the individual and group if they desire to see things in their truly realistic light and intent.

In America I have observed from a fairly close range the emergence of many castes and the nurture of deceptive values which can never lead to a truly relaxed peace of mind attitude or to a sense of honesty in personal and social relationships. Perhaps these things have always existed, perhaps everywhere. Perhaps they may continue to exist. But this doesn't mean that these are things which *must* be accepted, that we must finally resign ourselves to the decision that we can never possibly rise above these things. Such an attitude would be the final way of slaves, of living things that have given up the value of life, its meaning and its enjoyment, dooming themselves in their illusion of defeat and despair, when the affirmative and positive happiness of a true and honest life still breathes in the courage that pleads to express itself. Thomas Paine, Thomas Jefferson, Henry Thoreau, Abraham Lincoln, Walt Whitman had this courage; we now enjoy the freedoms they stood for, the freedoms they idealized, lived for, had the courage to acclaim to be the inalienable right of the living and of life itself.

In the American suburb we find the "farming" plebeian type of fellow who is usually wearing tan chinos and weeding his garden, fixing around the house, who more often than not would be found to be a "regular" guy, sometimes goes fishing or plays golf, reads the garden pages on Sunday and is attracted quickly to odd plants and bulbs or guaranteed-to-fruit trees and "nuclear" powered insecticides! While he is a more or less sociable chap, he would rather not be surprised by relatives or friends on his precious weekend but if he is, out comes the barbecue or the charcoal grill and he fusses all over the place to make everybody satisfied and happy. His life is pretty much a calendrical routine, week in and week out, month in and month out, doing the accustomed things and going to accustomed places according to an accustomed family plan or habit. If he genuinely loves this kind of life and is genuinely happy in it (and not too many of us are aware that we are not), all goes well and orderly, but if one should be subconsciously unhappy about it, we have what is called a case of *Suburbanitis,* a sudden heart attack, occlusion and death, often without any warning or previous symptoms, apparently in seemingly good condition—but the man died of *boredom* without ever knowing why! Routine has killed this man because routine can become an unnatural and anti-natural way of living, a chronic depressant, the invisible prison in which so many have succumbed and will continue to succumb. The free, uninhibited expression of the individual has been deceived and then suppressed.

In the cities the incidence of heart disease, ulcers, and many other ailments, is far worse because the life and artifacts of city existence are more artificial, more "civilized" than in the suburbs or countryside where at least the color of the environment—green—favors a more relaxed reception of the retina. "No wild animal suffers from coronary disease," writes Dr. A. T. W. Simeons, "though the hearts of higher mammals very closely resemble the human heart. Coronary disease is unknown among primitive races, but in modern urban man it is taking more lives than almost any other ailment." [34a] Gradually the primitive tree-top and cave human being is being imprisoned in his own made prisons of stone and steel. The *city*—which originally started as a walled-in fortress-like village to protect itself from the nomading invaders—has evolved to become a vast prison without walls or fortresses but still an unnatural prison, an artificial and depressing environment for the human being who evolved to walk on grass and whose lungs are not suited to breathe in poisonous gases, live in little boxes called apartments within bigger boxes and where the only exercise he gets is with his right hand which he uses to open and close doors. "The number of diseases to which a psychosomatic interpretation can be given is growing rapidly. Diseases which are not caused by external agencies, diseases which are neither congenital nor hereditary and those which do not occur in wild animals are all likely to have a psychosomatic background. We are witnessing an alarming increase in the frequency with which psychosomatic diseases, now recognized as such, are occurring in modern urban man, and there is good reason to be apprehensive about the sharp rise in the incidence of colitis, duodenal ulcers, high blood pressure, coronary diseases, diabetes, etc." [34b] "Psychosomatic ailments account for the bulk of urban man's ill-health and are the most frequent cause of his death. Man shares this kind of affliction with no other living creature . . . a disorder strictly comparable to a psychosomatic disease never occurs in any wild animal." [34c]

In my first published book, *Paradise Found* (1926), I wrote about what I called *metropolitanitis*—the city disease. If civilization is literally limited to its definition of the building of cities, then it has brought calamity upon the human being and unless he is able to escape from this entrapment he may smother under the mountainous heap of artifacts, inhibitions, and cultural impositions which he himself has manufactured through the centuries of cultural history. The meaning of *civilization* must be changed to interpret a process by which man pursues a healthful satisfaction in life, happiness, peace, a *naturalness* in which he can recover and rediscover his own self, his natural uninhibited personality, and to re-guide his own mind to serve a healthful, natural-expressive body that enjoys breathing oxygen and not gasoline, a body that has almost forgotten the sense and meaning of true freedom, a body that has by due course of civilization and culture become an artificial, thinkless (though calculating) robot in an environment which can only destroy him in time. "About a half million years ago, at the dawn of culture, our human ancestors began to use this unique brain to overcome their physical limitations. Then they rapidly refined abstract thought and built themselves an artificial environment to make up for their biological shortcomings. It was this trend towards an ever greater control over the world around him that led man into the appalling complexity of today's metropolitan life and into the psychosomatic diseases that are now threatening him with self-destruction." [34d]

The city is a machine to which man dedicates his life and his service. The machine must not stop for even a moment. The competition for wealth, power, prestige must not overlook the opportunities of each hour. Even though our shelves are overloaded with surpluses and our minds saturated with new whirls of desires and gadgetries, the sales graphs must show its lines upward even if we die in our tracks. "The tragedy of the rut into which man's brain has slipped is that his cortex cannot content itself with unproductivity. It is never satisfied with whatever security it achieves and must needs embark upon fresh intellectual adventures in the hope of achieving still greater security. In this process he is continually widening the gap between his diencephalic evolution and his cortical evolution." [34e] Thus man, in trying to reach the horizon only loses himself in the indeterminability of what he conceives as distance when in reality it actually lies at his very feet; in trying to achieve more and more without rational evaluation of his life and his Time, he gradually lives less and less; in his attempt to overtake time he leaves it behind. Thus man's ambitions become irrational, unnatural and anti-natural, cumbersome, unwieldly, overburdening, excessive, abnormal, the self-imprisonment within too many artificial and too-far-from-human mechanics, the dedication of man to the machine and to the "civilized" bastilles that came about with man's rise to power—it is these anti-natural excesses that may destroy him. The human being can still be man without being either beast or prey. He need not return to the jungle of his ancestors and he need not stay in the jungles of steel and concrete which are worse and more hardened. The world is beautiful and he can still enjoy its beauty so long as he allows it to stay beautiful. He will not find it in a billboard or on concrete. He must allow his mind to return to his body and his body to the world in which he has been born. Culture has "kidnapped" him away from his true self and put him in a gilded cage of wires, walls, and wedges. He must allow his mind and his mind through his body to reorientate culture so that the artifacts of his invention can serve him, not destroy him.

There is another typical class (which can exist anywhere and most probably goes on in most communities) which I like to call the caste of the "Aristocrat." This fellow walks about town in high black or red stockings, red shirt or shorts, and loves to show himself in places where people congregate; he just loves to be seen holding a conversation leaning on a Cadillac car, located as near as possible to any store or "hang-out" where the "nice people" of the town are apt to come around. He is also called the oversized "Little Lord" of the Suburb, and he feels, no doubt, as he struts along the village street that he is the reincarnated French noble from Versailles, that he is really "up-to-date" and should be easily identified as belonging to the "nicer" (wealthier), "higher" (the better located and more pretentious homes) and "finer" (belongs to a country club but never saw the Book Section of the *New York Times*) class of people. He would never walk this way in an ordinary street in Brooklyn or on Delancey Street but the "locale" of his home town is his proper stage and where the caste "understand" him. The children of these people must only go out with the other "nice" children, that is, children of parents who are just as wealthy, at least, but preferably more wealthy. The grown-ups leaving high school must only go to colleges or private schools where other such "fortunate" ones can be socially met. There is nothing wrong in giving the best, and wanting the finest for one's children, but there is a difference in cultivating them to worship the god of wealth as the spiritual land of human values, in habituating them to glorify and accept the god of Pretense and Show instead of trying to teach the children basic values, good common sense, a sense of fair play and realistic judgments, and a scale of genuinely human values. "Hypocrisy is often merely an unpleasant name for good breeding. In order to be polite one must be a hypocrite." [35]

Some of the women in town also have their "cultured" categories. The *sophisticated* type can only speak to you in a mumbling fashion with a smoking cigarette hanging limply from the lower lip. This means that she belongs to the modern school. This type also desires to rarely comb her hair and wears very tight slacks to outline more clearly her principal assets. She usually does not like to cook, as this is primitive, and her husband has already been accustomed to taking the family out too often for supper and if he isn't around she would take the kids out and stuff frankfurters down their throats with a bottle of coke and call it supper. She hopes to be considered an intellectual because she keeps the latest best seller where her visitors can easily see it and because she is really too lazy to go out and shop for breakfast on Sunday morning; let anyone go into any appetizer or delicatessen store early Sunday morning and you will see more or less ten men to one woman coming in to buy the victuals for the day. Of course, she feels that to be lazy and do very little is a sign of progress and the reward money for living in the "modern" age. She also likes to appear as "modern," anti-conventional and frets and pretends like a free spirit, so long as there are people to notice it. She usually believes in "progressive" education because she doesn't very often properly understand the term or the process, and will be usually found to be lazy sufficiently to let the schools take over the education of her children, lock-stock-and-barrel, because she just has so much to do and think about, which is, in actuality, very little. Whenever possible she will try to look Hepburn-like in order to give a more implied sophisticated portrait of sex-in-the-second-childhood version.

Then there is another kind of stylish women, the "charity" group; the woman who wouldn't miss being a hostess at any charitable, church, temple or hospital affair *so long as people can be there to see her* but otherwise she is really lazy and indifferent to "interfere" or "intrude" in order to give a sick neighbor a glass of water or mind their children while the neighbor is enroute to a doctor. This type wouldn't lift a finger unless she has some assurance of an audience and public approval of her "unselfish," "charming," and "motherly" personality. Besides, there is usually an acknowledgement of her noble work at the end of the year which she can pin on her blouse or frame to face you as you enter her home, or her name in the local newspaper which in itself is something to be proud of and gives some increase in the circulation of the paper. Silly people! It is good that people think good of us so long as we think good of ourselves; unless we *know* we are on the level with ourselves, what does it matter what other people think of us, seeing that we are prepared to fool ourselves by way of our own folly or ignorance? Admiring an illusion is admiring nothing. It is easier to *want* to be honest and good to ourselves and to others; to want to cultivate this principle as a way of life is much better than the expenditure of greater effort to defraud ourselves and others with pretenses and poor intent. This is part of the culture of deception, which lacks intelligence and good judgment, which lacks the courage to stand by whatever intelligence we have and whatever sense and good feeling we have in order to live the honest life, the *real* life, the life that acknowledges the best and most peaceful and the happiest in us.

There is still another class of women who are the patronesses of the arts or try to become artists themselves. This is a splendid thing and far better than playing poker, which may be good, too. Any effort to appreciate the arts in any form is a wholesome and commendable attitude, but the actual intent of many of these people is not directed really to the arts but the attempt of getting people to direct their attention to them by the pretense of trying to be considered differently from the usual run of other "ordinary" women. They, too, are usually mentally and physically lazy to read a competent book on the arts or make any serious attempt to study it in some acknowledged good art school but they group together and call some money-needing artist from New York or elsewhere who very often doesn't even know how to teach properly although he could be and often is a prominent and fine artist himself. These women who really do not take the arts very seriously, the intent of these women is to be considered apart from the "household" or "maid type" of women who haven't aspired to the fine arts or the more sensitive touch of artistic appreciation. Here we see the false concept of people trying to be looked "up at" from what they think are "lower" classes or groups of people. Art does not require this. Art belongs to the individual supreme; it is not an exhibition of pride, a public merit system, or a *soma* to self-inflation. Art, like a goddess, is the consort to the self and their communion is a sacred and conscious fusion that is intimate and free flowing; it transcends other things beyond the goddess and her lover, and unites them both before the mirror of their existences and gives birth to an *ideal*. There is no room for pretense here; the guards at this sanctuary will not allow it to enter except with honest and clean hands.

Now I do not wish to convey any wrong idea that anything a person does is not good, pretentious or stupid. I have no desire, fully conscious of the fallibility of human judgment and the old adage about throwing stones, to place myself on any pedestal supporting any unilateral attitude to please my own tastes. These people are really not bad people, and they are far from savage, cruel, or uncivilized. The contention here is not to make them appear ridiculous or as anything unworthy of society, nor is there any intention of setting standards or a scale of values based upon the concepts or acceptances of any individual or group. Our problem is to try to understand and to see through their actions and behaviors a self-deceptive pattern of what seems to us to be fictitious and superficial values which, in the present or the ultimate, is not conducive to a possibly fuller and finer enjoyment of life and the expenditure of Time in getting the most and best out of life. This is our critical base. As Havelock Ellis once said, "There are so many excellent people in the world whose hearts are in the right place. If only their heads were screwed on the right way!" [36]

A person has the right to wear high black stockings if he wants to and a wife has the right to be lazy and be addicted to frankfurters. We are not particularly concerned with what they do so long as no one else is the worse for it, but we are concerned with their attitudes which are directed to other people besides themselves and which are supposed to bring in and about, certain responses, and it is these *responses* which we have a right to be critical of, and we have the right to analyze them from the viewpoint of indicating to them the shallowness and the needlessness of such expressions regardless whether they are rambunctious, pretentious, hypocritical or just plain unadulteratedly stupid. It is the analysis of their behaviors to expose falsehood and pretense that is important in order to try to get them to understand their own behaviors and the responsibility of their relationships with others. Cicero once said, "There is none more base than that of the hypocrite, who, at the moment he is most false, takes care to appear most virtuous." [37] Locke states that "affectation is an awkward and forced imitation of what should be genuine and easy, wanting the beauty that accompanies what is natural." [38] A witness in court who tells the truth does not have to prepare for trial as much as the liar who needs rehearsal and prompting. To paraphrase Oscar Wilde's book, *"The Importance of Being Earnest,"* it is still more important to be honest. Robert Burns, in his *Epistle to the Rev. John M'Math,* so wonderfully muses:

> "God knows, I'm no the thing I should be,
> Nor am I even the thing I could be,
> But twenty times I rather would be
>   An atheist clean,
> Than under gospel colours hid be
>   Just for a screen."

Then there is another class of "nice" people, the *Entertainers.* This group is divided into subgroups like the ones who specialize in entertaining their business associates, buyers, possible new accounts, clients, etc., going to all extremes of lavish expense to give the most and the best, yet when an ordinary friend or relative comes around for a visit he is lucky if he gets a cup of tea and a cookie! Another subgroup entertains local people of their equal "status" in the community to maintain their own popularity and prestige and to more or less "restrict" their growing children to confine themselves within the "upper" classes. Money wants more money, but rarely is it willing to share itself with less fortunates. There is still another subgroup that necessitates a lot of entertaining because the occupants can't look at each other, are bored stiff with each other and its either a crowd, excitement and social activity or else a dose of milltowns. This class of people is always on the move; they can't stop. The moment they stop they take pills. They stay up late and hate to see their guests go because once the guests are gone the first thing that comes to their minds is to go to the shelf and take a pill in order to sleep. They fear to face themselves squarely and directly and call the shots. They are weak cowards, miserable and psychosomatically crippled, but before the world they appear as wonderful, warm and affectionate people. Behind all these people is the one possibility that keeps them from disintegration: *Pretension.* They are the members of the great parade of pretenders; you will find them in Hong Kong as well as in suburbia, in Zanzibar as well as in New York or Hollywood; they are everywhere. They *suffer needlessly,* if they are conscious of their pretensions, and in so doing they cannot possibly understand or be alerted to the value of Time in their lives because, if they knew or understood, they most certainly wouldn't behave the way they do.

It is an easy matter to agree with Bertrand Russell in his philosophy to oppose and possibly make unpopular any form of convention or custom or pretension which tends to keep people from being happy, natural, and from living as freely as possible to accomplish this. I join with Russell to expose social hypocrisy and to throw the religious "reformers" out of the homes of people, get them out of our minds, and keep free the hearts of human beings to express love, to further happiness, social and individual honesty and self-integrity, and to achieve an equitable and sound economy to live and let live. Russell writes these courageous words: "The day of nice people, I fear, is nearly over; two things are killing it. The first is the belief that there is no harm in being happy, provided no one else is the worse for it; the second is the dislike of humbug, a dislike which is quite as much esthetic as moral. . . . The essence of nice people is that they hate life. . . . Nice people are those who have nasty minds." [39] While these things may be true and the future encouraged I often wonder when the time may come when the majorities of humanity may really and fully understand what Mr. Russell is trying to convey. That is the problem, the perennial problem of human nature.

At the same time, from present observations, I also seem to discern that the general cultural trend in America is towards the nurture and emergence of more and better phonies. One might venture to call our present society the Great Order of Gay Deceivers. Even when one orders a corned-beef sandwich you can see this tendency operating when they pile high the meat in the center of the sandwich to give it the phony effect of a thick and generous portion. It's a great confidence game in which people really enjoy fooling others and getting fooled. Its like the Oriental custom of enjoying bargaining before making a deal; in America, this theatrical performance is translated in the desire to "kid" ourselves by so much make-believe. Of course, this is a generalization and there are many exceptions to any general rule. To accuse or condemn any community or group is unfair and unintended, as everywhere in any country or locale, I feel most certain, live *real* people who do not desire to or belong to the popular trend of deception and pretense. But there is no getting away from the reality that there is a cultural caste system emerging which is so definitely and observationally clear that it identifies a caste system primarily based on calculating prestige values on wealth or money values. This is not only the anti-intellectual upsurge of our times but it is a pre-indicator, among other things, of a general social attempt to replace or substitute former "moral" or "spiritual" values with the newer money-scale stratification values which are basically misleading, unfair, pretentious, and an unintelligent approach to the meaning and purpose of money itself.

It is axiomatic that everything is not gold that glitters. Judging people by their veneer brings veneer judgments. Brick, wood, and gold leaf, do not make anything sacred, wise, or holy. A library is a library and is no conclusive key to the mind of the owner. A degree or a professional standing is insufficient evidence of the competency or greatness of a person. A position is for what it is or should be, not a veranda to look down from. A house, small or big, plain or fancy, is still a house and what makes it a home or a palace depends upon the people who occupy it. Honor is something the public can never give a person; it is something the person builds up within himself. Honor can never be displayed or listed in a book, and the person who truly possesses it will not advertise it for public acclaim or even need the approval of people for his own acceptance of it. Wealth is good and poverty is bad, but these are the fortunes and misfortunes of living; neither can identify the honesty or the wisdom of either one. Prestige can never be judged by the clothes we wear or the club we belong to, or the section of the town we live in; that *may be* our good fortune but in itself does not really point at us, the recipients. Prestige can only be the person himself, not his drapery, or his art collection, or what may be painted on his office door. These tags are merely identities of a process, a clue to the tendencies of a person but never conclusive evidence as to his nature and worth. To identify people one must reduce the rich and the poor to a common level or denominator, that is, to judge them as individuals, as people, for their acts and thoughts and not for what they possess or could give you. It is more important to be a man today than to have "owned" treasures of yesterday. Precious and artistic things are good and they inspire people to appreciate the finer things and very often the right people do possess the things they really enjoy themselves and not because they desire to display them to the world as a prestige-getter. The point is that we shouldn't get into the habit of making judgments of people by what they possess alone; it is more important to judge people by what the people themselves contain. It is far greater to be tolerant, considerate, and compassionate, than to be the Chairman of the Board without it. It is far wiser to love life than to fear death. It is far more honest to be judged for what we are or could be than for what we have tried to appear to be or tried to make people believe we are. Let's stop kidding ourselves. Life and the Time it lives are just too precious and incomparably too valuable to make faces at it. If we really aim to feel the living pulse of democratic ideals, then let us live as democrats and make our judgments armed with democratic minds. There are too many people living in the dark ages of titles, prestiges and power psychoses, fearful neuroses, vanities and stupid show-offness. If we do live like democrats we may learn how to laugh heartily and we may be able to be really happy. Wisdom worships at the altar of Time devoted to the goddess of happiness; at least, this is what it should be, otherwise, why wisdom?

There are other caste classifications in our country and other

Westernized places such as the occupational castes by which people are judged, prestige-valued and rated according to the kind of work they do or the particular position they hold in a large company. Whether they are honest or not, fair or not, ethical, just, honorable, prudent, friendly, or not, doesn't seem to matter. What matters is the *rank* or *status* of their job, not how they got there. Success is not considered as a process of achievement but as a badge of identity. To judge a successful person properly one must take into account the events and circumstances by which his success was attained. Thieves, liars, grafters, and deceivers, are successful every day, but in my book I do not consider this success, but merely that they just got away with it.

Another caste in the country is the clan grouping by which people are judged by the lodge or club they belong to. Recently, I had an agreeable disagreement with one of my close friends who is a "high-degreed" top-ranking Mason, and we were discussing the subject of *friendship*. Incidentally, "the Catholics once believed that all Freemasons were the fruit of Eve's adultery with the Serpent." [40] Well, anyway, my friend claimed that the Masonic order is an ideal order that furthers and encourages friendship. I couldn't accept this claim because I replied that in order to become a Mason one has to *conform* to the requisites of admission and acceptance, or to be in agreement with them, and among other things one essential requisite is that an entrant or prospective member *must* believe in a Supreme Being. I explained that the principle of friendship, like love, cannot accept any form of dictation, regardless of the nature of the demands. In fact, the essence of friendship is the guardianship and the respect of the freedom of expression and thought that belongs to each individual and that a person who possesses this unobligated and unrestricted freedom and the similar concomitant respect of other individuals' freedoms, is capable of and worthy of becoming a friend and extending friendship to others moreso than one, regardless how fine he may be, who compromises his individuality to any extent and its inalienable right to free expression, to an order of rules and requisites, regardless whether they are right or wrong. To the extent of the compromise or submission to any set of rules, to this extent the principles and philosophy of friendship are hindered and not capable of fuller expression. What is not agreed upon is of secondary importance; this is the course of free expression in the first place, to agree or not to agree. What is most important is to uphold the equity of the process by which these forms of expression create themselves, and it is this process that is inviolate and essential to the spirit and essence of friendship. Like love, so in friendship, there can be no "riders" to group people together because of their adherence to any form of pre-required agreement. The philosophy of democracy and the government which it has constituted to make this philosophy a living process creates a method of free expression which can make, remake, modify, nullify, change the rules and ways of democratic society to meet the needs of the times and the maintenance of things by which this process can be continued without adulteration or the forfeit of its original premise. What is important, then, is that a person can be free to agree or disagree, that a person be not compelled to agree whether he believes in it or not, to a pre-requisite before being allowed to join a group or order, and it is this phase of requirement which is less conducive to the friendship which, in its greatest and most genuine fulfillment, requires a free society of free individuals, without bias or need of conformance. The only conformance that is justifiable is the respect of each person's freedom of conscience and voluntary action. When there *is* agreement among free individuals such agreement comes from the uninhibited and unsublimated emotions and chainless minds of people, not an agreement of expediency, conformance, and acculteration to ancestral custom and belief.

Seeking friendship by emblems or accepting *a priori* the existence of friendship because of a button or handshake, is primitive, clannish, immature, and inconclusive. The man who can stand alone if need be, and think for himself, is too good to be alone; this is the material for friendship. A millionaire never really knows who his friends are because so many people "agree" with him. Character requires honesty, independent judgment, courage in decisions and equity without bias and intolerance. This does not mean that one shouldn't give oneself the benefit of any doubt; it does mean that he shouldn't doubt that others deserve the same meaning. Honor, if present at all, will be found, I think, within the scope of truth of what a person feels or knows about himself, not what others think of him. What does it matter if the world applaud a man's honesty and integrity when that man knows these things are really absent? Even in success, the truer test of a man is not when he's down but when he is lifted high with power, fortune and praise. When one has to wear a badge or pin to gain the favor of others, it may be a sign that he lacks the merit to win without it. Regulations and rules, though they may be necessary like manners, are for those who unfortunately can't figure out for themselves what to do sensibly or how to think properly without having others to tell them what to do and what to believe.

Friendship is a philosophy that requires an *idealism,* that can be *within* a person, never found on a membership card identifying a group, never found in a set of rules which require conformance; conformance, itself, regardless of its nature, is the antithesis of free expression and without free expression there cannot be a genuine and honest-to-goodness friendship. Friendship owes nothing to anybody. It is without obligation or the necessity of agreement. It is part of the affection of human beings to *regard* each other on an equitable plane, and this regard is the sacred and respected freedom of conscience and thought that are the sole rights of each individual born into this world. I never saw a baby with a tag being born and I feel that when one is about to die he is not actually concerned with how many lodge buttons he is leaving behind. What he should mind is how much genuine friendship and love he is leaving behind in the hearts of people. It is in the warm hearts and in the free minds of people that we can possibly find the greatest happiness and glory in life and it is this that can keep alive the memories of those who were courageous, just and wise, to judge people as people and not as rules and buttons. *Time* always manages to change the rules, anyway, and it considers lodge buttons or fraternity pins no more valuable than the metal and jewels they are made of. We are not born with them and we certainly do not take them along with us to identify our status in heaven or hell. Nor do we need them to express ourselves while we live. Anything that requires a loss of individualism is a deterrent and not an encouragement of friendship. Friendship does not necessarily have to be amenable in order to be amiable.

As long as there are human beings and their evolved natures

there will be groups, statuses, categories, classes, societies, etc.; the sociologist, I think, realizes that these things cannot be eliminated any more than you can eliminate human nature without eliminating humans. Such is the nature of experience. It is natural for people to aspire to higher classes, greater prestige, more wealth, comforts, honors, etc., if these things can bring them happiness, peace of mind and life satisfaction; and it is rational if these traits are translated and transcended into realistic and unregretful concepts concerning them. These are understandable and historically very human. What we have to watch out for and contend with is the tendency to "get lost." That is, to lose our identities as individuals and the rights and freedoms which are naturally the life-given assets of an individual. Besides, it is more important to keep as a constant consciousness our own worth to ourselves, and to others, of course, and this worth should be translated into the fullest and finest expressions of experience. Among these we will not find phoniness or affectations but we will find honesty, character, wisdom and compassion, tolerance and equity, a sensible and humane attitude as much as possible in as many things and experiences. This is not an impossible task or something injurious to the happiness, comfort, and genuine prestige of people. With this realization a greater and wiser concept of the value and use of Time is attained, and this attainment is a fundamental factor in the "aging" of the individual and in the interests of the better acculteration and maturation of groups and classes.

The story of *Civilization* and the story of *Cultures* are intertwined and interrelated like the Siamese twin. Originally the meaning of Civilization indicated the process of building and emergence of cities, empires, etc. Today Civilization seems to be the art of having something done without doing it yourself. The higher the caste or rank the more are others expected to do things for you, and the lower the caste the greater the burden on you to serve others. The cities still exist, and with more concrete, steel and stone being constantly added to "hive" the millions of humans in them. A city is a large cemetery that contains live people. Everyone walks on stone, goes into stone, comes out of stone, and sleeps in stone. The bunions of city-living often grow into ulcers, for an ulcer is a protest against civilization. Were it possible for Moses to make a visit to Times Square and take a ride in the subway during the rush hour, no doubt he would rest in comfort for the next million years for having led his people into a wilderness. The Mexican peon would thank his Guadalupe with one look at the Big City, which he will no doubt consider as a vast insane asylum where the people are stricken with some unknown, mysterious, terrible malady that makes them always in a hurry. The cities do contain things of beauty and interest, and cities are probably necessary to fulfill the wants and satisfactions of people, but I fear the day when the last tree will suffocate and die and the last little patch of green grass will turn brown and turn into city dust. When this happens we shall all return to the Stone Age with new trimmings. It could be called the Neo-neolithic Period and it will be remembered millenia later as the period in which people gradually turned into stone, their minds became rigid, put into card files, and their hearts hardened into blocks of concrete. It will be remembered as the Age of Appreciated Land Values when space brought the highest prices until the next glacial coverage occurred and reclaimed the space again for trees and flowers and the people smiled again and came back to life and were happy to see things green and growing.

The more valuable property becomes so many more trees are cut down to make room for concrete blocks and higher go the stories into the sky so that no one is able to see the sun rise or set. So valuable has become space in the Florida resort areas that there was a report that a bird laid an egg in a solarium in Miami Beach because she couldn't find a tree! Gradually this "casting" and "termiting" of human beings may lead—I hope not—to the very reversal of what the founders of our country tried to maintain and hold sacred—the free individual and the importance and value of the individual as paramount and above the relative values of the social organism. Already it appears that individualism is declining in America as the caste systems emerge wider and deeper into the cultural maze of our country and people. It may be the tolling of the bell to mark the decline and decadence of the American civilization.

Let us stray away for a moment from the locale of our nation to the broader panorama of the world. In our survey of the historical stage of empires and nations we can fairly visualize the gradual growth of little tribal groups from the closing in of previously isolated or separated family groups or kinships, and from the tribal groups into the larger tribe or village and then into a confederacy or communion of villages with a more or less specific culture covering it. Later, by infiltration or conquest, these larger communities or kingdoms were incorporated into still larger nations and religions, military or imperial dominance in control. Empires were built and lost, and either the conquering or overwhelming culture took over or combined with the subjected culture, like the Toltecs and the Mayas, or else the conquered cultures absorbed the conquerors, such as occurred many times in China and India. Gradually the exploiting nations, by their greed and ambitions, may outdo themselves and lay the Dragon's Teeth for their own destruction. "Out of warlike peoples arose civilization, while the peaceful collectors and hunters were driven to the ends of the earth, where they are gradually being exterminated or absorbed with only the dubious satisfaction of observing the nations, which had wielded war so effectively to destroy them and to become great, now victimized by their own instrument." [41] For one example, witness the American Indian. There are many others. Slavery and colonialism have yet to draw their greatest dividends in reverse. The tragic story of the extermination of the "peaceful collectors and hunters," the *Ihalmiut* Eskimos of the Canadian Barrengrounds by the government forces and two churches, one Catholic, the other Protestant, is one of the most tearful, frightening accounts of the inhumanity of man, a story of a people being decimated and starved deliberately in concentration camps, exterminated by the "kindness" and "gentle guiding hand" of the two churches, with the "best of intentions," a story "so utterly disgusting that it comes very close to destroying any respect one may have remaining for our own race." [42]

The gradual integrating and disintegrating patterns of various cultures became so complex, interrelated and woven into the general experience of the peoples that the only example I could think of to portray it would be a heaping plateful of spaghetti with all kinds of sauces thrown in and mixed up into it. However, there are more or less separated or fairly isolated cultures existing today but with the intrusion of "strangers" and the shrinking of distances due to modern travel, many of these cultures are being exposed to, affected and influenced by, other cultures, such as the Westernization of most of the Orient, Asia Minor and Africa. In Europe the influence of the modern age of mechanics and the general material and scientific advance have affected the cultures of that continent. In America the spreading and infiltrating vanguard of Americanism and Western influences, together with its mechanical and material progress, are channeling themselves through North and South Americas. When I was fishing up in James Bay, Canada, among the Cree Indians, I noticed that the women didn't bother growing vegetables anymore, or weaving cloth or making shoes; they just buy their supplies, cans of food, cloth for clothes, rubber boots, shoes, tools, rope, etc., at the Hudson's Bay Company post at Rupert's. And if you should go up into the Eskimo country in Alaska you shouldn't be surprised to find belly-stoves, gas lamps, and cooking stoves, typewriters, tarpulins, hardware, outboard motors, fur coats, made in Detroit or New York, or lambskin-lined jackets from Oregon, and all the gadgets for living in that area that probably came from Chicago, Los Angeles, or Kalamazoo. It is reasonable to understand that these people will accept anything that will make their lives easier, more comfortable and secure; and this is good in many ways. If you go to Den Pesar in Bali you will see probably half the town riding on bicycles, and in Hong Kong and Singapore rickshaws are fast disappearing and little bicycle-tramcars are all over the place. Times do change and Time changes cultures.

Out of this general labyrinth of changing cultures evolved the fabric and texture of nationalisms. After all, *it is one world, humanity one race,* and all the peoples of the earth came from a common germinal root or nucleus, even though they spread all over the place and evolved separate identities, characteristics, colors, and cultures. As of today people still cling to their nationalistic cultures in spite of this, and this results in particularized ideologies and allegiances which, propagandized over many generations, become hindrances to tolerance, world unity, create dictators and power oligarchies for imperialistic ambitions, power blocs, police states, confusing politics, and the constant foment of hatreds, animosities, misunderstandings and jealousies. Man has manufactured the nations and now the nations have outgrown him. After all, the difference between a nationalist and an internationalist is a fence. The Communist hates the Capitalist, and vice versa. The religionists hate each other. The anarchist hates them all, including himself, and the only people who do not hate each other are the monkeys. The monkeys are not nationalists or partisans; every tree in the world could be their home, especially if it grows bananas. The deepest pity is the political impossibility of children running the world; only then could this big ball be what it should be, a playground, a kitchen, and a dormitory, three places of peace. As it is now, the Fascists wear black shirts, the Communists wear red shirts, the Phalangists wear striped shirts, the white supremacists wear white bathrobes, and the plain citizen is just glad he's got his pants on.

The great task of the wise leaders of the present and the future is to attempt the almost impossible—the gradual transference of national sovereignty into a world sovereignty. If human wisdom and fortitude do not accomplish this, perhaps the necessity of sheer survival may bring it about; or perhaps, after a terrible war in which all the nations will be destroyed and the poor remnants of the world will find themselves without sovereignties and thus create one new world nation, if there is any strength left. The long-range unity of the world into one sovereignty and with it the liquidation of any need of power, is the ultimate salvation for the human race if it is to survive.

The difficulty is, I think, in the confession that peoples are yet not so civilized as to realize the trust in such unity. Nationalism has been and still is the rage. But nationalism, we have seen, is a step ahead of the older oligarchical institutions. In its turn, the oligarchy was a step ahead of a still more despotical monarch who himself was an improvement upon the primitive tyrant who literally owned the people as cattle and goods. And yet, I hope, the time will come when nationalism will give rise to internationalism, as part of the ceaseless progress of the homogeneity of groups and general progress, as part of the newer necessities born out of the scientific age and its potential destructiveness as well as peaceful achievements. Perhaps the governments of the world will gradually change in principle, if not in name, into states, as in my own country, states of a central unified system, named, let us say, the *United States of Humanity.* And then, and only then, will the stigma of war, so attached to our history, subside and finally disappear save as a relic of record. The world, in a sense, is getting smaller and smaller and the peoples are getting closer and closer. We are beginning to understand each other much better and more intimately the relations between ourselves and the basic wants of peace and security and freedom for ourselves as people, and it is getting a little more difficult for the governments and special groups of the world to make people hate each other as they have done more easily before. This process is slow but there is progress. With understanding comes toleration, with toleration come trust and faith, and with these comes peace built upon the sacred respect of the individual, of his dignity, of his mind, and of liberty—the sacred respect of life itself and the little Time alloted to it. Communism is clearly the way of collectivization of humanity—human regimentation and the submergence of people into societal slavery. Democracy is the Savior, no doubt, if there ever could be one, but in sustaining and extending its premise of freedom within order it must clean its own stables of the same collectivizing and regimenting false concepts which is a growing danger to our own American culture.

The opposite of liberty is absolutism, dogmatic authority, slavery, and the dangers of power psychoses. Without liberty Time itself is enslaved. So important is it for the individual and the human race, as a whole, to safeguard this precious birthright of people that our brilliant Justice William O. Douglas strongly stresses this point when he wrote: "Where discretion is absolute, man has always suffered. At times it has been his property that has been invaded; at times, his privacy; at times, his liberty of movement; at times, his freedom of thought; at times, his life. Absolute discretion is a ruthless master. It is more destructive of freedom than any of man's other inventions. . . . So it has been from the beginning; and so it will be throughout time. The Framers of the Constitution knew human nature as well as we do. They too had lived in dangerous days; they too knew the suffocating influence of orthodoxy and standardized thought. They weighed the compulsions for restrained speech and thought against the abuses of liberty. They chose liberty."

I think it is fitting, to avoid a possible misinterpretation of meanings and intentions, to restate my thoughts regarding the content of society, culture, and the individual, and the relationships between them. The society and culture, because of the very nature of their being constantly subjected and allergic to change and mobility, have in the nature of themselves the potential and process of both heterogeneity and homogeneity, and even at times, through sudden mutations, a process of heteromorphosis. Strangely, the individual is no different, for he has within himself the potential of being alone if he desires to, yet not necessarily lonely. On the other hand, he can naturally desire to and enjoy fully the society of people or the "grouping" together with others and yet feel at all times that he is an individual and sincerely cherish his own status as a free individual. The fact that he is a *social* animal does not assume that he cannot be a *free* person, physically and intellectually free enough to understand for himself the maximum optimum of happiness that he is capable of according to his nature and by right of his very life. To the extent that the society and culture allow the free play of this process is also the extent of the value and meaning of society to him, of course, with the concomitant realization that *all* individuals are entitled to the same freedom, tolerance, and free play within and towards each other and the same obligations of regard and respect which can only make possible such a process of *social individualism.* A society progresses as it allows a greater and wider perimeter of expression, physical and intellectual movement, and the happiness potential for the individual. This process, by its *warming* nature, generates life and activity. Time moves with man as he goes along with the roll of the earth, of the moon and the sun. Each star, every new dawn, is approached and consumed with expectation, with a wholesome unshackled free spirit of courageous confidence and joyful living. The hour becomes precious because it is *lived.* Any social contract, be it cultural, political, economic or religious, that depresses, hinders, or misleads the individuals into a submissive state of intellectual coagulation, is not a progressive element and this process tends to *freeze* the individuals and holds in immobilized suspension the generating and free principles of living that comprise the content of a naturally normal and healthful sense of existence. The hour then becomes meaningless, and being bypassed it has died with the loss of its value.

Everything and anything the human being has been able to accomplish or evolve came out of individuals. Individuals built the nations and now the individuals have become the serfs of their own creations. Individuals have built great and gigantic machines and now the individuals have become the conscious slaves of these unconscious machines. The individuals have built religions and now they have become the dupes of their own fantasies and futile followers of their own dreams. The same with cultures. Individuals created them and all the rules and mores of every little society and every little custom. The intelligent person knows it is within his right to change them, modify them, destroy them if need be, reject or accept them, as they further or retard his pursuit of happiness, his general welfare and his life satisfaction. To the extent that one feels he *can* accept or reject any part of his culture or anyone else's culture, to this extent he remains more of an individual and a free person, and a more rational one. To the extent to which he subjects himself, submissively or ignorantly, to the "will" or absorptive, conforming tendencies of culture, irrespective of whether they are good or bad for him, to this extent he is a slave, a robot, less of an individual and more of a conformist than a rationalist. As stated before, conformance occurs when a person does not allow himself the freedom and natural right of making up his own mind.

What will the future society and culture consist of? I do not exactly know, nor does anyone else, but we can think, imagine, and hope. I fear that the religion of the future may be the attempt of the fewer intelligent but ruthless people to overwhelm and use the ignorant and misled masses of people, as they have done so consistently before. The intelligence of the human being may probably outdo itself in this direction most magnificently. The priestcraft might come from several principal groups: the high finance geniuses, the advertising agencies and promoters, and the power seekers; the last group including so-called idealists, self-called saviors of mankind, ideologists who in reality would seek the enslavement of the entire world if it meant that they could thus prove the egocentric success of their own illusions. Our own country may be overrun and overtaken with "bars and grills" and hot dog stands. As it is now the average growing child today is dietetically addicted to french fries, hot dogs, hamburgers, and pop! The oldfashioned kitchen with its enticing, tempting aromas of home cooking has now been transferred into shelves of superduper cans in the supermarket. Anybody interested in growth stocks should look into the container and can stocks for the future may, no doubt, become the *Canned Age*—and highly pressurized. Of course, as a result of this, there will be great need and expansion of the drug business because every new can will require an antidose or neutralizer until our people will become properly canned and drugged. Even the "detail men" of the "ethical" drug people will gradually, by "advising" everyone "to see your doctor," create the doctors into messenger boys for them and the doctors will eventually become expert catalogists and everything will be done for them by the drug manufacturers so that there will be little to do except write a number or a brand. It isn't that new drugs are bad or good; many of them are wonderful and have made possible the elimination of dreadful diseases, have eased pain considerably and lengthened the life of man. This is good, but the general importance of the drug industry to itself has become a more important factor than the genuine welfare of the people. Like any form of power it finally winds up only interested in itself and the profits of the drug industries indicate it; a recent author has written a book about the doctors and the drug-makers, titled *It's Cheaper to Die!*

A god in the heart is worth a thousand in the sky. The genuine sense of godliness (if it means anything good and wholesome) can come only from a free individual, one who is not chained to any belief, code, rule or culture. Something to expect in the future is much better than something to forget in the past. Let us look, then, hopefully at least, toward the hour and the coming hour as free men and women and join into the nature of Time, which is also, in itself, a free period. By enslaving ourselves we enslave the Time which we occupy or travel within. The importance of living is less important to those who take Time too lightly and the ways of acceptance and absorption as the "easy" way out. The only easy way is the free way.

A man is either a conforming robot with a robot brain, a live mannequin, or he is a free man possessing a free mind. Conforming people remind me so much of *Alice in Wonderland* in which Carroll so brilliantly writes: " 'You!' said the Caterpillar, contemptuously, 'Who are you?' "

Who am *I*? I am *Time,* pure as the next moment and as free as the coming day! Do not pollute me or waste me and in your *freedom* I may bring you paradise. In your *goodness* I may bring you heaven. In your *courage* I may give you the wonderful, ecstatic sense of self-respect incomparable with all other forms of respect. In your *honesty* of mind and purpose I may build you a home, within you, and around you, that will beautify you, give you greater solace and comfort than all the make-believe immortalities concocted out of your fearful dreams. In your *equity* I may bring you love and friendship which rules and conventions can never bring you. In the serenity and satisfaction of a *free mind* I may make you live like a song, full of rhyme and music, lightness, warmth and color. And the song of the bird, the echo of the forests, the murmur of the stream and the kiss of the wave upon the rock, the waltz of the wind and the holy peacefulness of the stillness beneath the tall pines, all these will tell you that you are great, wonderful, happy, because you have found it wonderful and great to be alive, because you have become part of that cosmic oneness, not in a dead Nirvana trying to escape from life, but in the living Nirvana of a precious living poem in which every moment of *Me,* Time, has become a precious moment with you! My recurring nature allows you to be free to accept me or reject me as you wish and when. And when you take me unto yourself in all my fullness, I *can* make you happy, peaceful, and serene. *Happiness* is the expression of freedom; *Serenity* is the expression of a free mind; *Peace* is their offspring; all born out of the courage, honesty, and intelligence of the individual who would rather be right than accepted, who would rather be free than a slave, who would rather be fair and equitable than a caterpillar, rich or poor, following mindlessly and blindly group judgments and social acceptances regardless of their nature or implications but only because it is "nice" and "more practical" and "proper" to be on their side. These slaves of submission are confessors of their own inferiorities and weaknesses, and they expose themselves as rigid little wooden soldiers which the invisible big ghosts of Culture and Status move about as suit their own so meaningless and nebulous natures.

To summarize and evaluate these two chapters on Cultures and Societies and their relation to our premise of Time, it seemed to me to be sufficiently important to review, in a critical as well as in an informative sense, a number of the salient cultures in the world in order to allow us an opportunity to understand and to reflect more lucidly upon the terrific and fastened impact that these cultures and their ideologies, their systems of worship and habituated and accumulated beliefs and customs, have upon the peoples of the earth.

So long as individuals are tied to their societies, cultures, and theologies, it would seem appropriate to examine them if the individuals are to be enabled to form a more rational approach and a more meaningful reevaluation of their own lives. This reevaluation or taking "inventory" of their life assets cannot be reasonably attained without using the Specific of Time as the basic essence of their own containment and actual being. If one can conclude, viewing this processional panorama of the historical pageantry of cultures and societies, that all these events are evolutional and their own natures as part of the general evolutional processes of the human being and his ideological products, then the individual may become fairly free and at least subjectively detached so that he can attempt to accept or deny any parts of it if such can make possible a more realizable happiness, peace, and general life satisfaction and a striving for the attainment, through this liberated rationalistic attitude, of a better and finer perception of the meaningfulness and appreciation of life to himself, and, thusly, a more meaningful concept of the nature, value, and the essence of Time within which all he is or could be expresses itself.

The individual cannot escape from the universal mold, nor should he desire to so long as he can feel that the universe is his to occupy and enjoy within the moments of Time he is more intelligently and physically aware of. Within this realization of the sanctity of life tied to the value of Time rests the happiness of the individual, the affirmative and positive approach to social ethics that may possibly bring about a peaceful humanity and an abatement, through the appreciation of and a more realistic analysis of life itself, of the cruelties and needless punishments which people for so many thousands of years have heaped upon themselves. If the individuals would cease creating more societies, but they will, nevertheless; and if the societies created more individuals, but they will not, nevertheless, perhaps both the individuals and the societies would become a more justifiable, harmonious and happy liaison of the individual and his environment. And in the meantime, like Candide, let's cultivate our own little garden and enjoy life—while we can—while we still have Time.

REFERENCE NOTES

CHAPTER I THE *Nature* OF TIME

1. Loren Eiseley, *Nature, Man, and Miracle,* Horizon, July, 1960, p. 29
2. *Ibid.,* p. 32
3. John Dewey, *Reconstruction in Philosophy,* p. 122
4. *Ibid.,* p. 117
5. Philip Eichler, *A Philosophy of Science,* p. 75
6. Miguel de Unamuno, *Tragic Sense of Life,* p. 97
7. Garrett Hardin, *Nature and Man's Fate,* p. 63
8. Antoine Nicolas de Condorcet, *The Historical Picture of the Progress of the Human Mind*
9. James Hutton, see *The Firmament of Time,* Eiseley, p. 25
10. Albert Schweitzer, *The Philosophy of Civilization,* p. 54
11. Bertrand Russell, *In Praise of Idleness,* pp. 99-100
12. Socrates, by Plato
13. John Dewey, *Nature and Experience,* preface p. ix
14. Bertrand Russell, *Our Knowledge of the External World,* p. 9
15. A. J. Carlson, *Scientific Monthly,* August, 1944
16. John Dewey, *Logic, The Theory of Inquiry*
17. William James, *The Meaning of Truth,* preface pp. xi-xii
18. John Dewey, in his Introduction to *Inside Experience,* by Joseph K. Hart, p. xxv

CHAPTER II TIME AND THE *Religious Factor*

1. John Dewey, *A Common Faith,* p. 32
2. Albert Einstein, *The World As I See It,* p. 26
3. Pope Pius V, *Bullarium Romanum,* ed. Gaude, Naples, 1882, tom. vii, pp. 430, 431. Also *see* A. D. White, *The History of the Warfare of Science with Theology,* p. 37, vol. ii
4. Alfred Weber, *History of Philosophy,* p. 18

5. James G. Frazer, *The Golden Bough,* Part I, The Magic Art, vol. I, pp. 222-225
6. Theophile James Meek, *Hebrew Origins,* pp. 83-84
7. John B. Watson, *Behaviorism,* p. 11
8. Theophile James Meek, *Hebrew Origins,* p. 87
9. Lucien Lévy-Bruhl, *The "Soul" of the Primitive,* p. 278
10. James G. Frazer, *The Belief in Immortality,* vol. I, pp. 84-86
11. Arthur Schopenhauer, *Die Welt als Wille und Vorstellung,* vol. II, p. 529
12. George W. Corner, director of Embryology Dept. of the Carnegie Institute, Washington, D.C., "Science and Sex Ethics," Saturday Evening Post, October 10, 1959
13. James G. Frazer, *The Golden Bough,* 1 vol. ed., p. 377
14. E. O. James, *Ancient Gods,* p. 293
15. Lévy-Bruhl, *The "Soul" of the Primitive,* p. 269
16. Joseph Campbell, *The Masks of God: Primitive Mythology,* p. 89
17. Ashley Montagu, *Immortality,* p. 32
18. E. O. James, *Ancient Gods,* p. 168
19. Edward B. Tylor, *Religion in Primitive Culture,* p. 70
20. *Ibid.,* pp. 71-78
21. Charles W. Mead, *Old Civilizations of Inca Land,* p. 77
22. Saint-Foix, *Essais Historiques sur Paris, Oeuvres Complètes,* vol. IV, p. 150. Maestricht, 1778
23. Edward B. Tylor, *Primitive Culture,* Ch. XI
24. Lillian Eichler, *The Customs of Mankind,* p. 586
25. *Ibid.,* p. 588
26. H. R. Hays, *From Ape to Angel,* p. 70
27. John Dewey, address delivered before the College of Physicians in St. Louis, April 21, 1937
28. W. Robertson Smith, *The Religion of the Semites,* p. 30
29. Edward B. Tylor, *Religion in Primitive Culture,* p. 196
30. E. P. Evans, *Evolutional Ethics and Animal Psychology,* p. 355
31. Theophile James Meek, *Hebrew Origins,* p. 101, states: "Yahweh was known as 'The Rider on the Clouds,' and, like most early gods, he was a god of war."
32. Edward B. Tylor, *Religion in Primitive Culture,* p. 88

33. Ashley Montagu, *Immortality,* p. 57
34. Llewelyn Powys, *Earth Memories,* p. 85
35. Edward B. Tylor, *Religion in Primitive Culture,* pp. 93-94
36. *Ibid.,* pp. 10-11
37. Aesop, The Fables of, *The Dog and the Shadow.*
38. Jawaharlal Nehru, "India," *New York Times,* September 7, 1958
39. Theophile James Meek, *Hebrew Origins,* p. 119
40. Joseph K. Hart, *Inside Experience,* p. 34
41. Barrows Dunham, *Man Against Myth,* pp. 14-15
42. Mark Graubard, *Man the Slave and Master,* p. 246
43. Benjamin C. Cardozo, *Notable Opinions of Mr. Justice Cardozo,* edited by Dr. A. L. Sainer, 1938.
44. Anatole France, *Penguin Island,* Introduction to
45. Ashley Montagu, *Immortality,* p. 42.
46. Bronislaw Malinowski, *A Scientific Theory of Culture,* p. 174
47. Bertrand Russell, *Why I Am Not a Christian,* p. 53
48. Plato, *The Symposium*
49. Ashley Montagu, *Immortality,* p. 37
50. *Ibid.,* pp. 39-40
51. A. Powell Davies, *The Mind and Faith of A. Powell Davies,* edited by William O. Douglas, p. 30
52. J. B. S. Haldane, *The Inequality of Man,* p. 205
53. Sébastien Faure, *Does God Exist?,* p. 14
54. *Ibid.,* p. 29
55. Joseph Campbell, *The Historical Development of Mythology,* from *Myth and Mythmaking,* edited by Henry A. Murray, p. 21
56. James G. Frazer, *The Golden Bough,* I vol. ed., p. 211
57. *Ibid.,* p. 209
58. *Ibid.,* p. 208
59. *Ibid.,* p. 211
60. *Ibid.,* p. 210
61. *Ibid.,* p. 216
62. *Ibid.,* pp. 268-269
63. Edward B. Tylor, *Religion in Primitive Culture,* p. 111
64. *Ibid.,* p. 36

65. *Ibid.*, pp. 58-59
66. *Ibid.*, pp. 85-86
67. H. R. Hays, *From Ape to Angel,* p. 144
68. E. P. Evans, *Evolutional Ethics and Animal Psychology,* p. 27
69. Bertrand Russell, *Sceptical Essays,* p. 211
70. James G. Frazer, *The Golden Bough,* 1 vol., ed., p. 29
71. Lester F. Ward, *Applied Sociology,* p. 88
72. Elie Metchnikoff, *The Nature of Man,* p. 287
73. Burnham P. Beckwith, *Religion, Philosophy and Science,* p. 57
74. James G. Frazer, *The Golden Bough,* Part I, The Magic Art, vol. i, pp. 373-374
75. James G. Frazer, *The Golden Bough,* Psyche's Task, pp. 111-113
76. Ashley Montagu, *Immortality,* p. 35
77. Morris R. Cohen, *The Meaning of Human History,* p. 278
78. Samuel F. Dunlap, *The Ghebers of Hebron,* preface p. iv
79. John Dewey, *A Common Faith,* p. 46
80. John Herman Randall, Jr., *Dualism in Metaphysics,* p. 307
81. Oscar Wilde, *The Picture of Dorian Gray,* p. 27
82. Bertrand Russell, *Why I Am Not a Christian,* p. 24
83. H. L. Mencken, *Living Philosophies,* p. 192
84. James Harvey Robinson, *Human Comedy,* p. 263
85. George A. Dorsey
86. Sigmund Freud, *The Future of an Illusion,* 1928
87. Albert Schweitzer, *The Philosophy of Civilization,* p. 139
88. Burnham P. Beckwith, *Religion, Philosophy and Science,* p. 62
89. A. D. White, *History of the Warfare of Science with Theology,* p. 325, vol. i
90. Burnham P. Beckwith, *Religion, Philosophy and Science,* p. 61
91. H. A. Overstreet, *The Enduring Quest,* p. 16
92. Arthur C. Clarke, *Horizon,* vol. 1, no. 3, January, 1959
93. Gaetano Mosca, *The Ruling Class,* Images of Man, edited by C. Wright Mills, p. 199
94. John Dewey, *A Common Faith,* p. 84
95. Winwood Reade, *The Martyrdom of Man,* p. 193
96. Friedrich Paulsen, *Introduction to Philosophy,* p. 4

97. A. D. White, *The History of the Warfare of Science with Theology,* p. 69, vol. II
98. Llewelyn Powys, *Earth Memories,* p. 70
99. Theodore Dreiser, *Living Philosophies,* p. 71
100. Morris R. Cohen, *The Meaning of Human History,* p. 278
101. J. H. Denison, *This Human Nature,* p. 240
102. Bertrand Russell, *Sceptical Essays,* p. 182
103. Robert Briffault, *Rational Evolution,* p. 40
104. William Graham Sumner, *Folkways,* p. 26
105. *Ibid.,* p. 24
106. Lee Alexander Stone, *The Power of a Symbol,* pp. 1-2
107. A. J. Carlson, Pres. Am. Assoc. Adv. Science, *Scientific Monthly,* August, 1944
108. H. A. Overstreet, *The Enduring Quest,* p. 79
109. A. D. White, *The History of the Warfare of Science with Theology,* p. 30, Vol. II
110. Charles Duff, *This Human Nature,* p. 49
111. Thomas Jefferson, *Jefferson Himself,* edited by Bernard Mayo, pp. 299, in a letter to Thomas Earle, of Monticello, September 24, 1823
112. Emmett McLoughlin, *American Culture and Catholic Schools,* pp. 35-37
113. *Ibid.,* p. 29
114. Theophile James Meek, *Hebrew Origins,* pp. 85-86
115. Bertrand Russell, *Sceptical Essays,* pp. 93-94
116. Philip Eichler, *A Philosophy of Science,* p. 107
117. Samuel Chugerman, *Lester F. Ward, Life of,* p. 134
118. L. C. Dunn, *Heredity and Variation,* p. 12
119. Elie Metchnikoff, *The Nature of Man,* p. 60
120. F. H. Shoosmith, *Life in the Animal World,* p. 220
121. George W. Beadle, *The Biological Sciences,* Frontiers of Science, p. 16
122. Roy Waldo Miner, *Fragile Creatures of the Deep, the Story of the Hydroids,* p. 246
123. George G. MacCurdy, *The Coming of Man,* the University Series,

Second Unit, Part IV; The University Society, New York, 1931, p. 4
124. Julian S. Huxley, *The Stream of Life,* p. 2
125. Donald Culross Peattie, *The Flowering Earth,* p. 87
126. Carleton Ray and Elgin Ciampi, *Marine Life,* p. 136
127. *Ibid.,* p. 138
128. Henry E. Crampton, *The Coming and Evolution of Life,* p. 53
129. A. I. Oparin, *Origin of Life,* p. 33
129a. A. T. W. Simeons, *Man's Presumptuous Brain,* p. 7
129b. *Ibid.,* pp. 25-26
130. Ivan P. Pavlov, *Lectures on Conditioned Reflexes,* p. 349
131. *Ibid.,* p. 114
132. Albert Schweitzer, *The Philosophy of Civilization,* p. 93
133. L. Adams Beck, *A Beginner's Book of Yoga*
134. George Bernard Shaw, *The Adventures of the Black Girl in Her Search for God,* pp. 88-89
135. Felice Belloti, *Fabulous Congo,* p. 58
136. Sir John Bland Sutton, famous English surgeon
137. James G. Frazer, *Belief in Immortality,* Vol. 1, p. vii-viii
138. Ashley Montagu, *Immortality,* p. 35
139. H. L. Mencken, *Living Philosophies,* p. 192
140. Samuel Chugerman, *Lester F. Ward, Life of,* p. 249
141. H. G. Wells, *First and Last Things,* p. 110
142. Bertrand Russell, *Value of Free Thought,* p. 93
143. Ashley Montagu, *Immortality,* p. 27
144. Elie Metchnikoff, *The Nature of Man,* p. 161
145. Sir Arthur Keith, *Living Philosophies,* p. 150
146. Fred Hoyle, *Nature of the Universe,* p. 141
147. F. J. Gould, *A Short History of Religion*
148. James G. Frazer, *The Golden Bough,* 1 vol. ed., p. 107
149. James G. Frazer, *Belief in Immortality,* Vol. 1, pp. 11-23
150. Edward B. Tylor, *Religion in Primitive Culture,* p. 461
151. Albert Einstein, *The World As I See It,* p. 27
152. Morris R. Cohen, *The Meaning of Human History,* p. 283
153. *Ibid.,* p. 285
154. Auguste Forel, *The Social World of the Ants,* Vol. II, p. 352

155. Llewelyn Powys, *Earth Memories,* p. 86
156. Alfred Weber, *History of Philosophy,* p. 194
157. John Stuart Mill, *Autobiography*
158. George Jean Nathan, *Living Philosophies,* p. 231
159. Elie Faure, *History of Art,* Vol. Spirit of Forms, Introduction xv
160. Albert Schweitzer, *The Philosophy of Civilization,* p. 309
161. Frederick Barnard, quoted in *Immortality,* by Ashley Montagu, pp. 22-23
162. Miguel de Unamuno, *Tragic Sense of Life,* p. 45
163. *Ibid.,* p. 57
164. John Dewey, *A Common Faith,* p. 26
165. Albert Schweitzer, *The Philosophy of Civilization,* pp. 63-64

CHAPTER III TIME AND THE *Power Factor*

1. Samuel F. Dunlap, *The Ghebers of Hebron,* p. 998
2. Philip Eichler, *A Philosophy of Science,* p. 46
3. Aesop, The Fables of, *The Miser*
4. Donald Culross Peattie, *The Flowering Earth,* p. 24
5. Lewis Mumford, *Living Philosophies,* p. 209
6. Frazier Hunt, *This Bewildered World,* p. 246
7. Mark Graubard, *Man the Slave and Master,* p. 348
8. E. L. Grant Watson, *Mysteries of Natural History,* pp. 148-149
9. John Stuart Mill, *The Idea of God in Nature*
10. J. B. S. Haldane, *The Inequality of Man,* p. 143
11. Elie Metchnikoff, *The Nature of Man,* p. 253
12. *Ibid.,* p. 31
13. Henry E. Crampton, *The Coming and Evolution of Life,* p. 81
14. Forest Ray Moulton, *Astronomy, The Nature of the World and of Man,* p. 17
15. J. H. Fabre, *Social Life in the Insect World,* p. 91
16. Donald Culross Peattie, *This is Living*
17. W. C. Allee, *Social Life of Animals,* pp. 241-242
18. Theodore Dreiser, *Living Philosophies,* p. 58
19. Irving Adler, *How Life Began,* p. 21
20. Leon Tolstoi, *Les Confessions,* Paris, 1891, p. 49

21. Eugene N. Marais, *The Soul of the White Ant,* p. 75
22. Mark Graubard, *Man the Slave and Master,* p. 76
23. Joseph Needham, *Man a Machine,* p. 95
24. Donald Culross Peattie, *The Flowering Earth,* p. 203
25. Joseph Ratner, *Introduction to Intelligence in the Modern World,* p. 14
26. Eugene N. Marais, *The Soul of the White Ant,* p. 106
27. *Ibid.,* p. 111
28. John Dewey, *Individualism Old and New,* p. 49
29. Walter Russell, *The Universal One,* p. 61, states: "Matter, in its form dimensions, is an appearance which is the result of motion."
30. John Dewey, *Nature and Experience,* p. 4a
31. *Ibid.,* p. 1
32. William H. Kilpatrick, *The Educational Frontier,* p. 296
33. John Dewey, *Human Nature and Conduct,* p. 165
34. Thomas Gann and J. Erick Thompson, *The History of the Maya,* pp. 130-131
35. J. H. Denison, *This Human Nature,* p. 293
36. Bertrand Russell, *Sceptical Essays,* p. 108
37. Hendrik De Leeuw, *Sinful Cities of the Western World,* p. 223
38. H. C. Thomas and W. A. Hamm, *The Foundations of Modern Civilization,* p. 257
39. J. H. Fabre, *Social Life in the Insect World,* p. 87
40. Quincy Wright, *A Study of War,* Vol. 1, p. 78
41. *Ibid.,* p. 45. Also *see* W. M. Wheeler, *Social Life Among Insects,* p. 7
42. Herbert S. Dickey, *My Jungle Book,* p. 213
43. James Harvey Robinson, *Human Comedy,* p. 98
44. F. Max Müller, *My Autobiography,* p. 316
44a. Joost A. M. Meerloo, foreword to *Man's Presumptuous Brain,* pp. vii-viii
44b. A. T. W. Simeons, *Man's Presumptuous Brain,* p. 281
45. Jerome Davis, *Capitalism and its Culture,* p. 516
46. Edmond N. Cahn, *The Sense of Injustice,* p. 49
47. Robert Briffault, *Rational Evolution,* p. 208

48. Carl L. Becker, *How New Will the Better World Be?*, p. 83
49. *Jefferson Himself*, edited by Bernard Mayo, p. 240
50. Mark Graubard, *Man the Slave and Master*, p. 36
51. Winwood Reade, *The Martyrdom of Man*, p. 447
52. Bertrand Russell, *Icarus or The Future of Science*, p. 5
53. Carl L. Becker, *How New Will the Better World Be?*, p. 84
54. Thorstein Veblen, *The Theory of the Leisure Class*, p. 1
55. Pierre Bovet, *The Fighting Instinct*, p. 225
56. Benjamin N. Cardozo, *Law and Literature*, p. 170
57. James Harvey Robinson, *Human Comedy*, p. 265
57a. A. T. W. Simeons, *Man's Presumptuous Brain*, p. 278
58. Charles Duff, *This Human Nature*, p. 57
59. Pierre Bovet, *The Fighting Instinct*, p. 156
60. Bertrand Russell, *Why I Am Not a Christian*, p. 91
61. Charles Duff, *This Human Nature*, p. 77
62. J. H. Fabre, *Social Life in the Insect World*, p. 86
63. Frederick Palmer, *Our Gallant Madness*, p. 320
64. John Dewey, Antinaturalism in Extremis, from *Naturalism and the Human Spirit*, p. 3
65. James Jeans, *Living Philosophies*, p. 113
66. John Dewey, *Freedom and Culture*, p. 71
67. Carl L. Becker, *How New Will the Better World Be?*, p. 71
68. Philip C. Jessup, *A Modern Law of Nations*, p. 2

CHAPTER IV TIME AND THE *Social Factor*

1. James G. Frazer, *Folk-lore in the Old Testament*, Vol. III, pp. 93-94
2. John Dewey, *The Public and Its Problems*, p. 70
3. Philip Eichler, *A Philosophy of Science*, p. 12
4. Edward B. Tylor, *Anthropology*, pp. 54-55
5. Lester F. Ward, *Glimpses of the Cosmos*, p. 264
6. Auguste Forel, *Social World of the Ants*, Vol. II, p. 338
7. John Dewey, *Liberalism and Social Action*, pp. 74-75
8. Winwood Reade, *Martyrdom of Man*, p. 25
9. Julian Huxley, *Ants*, p. 100
10. Donald Culross Peattie, *This is Living*

11. Otto Klineberg, *Social Psychology*, p. 32
12. H. A. Junod, *The Life of a South African Tribe*, I, pp. 121-122
13. "Meet Mister Porpoise," *Natural History Magazine*, January, 1940
14. E. G. Boulenger, *Apes and Monkeys*, p. 54
15. *Ibid.*, p. 62
16. Lucien Lévy-Bruhl, *The "Soul" of the Primitive*, p. 103
17. *Ibid.*, p. 87
18. José Ortego y Gasset, *Revolt of the Masses*, p. 134
19. Anthony Eden, during his visit to America, December, 1938
20. Albert Schweitzer, *The Philosophy of Civilization*, pp. 293-294
21. Danielle Hunebelle, "The Endless Crucifixion of Spain," *Réalités* Magazine, Paris, France, 1959
22. John Dewey, *Experience and Education*, p. 24
23. Eric Larrabee, "After Abundance, What?," *Horizon*, July, 1960, p. 72
24. John Dewey, *Ethics*, revised ed., Dewey and Tufts, p. 366
25. Matilde Castro Tufts, *Looking to Philosophy*, p. 403
26. John Dewey, *Experience and Nature*, frontispiece
27. William Graham Sumner, *Folkways*, pp. 29-30
28. Miguel de Unamuno, *Tragic Sense of Life*, p. 16
29. William O. Douglas, *America Challenged*, pp. 4 and 14
30. Llewelyn Powys, *Earth Memories*, p. 51
31. Elie Metchnikoff, *The Nature of Man*, p. 290
32. Benjamin N. Cardozo, *The Notable Opinions of Mr. Justice Cardozo*, edited by A. L. Sainer, p. 431
33. Oscar Wilde, *Picture of Dorian Gray*, p. 93
34. George Jean Nathan, *Living Philosophies*, p. 233
35. Friedrich Paulsen, *Introduction to Philosophy*, p. 429
36. John Dewey, *Human Nature and Conduct*, pp. 30-31
37. Philip Eichler, *A Philosophy of Science*, p. 105
38. Albert Einstein, *The World As I See It*, p. 29
39. Sir Leslie Stephen, *An Agnostic's Apology*, p. 50
40. Joseph K. Hart, *Inside Experience*, p. 108
41. Konrad Z. Lorenz, *King Solomon's Ring*, p. 18
42. Raymond A. Dart, *Adventures with the Missing Link*, p. 238
43. Frank Gibney, *The Operators*, p. 163, states: "The annual sales volume of American department stores is now (1960)

roughly $16 billion. Shortages come to $200 million annually—more than half the amount of net profits. In 1959 the country's supermarkets sustained $100 million loss from employee thefts, which cancelled out profits on $5 billion worth of sales. Authorities in the field agree that their statistical embezzler archetype is a respected employee. . . ." This is noted from just two channels of distribution. What total thievery by employees amounts to in all fields of activity is anyone's guess, but it would not surprise any student on public and private morals to be staggering and shocking.

44. H. R. Hays, *From Ape to Angel*, p. 84
45. Samuel Chugerman, *Lester F. Ward*, pp. 521-522
46. Felice Belloti, *Fabulous Congo*, pp. 61-62
47. John Dewey, *Freedom and Culture*, p. 15
48. LaForest Potter, *Strange Loves*, p. 12
49. *Ibid.*, p. 14
50. Allen Edwardes, *The Jewel in the Lotus*, p. 21
51. *Ibid.*, p. 250
52. Albert Einstein, *The World As I See It*, p. 27
53. Phillip C. Jessup, *A Modern Law of Nations*, p. 3
54. Benjamin Cardozo, *The Nature of the Judicial Process*, p. 127
55. Lord Josiah Stamp, "Essential Characteristics of Democracy," p. 4, from proceedings of Congress on *Education for Democracy*, Columbia University, August, 1939
56. Erwin Schrödinger, *Science and the Human Temperament*, p. 132
57. Loren Eiseley, *The Firmament of Time*, pp. 4-5
58. Elie Metchnikoff, *The Nature of Man*, p. 110
59. Joseph K. Hart, *Inside Experience*, p. 232
60. Albert Schweitzer, *The Philosophy of Civilization*, p. 84
61. Loren Eiseley, *The Firmament of Time*, p. 146
62. Albert Schweitzer, *The Philosophy of Civilization*, p. 79
63. Sir Arthur Kennedy, *African Sketch Book*, Vol. II, p. 449
64. Ralph Waldo Emerson, essay on *Experience*.
65. Lin Yutang, *The Importance of Living*, p. 10
66. Epictetus, *Discourses of Epictetus*, translated by George Long, p. 427
67. Vance Packard, *The Wastemakers*, p. 223

68. *Ibid.*, p. 204
69. *Ibid.*, p. 205
70. Albert Schweitzer, *The Philosophy of Civilization,* p. 46

CHAPTER V TIME AND THE *Ego Factor*

1. Joseph K. Hart, *Inside Experience*, p. 80
2. Irving Adler, *How Life Began,* p. 51
3. Oliver L. Reiser, *The Alchemy of Light and Color,* p. 65
4. Wen Kwei Liao, *The Individual and the Community,* pp. 298-299
4a. A. T. W. Simeons, *Man's Presumptuous Brain,* p. 76
5. Irving Adler, *How Life Began,* p. 121
6. *Ibid.*, p. 58
6a. Richard Carrington, *A Biography of the Sea,* pp. 3 and 66
7. Earl A. Evans, Jr., "How Life Began," *Saturday Evening Post,* Nov. 26, 1960, pp. 24, 25, 53, 57
8. Loren Eiseley, "Nature, Man, and Miracle," *Horizon,* July, 1960, p. 30
9. A. Powell Davies, *The Mind and Faith of A. Powell Davies,* edited by William O. Douglas, p. 45
10. Albert Schweitzer, *The Philosophy of Civilization,* p. 306
11. Werner Heisenberg, "From Plato to Max Planck, the philosophical problems of Atomic Physics," *Atlantic Monthly,* Anniversary Issue, 1959, p. 112
12. Raymond A. Dart, *Adventures with the Missing Link,* p. 235
13. Sir Arthur Keith, *Living Philosophies,* p. 142
14. Hanns Heinz Ewers, *The Ant People,* p. 293
15. L. C. Dunn, *Heredity and Variation,* The University Series, Second Unit, Part III, p. 5
16. *Ibid.*, p. 7
17. Sir Arthur Keith, *Living Philosophies,* p. 142
18. Donald Culross Peattie, *Flowering Earth,* p. 47
19. Sir Arthur Keith, *Living Philosophies,* p. 140
20. John Dewey, *Freedom and Culture,* p. 102
21. *Jefferson Himself,* edited by Bernard Mayo, pp. 344-345
22. Gordon W. Allport, *Social Frontier,* June 1939, Vol. V, No. 46, quotation on cover

23. H. R. Hays, *From Ape to Angel,* p. 368
24. Charles Duff, *This Human Nature*, p. 55
25. Joseph Ratner, *Introduction to Intelligence in the Modern World,* pp. 24-25
26. Bertrand Russell, *Why I Am Not a Christian,* p. 191
27. Henry C. Link, *The Rediscovery of Man*
28. John Dewey, *Human Nature and Conduct,* pp. 87-88
29. Clyde Kluckhorn, *Mirror for Man,* p. 223
30. John Dewey, *Ethics;* Dewey and Tufts, revised ed., p. 347
31. *Ibid.,* 358
32. J. H. Denison, *Emotion As the Basis of Civilization,* p. 160
33. Ralph Linton, *The Tree of Culture,* p. 520
34. Albert Schweitzer, *The Philosophy of Civilization,* p. 86
35. Eric Larrabee, "After Abundance, What?," *Horizon,* July, 1960, p. 71
36. Harry Golden, *2¢ Plain,* p. 21
37. James Harvey Robinson, *The Human Comedy,* p. 47
38. Robert Briffault, *Rational Evolution,* p. 3
39. Raymond A. Dart, *Adventures with the Missing Link,* p. 225
40. Herbert S. Dickey, *My Jungle Book,* p. 71
41. Bertrand Russell, *Why Men Fight,* p. 153
42. Albert G. A. Balz, *The Basis of Social Theory,* p. 249

CHAPTER VI TIME AND THE *Culture Factor,* PART I

1. Donald Culross Peattie, *This is Living; see* Irving Adler, *How Life Began,* p. 55: "Water is especially important for life. Life began in water, probably over two billion years ago. As a result, there are living things in water, and there is water in all living things."
2. A. D. White, *History of the Warfare of Science with Theology,* pp. 98-99, Vol. I
3. James Harvey Robinson, Columbia University, N. Y.
4. Sir James Jeans, *Living Philosophies,* p. 108
5. Joseph K. Hart, *Inside Experience,* p. 151
6. Edward B. Tylor, *Anthropology,* p. 133. *See also* page 125, relationship of language to sound expression of natural acts: In

the Tecuna language of Brazil, the verb to sneeze is *haitschu;* the Galla of Abysinnia, to express "the smith blows the bellows," say *tumtun bufa bufti,* or "the tumtum puffs the puffer"; the Japanese say *pata-pata* to denote flapping or clapping; the Yoruba Negroes say *gbang* to denote "to beat."

7. John Dewey, *Human Nature and Conduct,* p. 69
8. Morris R. Cohen, *The Meaning of Human History,* p. 171
9. LaForest Potter, *Strange Loves,* pp. 13-14
10. W. Lavallin Puxley, *Deep Seas and Lonely Shores,* in which the author illustrates the life of the sperm-whale, a very peaceful animal, which is being slaughtered by man.
11. Herbert S. Dickey, *My Jungle Book,* p. 88
12. Clyde Kluckhorn, *Mirror for Man,* p. 85
13. Fred Hoyle, "Forecasting the Future"; *Frontiers of Science,* p. 118
14. Basil Davidson, *The Lost Cities of Africa,* p. 6
15. L. S. B. Leakey, "Finding the World's Earliest Man," *National Geographic* Magazine, September, 1960, pp. 420-435
16. Ralph Linton, *The Tree of Culture,* pp. 523-524
17. Frederick Starr, *Confucianism,* p. 7
18. Fung Yu-Lan, *A Short History of Chinese Philosophy,* pp. 1-3
19. Frederick Starr, *Confucianism,* p. 43
20. J. E. Ellam, *Religion of Tibet,* p. 20
21. *Ibid.,* p. 35
22. *Ibid.,* p. 24
23. *Ibid.,* p. 26
24. *Ibid.,* p. 27
25. *Ibid.,* p. 98
26. James G. Frazer, *The Golden Bough,* 1 vol. ed., p. 60
27. Edward B. Tylor, *Religion in Primitive Culture,* p. 79
28. Albert Schweitzer, *Indian Thought and Its Development,* p. 42
29. Edward B. Tylor, *Religion in Primitive Culture,* p. 96
30. Ernest Crawley, *The Mystic Rose,* Vol. 1, p. 124. *See also* Allen Edwardes, *The Jewel in the Lotus,* p. 60, who states regarding the origin of castes in India: "Indian castes were originated not only out of occupation and wealth but in relation to bodily charm. Genital proportions were also a basis for

classification and men were rated by the size of their *lingams,* from insignificance (*Tussoo,* inchlet) to greater length and thickness (*Lumbah,* yard)."

31. Edward B. Tylor, *Religion in Primitive Culture,* p. 97
32. James G. Frazer, *Golden Bough,* 1 vol. ed., p. 209
33. Miguel Covarrubias, *Island of Bali,* p. 123
34. Richard Lewinsohn, *A History of Sexual Customs,* p. 34; Also, *see* R. W. Frazer, *Sati,* Encyclopedia of Religion and Ethics, Vol. XI, p. 207
35. Miguel Covarrubias, *Island of Bali,* p. 196
36. *Ibid.,* pp. 358-359
37. Allen Edwardes, *The Jewel in the Lotus,* pp. 49-50
38. *Ibid.,* pp. 62-63
39. Max-Pol Fouchet, *The Erotic Sculpture of India,* p. 8
40. Allen Edwardes, *The Jewel in the Lotus,* p. 42
41. Abbé J. A. DuBois, *Hindu Manners, Customs, and Ceremonies.* Also, *see* Allen Edwardes, *The Jewel in the Lotus,* p. 18
42. James G. Frazer, *The Golden Bough,* Part IV, Adonis, Attis, Osiris, vol. ii, pp. 209-212
43. Allen Edwardes, *The Jewel in the Lotus,* pp. 264-265
44. D. Chowry Muthu, *The Antiquity of Hindu Medicine,* pp. 42-43
45. J. H. Denison, *Emotion as the Basis of Civilization,* p. 56
46. James G. Frazer, *The Golden Bough,* 1 vol. ed., p. 104
47. *Ibid.,* p. 61
48. James G. Frazer, *The Golden Bough,* Part I, Magic Art, Vol. I, pp. 233-234
49. James G. Frazer, *The Golden Bough,* 1 vol. ed., p. 53
50. Wallis Budge, *Egyptian Magic,* p. xi, preface
51. *Ibid.,* 28
52. Ralph Linton, *The Tree of Culture,* pp. 409 and 423
53. Bronislaw Malinowski, *Myth in Primitive Psychology,* p. 89
54. W. Oldfield Howey, *The Cat in the Mysteries of Religion and Magic,* p. 52
55. James G. Frazer, *The Golden Bough,* 1 vol. ed., pp. 308-309
56. Robert Briffault, *Rational Evolution,* p. 23
57. John P. Mahaffy, *Prolegomena to Ancient History,* p. 416

58. Edward B. Tylor, *Anthropology,* p. 21
59. Winwood Reade, *The Martyrdom of Man,* p. 27
60. Allen Edwardes, *The Jewel in the Lotus,* p. 91
61. Ralph Linton, *The Tree of Culture,* p. 288
62. H. R. Hays, *From Ape to Angel,* p. 127
63. Clyde Kluckhorn, *Mirror for Man,* p. 46
64. E. O. James, *Ancient Gods,* p. 272
65. Theophile James Meek, *Hebrew Origins,* p. 217
66. E. O. James, *Ancient Gods,* p. 309
67. *Ibid.,* p. 148
68. *Ibid.,* p. 222
69. Theophile James Meek, *Hebrew Origins,* p. 107
70. W. Robertson Smith, *The Religion of the Semites,* p. 74
71. E. W. Lane, *An Account of the Manners and Customs of the Modern Egyptians,* ii, 346.
72. William Graham Sumner, *Folkways,* p. 39
73. M. Rudwin, *The Devil in Legend and Literature,* p. 231
74. E. O. James, *Ancient Gods,* p. 187
75. Clyde Kluckhorn, *Mirror for Man,* p. 110
76. William Graham Sumner, *Folkways,* p. 323
77. Richard Lewinsohn, *A History of Sexual Customs,* p. 97
78. A Jewish Reader: *In Time and Eternity,* edited by Nahum N. Glatzer, p. 62
79. Maimonides, *In Time and Eternity,* edited by Nahum N. Glatzer, pp. 65-66
80. Clyde Kluckhorn, *Mirror for Man,* p. 278
81. Bertrand Russell, *Why I Am Not a Christian,* p. 202
82. D. T. Atkinson, *Magic, Myth and Medicine,* p. 71
83. Walter Libby, *The History of Medicine,* pp. 14-15
84. Richard Lewinsohn, *A History of Sexual Customs,* pp. 25-26
85. Edward B. Tylor, *Anthropology,* p. 12
86. James G. Frazer, *The Golden Bough,* Vol. IX, The Scapegoat, p. 400
87. C. W. Ceram, *The Secret of the Hittites,* p. 209
88. James G. Frazer, *The Golden Bough,* 1 vol. ed., p. 302
89. J. H. Denison, *Emotion As the Basis of Civilization,* p. 225
90. Ralph Linton, *The Tree of Culture,* p. 368

91. *Ibid.*, p. 379
92. Edward B. Tylor, *Religion in Primitive Culture,* pp. 163-4
93. *Ibid.*, p. 156
94. *Ibid.*, p. 157
95. Wallis Budge, *Egyptian Magic,* p. xiii, preface
96. Sylvanus G. Morley, *Ancient Maya,* pp. 213-214
97. George C. Vaillant, *Aztecs of Mexico*, p. 53
98. Charles W. Mead, *Old Civilizations of Inca Land,* p. 86
99. Winwood Reade, *Martyrdom of Man,* p. 207
100. J. H. Denison, *Emotion As the Basis of Civilization,* p. 57
101. *Ibid.*, p. 189
102. David Forsyth, *Psychology and Religion,* p. 97
103. James G. Frazer, *The Golden Bough,* Part VI, The Scapegoat, preface, pp. v-vi
104. James G. Frazer, *The Golden Bough,* 1 vol. ed., p. 566
105. *Ibid.*, p. 11
106. *Ibid.*, p. 686
107. Wallis Budge, *The Gods of the Egyptians,* preface
108. James Harvey Robinson, *Human Comedy,* p. 165
109. Geoffrey Bibby, *The Testimony of the Spade,* p. 341
110. James G. Frazer, *The Golden Bough,* 1 vol. ed., p. 426
111. *Ibid.*, pp. 667-668
112. A. Powell Davies, *The Mind and Faith of A. Powell Davies,* edited by William O. Douglas, p. 273
113. James G. Frazer, *The Golden Bough,* 1 vol. ed., p. 401
114. Emmett McLoughlin, *American Culture and Catholic Schools,* p. 81
115. *Ibid.*, p. 84
116. Joseph Campbell, *The Historical Development of Mythology,* from *Myth and Mythmaking,* edited by Henry A. Murray, p. 19
117. Lee Alexander Stone, *The Power of a Symbol,* p. 49
118. B. Z. Goldberg, *The Sacred Fire*, p. 69
119. Lee Alexander Stone, *The Power of a Symbol*, p. 56
120. H. R. Hays, *From Ape to Angel*, p. 67: "Primitive Catholic peasants beat the images of the saints who failed to answer their prayers."
121. James Harvey Robinson, *Human Comedy,* p. 117

122. Edward B. Tylor, *Religion in Primitive Culture,* pp. 383-384
123. Charles Duff, *This Human Nature,* p. 133: "Faith seems to flourish better in filth."
124. D. T. Atkinson, *Magic, Myth and Medicine*, p. 72
125. Ralph Linton, *The Tree of Culture,* p. 375
126. E. O. James, *Ancient Gods,* p. 281
127. Maximilian Rudwin, *The Devil in Legend and Literature,* p. 1
128. St. Thomas Aquinas, *The Summa*, Part III
129. William James, *The Will to Believe,* p. 218
130. John Langdon Davies, *A Short History of Women,* p. 299, states: "Martin Luther was perfectly certain that witches had intercourse with the Devil, and his enemies, the Catholics, preached sermons proving that he himself was the result of such intercourse."
131. W. R. Alger, *A Critical History of the Doctrine of a Future Life,* p. 76
132. J. H. Denison, *This Human Nature,* p. 215
133. Pedro de Cieza de Leon, *The Incas,* translated by Harriet de Onis, edited by Victor Wolfang von Hagen, pp. 358-359
134. Bertrand Russell, *Value of Free Thought,* p. 77
135. James G. Frazer, *The Golden Bough,* Part II, Taboo and the Perils of the Soul, pp. 214-215; 217-218
136. Edward B. Tylor, *Religion in Primitive Culture,* p. 179
137. Lin Yutang, *The Importance of Living,* pp. 16-17
138. Maximilian Rudwin, *The Devil in Legend and Literature,* p. 263
139. *Ibid.*, p. 14
140. Philip Wylie, *Innocent Ambassadors,* p. 104
141. John Dewey, *Human Nature and Conduct*, p. 55
142. E. Martin, *Farewell to Revolution,* p. 113
143. Joseph Wheless, *Forgery in Christianity,* p. xix, foreword
144. Philip Wylie, *Innocent Ambassadors,* p. 104
145. St. Fulgentius, *De Fide*
146. Hoenus, *Visit to Hell: The Christian Hell,* by Hypatia Bradlaugh Bonner, p. 58
147. Father Furniss, *Books for Children,* first published 1861
148. *Ibid.*

149. Emmett McLoughlin, *American Culture and Catholic Schools,* p. 48
150. Charles Duff, *This Human Nature,* pp. 203-204. Taken from the Vision of Tundale and others in Delepierre's *L'Enfer décrit par ceux qui l'ont vu.*
151. Maximilian Rudwin, *The Devil in Legend and Literature,* p. 25
152. A. D. White, *The History of the Warfare of Science with Theology,* pp. 27-28, Vol. II
153. *Ibid.,* pp. 40-41, Vol. II
154. *Ibid.,* p. 113, Vol. II
155. *Ibid.,* pp. 120-121, Vol. II
156. J. A. MacCulloch, *Medieval Faith and Fable,* pp. 86-87
157. G. G. Coulton, *The Inquisition*
158. Jacobus de Voragine, Archbishop of Genoa, *The Golden Legend, (Legenda aurea sive historia Lombardica),* a collection of stories of the saints compiled about 1275.
159. Maximilian Rudwin, *The Devil in Legend and Literature,* p. 250. He also states that "the invention of paper money was attributed to the Devil by Gérard de Nerval in *L'Imagier de Harlem ou la Découverte de l'imprimerie* (1851)."
160. Edmond N. Cahn, *The Sense of Injustice,* p. 10
161. W. Oldfield Howey, *The Cat in the Mysteries of Religion and Magic,* pp. 64-65
162. Fridtjof Nansen, *Living Philosophies,* pp. 98-99
163. Spinoza, *Tracticus Theologicus*
164. L. Guy Brown, *Problems of Social Psychiatry; Fields and Methods of Sociology,* edited by L. L. Bernard, pp. 130-131
165. Emmett McLoughlin, *American Culture and Catholic Schools,* p. 120
166. Maximilian Rudwin, *The Devil in Legend and Literature,* p. 247
167. Walter Kaufman, *Critique of Religion and Philosophy,* pp. 130-131
168. Richard Lewinsohn, *A History of Sexual Customs,* p. 135
169. Henry Charles Lea, *History of the Inquisition of the Middle Ages,* Vol. III, p. 549
170. Edward B. Tylor, *Origins of Culture,* p. 139

171. Elie Metchnikoff, *The Nature of Man,* p. 14
172. Burnham P. Beckwith, *Religion, Philosophy and Science*, p. 40
173. D. T. Atkinson, *Magic, Myth and Medicine,* pp. 211-212
174. Bertrand Russell, *Why I Am Not a Christian*, p. 37
175. Emmett McLoughlin, *American Culture and Catholic Schools,* p. 59
176. Everett Dean Martin, *Farewell to Revolution,* p. 120
177. J. H. Denison, *Emotion As the Basis of Civilization,* p. 294
178. Bertrand Russell, *Why I Am Not a Christian*, p. 16
179. James G. Frazer, *Folk-lore in the Old Testament,* Vol. I, pp. 76-77
180. Maximilian Rudwin, *The Devil in Legend and Literature,* p. 139
181. Richard Lewinsohn, *A History of Sexual Customs*, p. 96
182. *Ibid.*, p. 98
183. *Ibid.,* pp. 98-99
184. Anselm of Canterbury, *De Contemptu Mundi*
185. Gibbon, Edward, *Decline and Fall of the Roman Empire*
186. Richard Lewinsohn, *A History of Sexual Customs,* p. 172
187. J. H. Denison, *This Human Nature,* p. 298
188. *Ibid.,* p. 300
189. Hendrik De Leeuw, *Sinful Cities of the Western World*
190. Charles Duff, *This Human Nature,* p. 103
191. Bertrand Russell, *Sceptical Essays,* p. 109
192. John Dewey, *Human Nature and Conduct*, p. 75
193. Emmett McLoughlin, *American Culture and Catholic Schools,* p. 250
194. Louise Marie Spaeth, *Marriage and Family Life Among Strange Peoples,* p. 18: "Monogamy prevails in the majority of primitive people."
195. Albert Schweitzer, *Indian Thought and Its Development,* p. 4
196. William J. Fielding, *The Shackles of the Supernatural,* p. 125
197. Louise Marie Spaeth, *Marriage and Family Life Among Strange Peoples,* pp. 25-26
198. B. Z. Goldberg, *The Sacred Fire,* p. 48
199. Felice Belloti, *The Fabulous Congo,* p. 77
200. Bengt Danielsson, *Love in the South Seas,* p. 74
201. *Ibid.,* pp. 56-57

202. Philip Wylie, *Innocent Ambassadors*, p. 54
203. Bertrand Russell, *Why I Am Not a Christian*, p. 47
204. Benjamin N. Cardozo, *The Notable Opinions of Mr. Justice Cardozo*, edited by A. L. Sainer
205. Herbert W. Schneider, *Radical Empiricism and Religion*, From *Essays in Honor of John Dewey*, p. 348
206. George Santayana, *Winds of Doctrine*, p. 1
207. George Bernard Shaw, *The Adventures of the Black Girl in Her Search for God*, p. 85
208. Samuel Miller, Harvard Divinity School, "Evolution of Religion," *Saturday Review of Literature*, November 14, 1959, p. 70
209. George C. Vaillant, *Aztecs of Mexico*, pp. 201-202
210. Edward B. Tylor, quoted in H. R. Hays' *From Ape to Angel*, p. 62
211. Sylvanus G. Morley, *The Ancient Maya*, p. 181
212. *Ibid.*, p. 210
213. Ralph Linton, *The Tree of Culture*, pp. 638-639
214. Charles Gallenkamp, *Maya*, p. 92
215. *Ibid.*, p. 96
216. Thomas Gann and J. Eric Thompson, *The History of the Maya*, p. 249
217. James G. Frazer, *The Golden Bough*, 1 vol. ed., pp. 500-501
218. Felice Belloti, *The Fabulous Congo*, p. 44
219. *Ibid.*, p. 46
220. *Ibid.*, p. 61
221. Charles Gallenkamp, *Maya*, p. 21
222. *Ibid.*, p. 20
223. *Ibid.*, p. 8
224. Sylvanus G. Morley, *Ancient Maya*, p. 212
225. *Ibid.*, p. 213. The diffusion and infusion of religious symbols and deities are an endless chain in the story of man's religions. This process of religious continuity with new names is clarified, philosophically, by the brilliant anthropologist, Ralph Linton, who states: "The changes in official religion by which, let us say, the local baal of a Palestinian village later became a Roman deity, and then a Christian saint, and is now a Mohammedan saint—I don't know what he will be after the

new state of Israel is established, but I will bet he will still be around—these changes are only superficial."—*Tree of Culture,* p. 662

226. Pedro de Cieza de Leon, *Incas,* translated by Harriet de Onis, edited by V. Wolfang von Hagen, p. 53
227. *Ibid.,* p. 110
228. *Ibid.,* pp. 274-275
229. *Ibid.,* p. 310
230. Richard Lewinsohn, *A History of Sexual Customs,* p. 43
231. Allen Edwardes, *The Jewel in the Lotus,* p. 47
232. *Ibid.,* p. 76
233. Enrique Cardinal Pla y Denseel, Cardinal of Toledo, Spain, Associated Press report, *The New York Times,* July 11, 1959.

CHAPTER VII TIME AND THE *Culture Factor,* PART II

1. Mark Graubard, *Man the Slave and Master,* p. 20
2. Walter Kaufmann, *Critique of Religion and Philosophy,* p. 46
3. Art Young, *On My Way,* p. 188
4. Everett Dean Martin, *Farewell to Revolution,* p. 371
5. Erich Kahler, *Man the Measure,* pp. 610-611
6. Albert Schweitzer, *The Philosophy of Civilization,* p. 14
7. Loren Eiseley, *The Firmament of Time,* p. 57
8. August Bebel, *Woman and Socialism,* p. 455
9. Read Bain, *The Fields and Methods of Biological Sociology,* p. 42, from *Field and Methods of Sociology,* edited by L. L. Bernard
10. James Harvey Robinson, *Human Comedy,* p. 74
11. Erich Kahler, *Man the Measure,* pp. 608-609
12. *Ibid.,* pp. 616-617
13. William O. Douglas, *America Challenged,* p. 16
14. John Dewey, *Freedom and Culture,* p. 10
15. William O. Douglas, *America Challenged,* p. 3
16. Albert Schweitzer, *The Philosophy of Civilization,* pp. 20, 45-46
17. Loren Eiseley, *The Firmament of Time,* p. 136
18. Albert Schweitzer, *The Philosophy of Civilization,* p. 18
19. Erich Kahler, *Man the Measure,* pp. 613 and 618

20. Lin Yutang, *N.Y. Times Magazine*, Nov. 12, 1939
21. Woodrow Wilson, *Memoirs*, on Government
22. John Dewey, *Authority and Social Change*, paper read at Harvard Tercentenary Conference of Arts and Sciences, September, 1936
23. Declaration of the *Rights of Man*, Article I
24. Walter Kaufmann, *Critique of Religion and Philosophy*, pp. 70-71
25. Albert Schweitzer, *The Philosophy of Civilization*, pp. 9-10
26. John Dewey, *The Public and Its Problems*, p. xx
27. John Langdon-Davies, *A Short History of Women*, p. 168
28. James G. Frazer, *The Scope of Social Anthropology*, pp. 166-167
29. Mark Graubard, *Man the Slave and Master*, p. 203
30. Clyde Kluckhorn, *Mirror for Man*, p. 112
31. H. C. Thomas and W. A. Hamm, *Foundations of Modern Civilization*, p. 21
32. Bertrand Russell, *In Praise of Idleness*, p. 225
33. Letters of Charles Darwin; *see* Garrett Hardin's *Nature and Man's Fate*, p. 19
34. Burnham P. Beckwith, *Religion, Philosophy and Science*, pp. 59-60
34a. A. T. W. Simeons, *Man's Presumptuous Brain*, p. 161
34b. *Ibid.*, p. 3
34c. *Ibid.*, p. 1
34d. *Ibid.*, p. 4
34e. *Ibid.*, p. 55
35. Charles Duff, *This Human Nature*, p. 51
36. Havelock Ellis, *Questions of Our Day*, p. 15
37. Cicero, *De Officiis*, Bk. 1, Ch. 13, Sec. 41
38. Locke, *On Education*, Sec. 66
39. Bertrand Russell, *Why I Am Not a Christian*, p. 156
40. Maximilian Rudwin, *The Devil in Legend and Literature*, p. 230
41. Quincy Wright, *A Study of War*, Vol. I, p. 100
42. Ivan Sanderson, reviewing *The Desperate People*, by Farley Mowatt, *Saturday Review of Literature*, November 28, 1959

BIBLIOGRAPHY

Adler, Irving, *How Life Began,* New York, John Day, 1957.

Alger, W. R., *A Critical History of the Doctrine of a Future Life,* Philadelphia, G. W. Childs, 1864.

Allee, W. C., *Social Life of Animals,* New York, W. W. Norton, 1938.

Aquinas, St. Thomas, *Summa Theologicus,* Vatican Library, Rome.

Atkinson, D. T., *Magic, Myth and Medicine,* Cleveland, World Publishing Co., 1956.

Bain, *The Fields and Methods of Biological Sociology,* from *Fields and Methods of Sociology,* New York, Ray Long & R. R. Smith, 1934.

Balz, Albert G. A., *The Basis of Social Theory,* New York, Knopf, 1924.

Beadle, George W., *The Biological Sciences,* from *Frontiers of Science,* New York, Basic Books, 1958.

Bebel, August, *Woman and Socialism,* New York, Socialist Literature, 1910.

Beck, L. Adams, *A Beginner's Book of Yoga,* New York, Farrar & Rinehart, 1937.

Becker, Carl L., *How New Will the Better World Be?,* New York, Knopf, 1944.

Beckwith, Burnham P., *Religion, Philosophy and Science,* New York, Philosophical Library, 1957.

Belloti, Felice, *Fabulous Congo,* London, A. Dakers.

Bibby, Geoffry, *The Testimony of the Spade,* New York, Knopf, 1956.

Boulenger, E. G., *Apes and Monkeys,* New York, McBride, no date.

Bovet, Pierre, *The Fighting Instinct,* London, Allen & Unwin, 1923.

Briffault, Robert, *Rational Evolution,* New York, MacMillan, 1930.

Brown, L. Guy, *Problems of Social Psychiatry,* from *Fields and Methods of Sociology,* New York, Ray Long & R. R. Smith, 1934.

Budge, Sir Wallis, *Egyptian Magic,* Evanston, Ill., University Books, no date.

——— *The Gods of the Egyptians,* London, Metheuen & Co., 1904.

Cahn, Edmond N., *The Sense of Injustice,* New York, University Press, 1949.

Campbell, Joseph, *The Masks of God: Primitive Mythology,* New York, Viking Press, 1959.

——— *The Historical Development of Mythology,* from *Myth and Mythmaking,* New York, Braziller, 1960.

Cardozo, Benjamin N., *Notable Opinions of Mr. Justice Cardozo,* New York, Ad Press, 1938.

——— *Law and Literature,* New York, Harcourt, Brace & Co., 1931.
——— *The Nature of the Judicial Process,* New Haven, Yale University Press, 1946.
Carrington, Richard, *A Biography of the Sea,* New York, Basic Books, 1960.
Ceram, C. W., *The Secret of the Hittites,* New York, Knopf, 1956.
Chugerman, Samuel, *Life of Lester F. Ward,* Durham, N.C., Duke University Press, 1939.
Cieza de Leon, Pedro de, *The Incas,* Norman, Okla., University of Oklahoma Press, 1959.
Clarke, Arthur C., *Horizon,* New York, January, 1959.
Cohen, Morris R., *The Meaning of Human History,* La Salle, Ill., Open Court Pub., 1947.
Covarrubias, Miguel, *Island of Bali,* New York, Knopf, 1937.
Crampton, Henry E., *The Coming and Evolution of Life,* New York, University Society, 1931.
Crawley, Ernst, *The Mystic Rose,* 2 vol., New York, Boni & Liveright, 1927.

Danielsson, Bengt, *Love in the South Seas,* New York, Reynal & Co., 1956.
Dart, Raymond A., *Adventures with the Missing Link,* New York, Harper, 1959.
Davidson, Basil, *The Lost Cities of Africa,* Boston, Little, Brown & Co., 1959.
Davies, A. Powell, *The Mind and Faith of A. Powell Davies,* Garden City, N.Y., Doubleday, 1959.
Davies, John Langdon, *A Short History of Women,* New York, Viking, 1927.
Davis, Jerome, *Capitalism and Its Culture,* New York, Farrar & Rinehart, 1935.
Denison, John Hopkins, *Emotion As the Basis of Civilization,* New York, Scribner's, 1928.
Dewey, John, *Reconstruction in Philosophy,* New York, Henry Holt, 1920.
——— *Nature and Experience,* New York, W. W. Norton, 1929.
——— *Logic, The Theory of Inquiry,* New York, Henry Holt, 1938.
——— *A Common Faith,* New Haven, Conn., Yale University Press, 1934.
——— *Individualism Old and New,* New York, Milton Balch Co., 1930.
——— *Human Nature and Conduct,* New York, Henry Holt, 1922.
——— *Naturalism and the Human Spirit,* New York, Columbia University Press, 1944.
——— *Freedom and Culture,* New York, G. P. Putnam's, 1939.
——— *The Public and Its Problems,* New York, Henry Holt, 1927.
——— *Liberalism and Social Action,* New York, G. P. Putnam's Sons, 1935.
——— *Experience and Education,* New York, MacMillan, 1938.
——— *Ethics,* Revised edition, Dewey and Tufts, New York, Henry Holt, 1926.

Dickey, Herbert S., *My Jungle Book,* Boston, Little Brown & Co., 1932.
Dorsey, George A., *Why We Behave Like Human Beings,* New York, Harper, 1926.
Douglas, William O., *America Challenged,* Princeton, N.J., Princeton University Press, 1960.
DuBois, Abbé J. A., *Hindu Customs, Manners and Ceremonies,* Oxford, Oxford University Press, 1897.
Duff, Charles, *This Human Nature,* New York, Cosmopolitan, 1930.
Dunham, Barrows, *Man Against Myth,* Boston, Little Brown & Co., 1947.
Dunlap, Samuel F., *The Ghebers of Hebron,* New York, J. W. Bouton, 1898.
Dunn, L. C., *Heredity and Variation,* New York, University Society, 1932.

Edwardes, Allen, *The Jewel in the Lotus,* New York, Julian, 1959.
Eichler, Lillian, *The Customs of Mankind,* Garden City, N.Y. Garden City Publishing Co., 1924.
Eichler, Philip, *A Philosophy of Science,* New York, G. P. Putnam's Sons, 1936.
Einstein, Albert, *The World As I See It,* New York, Philosophical Library, 1949.
Eiseley, Loren, *Nature, Man, and Miracle,* New York, Horizon, July, 1960.
——— *The Firmament of Time,* New York, Atheneum, 1960.
Ellam, J. E., *Religion of Tibet,* New York, Dutton, 1927.
Ellis, Havelock, *Questions of Our Day,* London, Lane, 1936.
Epictetus, *Discourses of Epictetus,* translated by George Long, New York, A. L. Burt & Co., no date.
Evans, E. P., *Evolutional Ethics and Animal Psychology,* London, Heinemann, 1898.
Ewers, Hanns Heinz, *The Ant People,* New York, Dodd, Mead, 1943.

Fabre, J. H., *Social Life in the Insect World,* London, Penguin Books, 1911.
Faure, Elie, *History of Art,* Garden City, N.Y., Garden City Publishing Co., 1937.
Faure, Sébastien, *Does God Exist?,* Stelton, N.J., Kropotkin Library.
Fielding, William J., *Shackles of the Supernatural,* Girard, Kan., Haldeman-Julius, 1938.
Forel, Auguste, *The Social World of the Ants,* 2 vol., New York, Albert & Charles Boni, 1929.
Forsyth, David, *Psychology and Religion,* London, Watts & Co., 1935.
Fouchet, Max-Pol, *The Erotic Sculpture of India,* New York, Criterion, 1959.
France, Anatole, *Penguin Island,* London, Lane, 1909.
Frazer, Sir James G., *The Golden Bough,* 13 vol., New York, MacMillan, 1935.

——— *The Golden Bough,* 1 vol., edition, New York, MacMillan, 1922.
——— *Folk-lore in the Old Testament,* New York, MacMillan, 1935.
——— *The Scope of Social Anthropology,* London, P. Lund Humphries & Co., 1938.

Freud, Sigmund, *The Future of an Illusion,* London, Hogarth Press, 1949.
Fulgentius, Saint, *De Fide,* London, Watts & Co., 1913.
Furniss, Father, *Books for Children,* London, Watts & Co., 1913.

Gallenkamp, Charles, *Maya,* New York, David McKay, 1959.
Gann, Thomas, and J. Erick Thompson, *The History of the Maya,* New York, Scribner's, 1931.
Gibney, Frank, *The Operators,* New York, Harper, 1960.
Glatzer, Nahum N., editor, *A Jewish Reader: In Time and Eternity,* New York, Schocken Books, 1946.
Goldberg, B. Z., *The Sacred Fire,* New York, Horace Liveright, 1932.
Golden, Harry, *2¢ Plain,* Cleveland, World Publishing Co., 1959.
Gould, F. J., *A Short History of Religion,* London, Watts & Co., 1903.
Graubard, Mark, *Man the Slave and Master,* New York, Covici-Friede, 1938.

Haldane, J. B. S., *The Inequality of Man,* London, Penguin Books, 1932.
Hardin, Garrett, *Nature and Man's Fate,* New York, Rinehart, 1959.
Hart, Joseph K., *Inside Experience,* New York, Longmans, Green & Co., 1927.
Hays, H. R., *From Ape to Angel,* New York, Knopf, 1958.
Hoenus, *Visit to Hell: The Christian Hell,* by Hypatia Bradlaugh, London, Bonner, Watts & Co., 1913
Howey, W. Oldfield, *The Cat in the Mysteries of Religion and Magic,* New York, Castle Books, 1956.
Hoyle, Fred, *Nature of the Universe,* New York, Harper, 1950.
——— *Forecasting the Future* from *Frontiers of Science,* New York, Basic Books, 1958.
Hunt, Frazier, *This Bewildered World,* New York, Stokes, 1937.
Huxley, Julian S., *The Stream of Life,* New York, Harper, 1927.
——— *Ants,* New York, Jonathan Cape & R. Ballou, 1930.

James, E. O., *Ancient Gods,* New York, G. P. Putnam's Sons, 1960.
James, William, *The Meaning of Truth,* New York, Longmans, Green & Co., 1909.
——— *The Will to Believe,* New York, Longmans, Green & Co., 1917.
Jefferson, Thomas, *Jefferson Himself,* edited by Bernard Mayo, Boston, Houghton, Mifflin Co., 1937.
Jessup, Philip C., *A Modern Law of Nations,* New York, MacMillan, 1948.

Junod, H. A., *The Life of a South African Tribe,* London, MacMillan, 1927.
Kahler, Erich, *Man the Measure,* New York, Braziller, 1961.
Kaufmann, Walter, *Critique of Religion and Philosophy,* New York, Harper, 1958.
Kilpatrick, William H., *The Educational Frontier,* New York, Appleton-Century, 1933.
Klineberg, Otto, *Social Psychology,* New York, Henry Holt, 1940.
Kluckhorn, Clyde, *Mirror for Man,* New York, McGraw-Hill, 1949.

Lane, E. W., *An Acount of the Manners and Customs of the Modern Egyptians,* London, A. Gardner, 1898.
Larrabee, Eric, *After Abundance, What?,* New York, Horizon, July, 1960.
Lea, Henry Charles, *History of the Inquisition of the Middle Ages,* New York, Harper, 1887.
Leeuw, Hendrik de, *Sinful Cities of the Western World,* New York, Julian Messner, 1934.
Lévy-Bruhl, Lucien, *The "Soul" of the Primitive,* New York, MacMillan, 1928.
Lewisohn, Richard, *A History of Sexual Customs,* New York, Harper, 1958.
Liao, Wen Kwai, *The Individual and the Community,* New York, Harcourt, Brace & Co., 1933.
Libby, Walter, *The History of Medicine,* Boston, Houghton, Mifflin Co., 1922.
Link, Henry C., *The Rediscovery of Man,* New York, MacMillan, 1939.
Linton, Ralph, *The Tree of Culture,* New York, Knopf, 1955.
Living Philosophies, A Symposium, New York, Simon & Schuster, 1931.
Lorenz, Konrad Z., *King Solomon's Ring,* New York, Thomas Y. Crowell, 1952.

MacCulloch, J. A., *Medieval Faith and Fable,* London, G. G. Harrap & Co., 1932.
MacCurdy, George G., *The Coming of Man,* University Series, New York, University Society, 1931.
Mahaffy, John P., *Prolegomena to Ancient History,* London, MacMillan, 1871.
Malinowski, Bronislaw, *A Scientific Theory of Culture,* Chapel Hill, University of North Carolina Press, 1944.
——— *Myth in Primitive Psychology,* New York, W. W. Norton, 1929.
Marais, Eugene N., *The Soul of the White Ant,* New York, Dodd-Mead, 1937.
Martin, Everett Dean, *Farewell to Revolution,* New York, W. W. Norton, 1935.
McLoughlin, Emmett, *American Culture and Catholic Schools,* New York, Lyle Stuart, 1960.

Mead, Charles W., *Old Civilizations of Inca Land,* Hand Book Series II, New York, American Museum of Natural History, 1935.

Meek, Theophile James, *Hebrew Origins,* New York, Harper, 1936.

Meerloo, Joost A. M., foreword to *Man's Presumptuous Brain,* by A. T. W. Simeons, New York, E. P. Dutton, 1961.

Metchnikoff, Elie, *The Nature of Man,* New York, G. P. Putnam's Sons, 1906.

Mill, John Stuart, *Autobiography,* Garden City, N.Y., Dolphin Books.

——— *The Idea of God in Nature,* London, Watts & Co., 1914.

Miner, Roy Waldo, *Fragile Creatures of the Deep, the Story of the Hydroids,* Guide Leaflet no. 98, American Museum of Natural History, *Natural History Magazine,* November, 1938.

Montagu, Ashley, *Immortality,* New York, Grove Press, 1955.

Morley, Sylvanus G., *The Ancient Maya,* Stanford, Calif., Stanford University Press, 1946.

Mosca, Gaetano, *The Ruling Class,* from *Images of Man,* edited by C. Wright Mills, New York, Braziller, 1960.

Moulton, Forest Ray, *Astronomy* from *The Nature of the World and of Man,* Garden City, N.Y., Doubleday, 1937.

Moulton, G. G., *The Inquisition,* New York, J. Cape & H. Smith, 1929.

Müller, F. Max, *My Autobiography,* New York, Scribner's, 1901.

Muthu, D. Chowry, *Antiquity of Hindu Medicine,* New York, P. B. Hoeber, 1931.

Needham, Joseph, *Man a Machine,* New York, W. W. Norton, 1928.

Oparin, A. I., *Origin of Life,* New York, MacMillan, 1938.

Ortega y Gasset, José, *The Revolt of the Masses,* New York, W. W. Norton, 1932.

Overstreet, H. A., *The Enduring Quest,* Chautauqua, N.Y., Chautauqua Press, 1931.

Packard, Vance, *The Wastemakers,* New York, David McKay Co., 1959.

Palmer, Frederick, *Our Gallant Madness,* Garden City, N.Y., Doubleday, 1937.

Paulsen, Friedrich, *Introduction to Philosophy,* New York, Henry Holt, 1907.

Pavlov, Ivan P., *Lectures on Conditioned Reflexes,* New York, Liveright, 1928.

Peattie, Donald Culross, *The Flowering Earth,* New York, G. P. Putnam's Sons, 1939.

——— *This is Living,* New York, Dodd-Mead & Co., 1938.

Potter, LaForest, *Strange Loves,* New York, National Library Press, 1933.

Powys, Llewelyn, *Earth Memories,* New York, W. W. Norton, 1938.

Puxley, W. Lavallin, *Deep Seas and Lonely Shores,* New York, E. P. Dutton, 1936.

Randall, Jr., John Herman, *Dualism in Metaphysics,* from *Essays in Honor of John Dewey,* New York, Henry Holt, 1929.

Ratner, Joseph, *Introduction to Intelligence in the Modern World,* New York, Modern Library, 1939.

Ray, Carleton, and Ciampi, Elgin, *Marine Life,* New York, A. S. Barnes, 1956.

Reade, Winwood, *The Martyrdom of Man,* London, Watts & Co., 1924.

Reiser, Oliver L., *Alchemy of Light and Color,* New York, W. W. Norton, 1928.

Robinson, James Harvey, *Human Comedy,* New York, Harper, 1937.

Rudwin, Maxmillian, *The Devil in Legend and Literature,* La Salle, Ill., Open Court Pub. Co., 1931.

Russell, Bertrand, *In Praise of Idleness,* New York, W. W. Norton Co., 1935.

——— *Our Knowledge of the External World,* New York, W. W. Norton Co., 1929.

——— *Why I Am Not a Christian,* New York, Simon & Schuster, 1957.

——— *Sceptical Essays,* New York, W. W. Norton Co., 1928.

——— *Value of Free Thought,* Girard, Kan., Haldeman-Julius, 1944.

——— *Icarus or the Future of Science,* New York, E. P. Dutton, 1924.

——— *Why Men Fight,* New York, Century, 1916.

Santayana, George, *Winds of Doctrine,* New York, Scribner's, 1913.

Schneider, Herbert W., *Radical Empiricism and Religion* from *Essays in Honor of John Dewey,* New York, Henry Holt, 1929.

Schrödinger, Erwin, *Science and the Human Temperament,* New York, W. W. Norton & Co., 1935.

Schweitzer, Albert, *The Philosophy of Civilization,* New York, MacMillan, 1960.

——— *Indian Thought and Its Development,* New York, Henry Holt, 1936.

Shaw, George Bernard, *The Adventures of the Black Girl in Her Search for God,* New York, Dodd-Mead, 1933.

Shoosmith, F. H., *Life in the Animal World,* New York, McBride.

Simeons, A. T. W., *Man's Presumptuous Brain,* New York, E. P. Dutton, 1961.

Smith, W. Robertson, *The Religion of the Semites,* New York, Meridian Library, 1956.

Spaeth, Louise Marie, *Marriage and Family Life Among Strange Peoples,* Chicago, T. S. Rockwell, 1931.

Starr, Frederick, *Confucianism,* New York, Covici-Friede, 1930.

Stephen, Sir Leslie, *An Agnostic's Apology,* London, Watts & Co., 1931.
Stone, Lee Alexander, *The Power of a Symbol,* Chicago, Pascal-Covici, 1925.
Sumner, William Graham, *Folkways,* Boston, Ginn & Co., 1906; New York, Dover, 1959.

Thomas, H. C., and Hamm, W. A., *The Foundations of Modern Civilization,* New York, Vanguard Press, 1927.
Tufts, Matilde Castro, *Looking to Philosophy* from *Essays in Honor of John Dewey,* New York, Henry Holt, 1929.
Tylor, Edward B., *Religion in Primitive Culture,* London, John Murray, 1900; New York, Harper, 1958.
——— *Primitive Culture,* as above.
——— *Anthropology,* New York, D. Appleton & Co., 1897.
——— *Origins of Culture,* New York, Harper, 1958.

Unamuno, Miguel de, *Tragic Sense of Life,* New York, Dover, 1954.

Vaillant, George C., *Aztecs of Mexico,* Garden City, N.Y., Doubleday, 1944.
Veblen, Thorstein, *The Theory of the Leisure Class,* New York, Vanguard Press, 1922.

Ward, Lester F., *Applied Sociology,* New York, D. Appleton & Co., 1902.
——— *Glimpses of the Cosmos,* New York, G. P. Putnam's Sons, 1913.
Watson, E. L. Grant, *Mysteries of Natural History,* New York, Stokes, 1937.
Watson, John B., *Behaviorism,* New York, W. W. Norton & Co., 1925.
Weber, Alfred, *History of Philosophy,* New York, Scribner's, 1899.
Wells, H. G., *First and Last Things,* New York, G. P. Putnam's Sons, 1908.
Wheless, Joseph, *Forgery in Christianity,* New York, Knopf, 1930.
White, A. D., *History of the Warfare of Science with Theology,* New York, Dover, 1960; originally published 1896.
Wilde, Oscar, *The Picture of Dorian Gray,* New York, Modern Library, 1925.
Wright, Quincy, *A Study of War,* 2 vol., Chicago, University of Chicago Press, 1942.
Wylie, Philip, *Innocent Ambassadors,* New York, Rinehart, 1957.

Young, Art, *On My Way,* New York, Horace Liveright, 1928.
Yu-Lan, Fung, *A Short History of Chinese Philosophy,* New York, MacMillan, 1948.
Yutang, Lin, *The Importance of Living,* New York, Reynal & Hitchcock, 1937.

INDEX

INDEX

INDEX

INDEX